SPACE SCIENCE, EXPLORATION AND POLICIES

PLASMA OF METEORITE IMPACT AND PREHISTORY OF LIFE

SPACE SCIENCE, EXPLORATION AND POLICIES

Additional books in this series can be found on Nova's website under the Series tab.

Additional E-books in this series can be found on Nova's website under the E-books tab.

SPACE SCIENCE, EXPLORATION AND POLICIES

PLASMA OF METEORITE IMPACT AND PREHISTORY OF LIFE

GEORGE MANAGADZE

Nova Science Publishers, Inc.
New York

Copyright © 2011 by Nova Science Publishers, Inc.

All rights reserved. No part of this book may be reproduced, stored in a retrieval system or transmitted in any form or by any means: electronic, electrostatic, magnetic, tape, mechanical photocopying, recording or otherwise without the written permission of the Publisher.

For permission to use material from this book please contact us:
Telephone 631-231-7269; Fax 631-231-8175
Web Site: http://www.novapublishers.com

NOTICE TO THE READER

The Publisher has taken reasonable care in the preparation of this book, but makes no expressed or implied warranty of any kind and assumes no responsibility for any errors or omissions. No liability is assumed for incidental or consequential damages in connection with or arising out of information contained in this book. The Publisher shall not be liable for any special, consequential, or exemplary damages resulting, in whole or in part, from the readers' use of, or reliance upon, this material. Any parts of this book based on government reports are so indicated and copyright is claimed for those parts to the extent applicable to compilations of such works.

Independent verification should be sought for any data, advice or recommendations contained in this book. In addition, no responsibility is assumed by the publisher for any injury and/or damage to persons or property arising from any methods, products, instructions, ideas or otherwise contained in this publication.

This publication is designed to provide accurate and authoritative information with regard to the subject matter covered herein. It is sold with the clear understanding that the Publisher is not engaged in rendering legal or any other professional services. If legal or any other expert assistance is required, the services of a competent person should be sought. FROM A DECLARATION OF PARTICIPANTS JOINTLY ADOPTED BY A COMMITTEE OF THE AMERICAN BAR ASSOCIATION AND A COMMITTEE OF PUBLISHERS.

Additional color graphics may be available in the e-book version of this book.

Library of Congress Cataloging-in-Publication Data

Managadze, G. G. (Georgii Georgievich), 1936-
[Plazma meteoritnogo udara i dobiologicheskaia evoliutsiia. English]
Plasma of meteorite impact and prehistory of life / George Managadze.
p. cm.
Includes bibliographical references and index.
ISBN 978-1-61324-510-1 (hardcover)
1. Meteoritic hypothesis. 2. Plasma jets. 3. Organic compounds--Synthesis. 4. Life--Origin. I. Title.
QB981.M256613 2011
576.8'3--dc23

2011015639

Published by Nova Science Publishers, Inc. † New York

CONTENTS

PREFACE

The book is intended for researchers and specialists, students and post-graduate students engaged in the problem of the origin of life on Earth, as well as any interested laymen.

The author proposes a new concept of the prehistory of life based on the possibility of the origin of primeval forms of living matter in the processes that accompany a hypervelocity impact of a meteorite onto the surface of a planet.

It was experimentally shown that the plasma torch produced by a meteorite impact has exceptional and hitherto unknown properties. In the process of its explosive expansion-away the plasma medium of the torch ensures the synthesis of organic compounds, their assembly, and ordering. The electromagnetic fields of the torch plasma make possible the breaking of the mirror symmetry of the synthesis products, which could have determined the «sign» of the asymmetry of the bioorganic world. The author shows that the combination of these extremely important properties of the torch plasma could have favored the formation of the first primitive forms of living matter, which were not only capable of replication, but also possessed a primitive genetic code.

It is shown that an impact of a large meteorite could have produced in its crater the initial conditions (temperature, humidity, concentration of organic compounds) required for the survival and evolutionary development of primitive forms of life for many thousands and million years.

Evidence is provided that supports the hypothesis that the synthesis of simple organic compounds in interstellar gas and dust clouds may also occur in the plasma torch produced as a result of hypervelocity collisions of dust particles. The proposed concept is based on real processes and on the current views on the conditions that reigned on the early Earth. Moreover, the concept is also supported by material evidence found in the past 10-15 years.

The new concept of the origin of primeval forms of living matter is mostly based on plasmachemical processes in the torch, which are characterized by extremely high catalytic activity. These processes could have reduced substantially the time required for the emergence of a primitive living being and, what is important, without involving the atmosphere or solar energy.

The main ideas of the new concept have been corroborated by the results of laboratory experiments where laser radiation was used to simulate the processes that occur in the impact-produced plasma torch.

The torch produced by laser radiation was shown experimentally to be a full analog of the torch produced in the process of a hypervelocity impact. The plasma torch was shown to be exactly and correctly reproduced in laboratory experiments, which allow it to be thoroughly studied.

The book is original in that its author was the person who proposed the core ideas on the development of the new concept and creation of the dedicated scientific instruments for model experiments involving mass-spectrometric measurements of the synthesis products. The book gives a detailed description of the results of these experiments and their interpretation.

The author analyzes special configurations of meteorite impacts, which could ensure the introduction of complex organic compounds synthesized in the torch plasma into deep-seated rock layers or into underground water reservoirs on the Earth and planetary satellites with extremely low or high surface temperatures. Novel techniques and original analytical onboard instruments are proposed, which should be capable of detecting signs of extraterrestrial life (if any) on the satellites of the planets of the Solar System. The author proposes a novel technique for the experiments in the near magnetosphere, which allows reproducing hypervelocity impacts with relative velocities of about 16 km/s, which are impossible to simulate under terrestrial conditions.

AUTHOR'S FOREWORD TO THE ENGLISH EDITION

Man is the dominating species in the fantastic variety of forms of life on modern Earth.

It is therefore for Man to sort out how and under what conditions the primary form of living matter could have developed from nonliving and inorganic substance, which, evolving, made possible the emergence of Civilization or the community of humans capable of comprehending how life came into being.

The assumption that life was imported to Earth from space only further complicates the issue. In this case we must know where and how it could have originated and why the ideal conditions at the surface and in the interiors of our planet were not suitable enough to this end.

Today we should frankly admit that we do not know how nonliving matter became living. However, we know that the primary form of living matter originated in multiple-factor natural processes of extreme complexity, which so far remain beyond our understanding.

There is currently a consensus of opinion that an environment meeting certain conditions was necessary for primary forms of life to originate and evolve. The first and simplest of these conditions consist in the availability of water and a certain temperature to ensure that this liquid would not freeze. Another important property that the environment must possess is that it must contain high concentration of simple organic compounds. However, these were the simplest and most obvious requirements to the properties of the natural environment that could meet the initial conditions necessary for the origination of living matter. There were, however, other, more complex and difficult-to-ensure conditions to the environment – it had to be homochiralic. We can neither rule out the need for and the existence of other properties, which are currently little known or even completely unknown.

Hence a more efficient approach toward solving the problem of the origin of living matter consisting in the successful search involving the discovery, identification, and analysis of a natural environment capable of ensuring the development of the key properties of living matter could evidently have brought us closer to solving the Eternal Problem.

In the case of a discovery of a natural environment where relatively simple conditions are realized, when one searches for such an environment in natural conditions, the presence of mechanisms ensuring the homochirality of this environment should be viewed as one of the key criteria for its identification. However, this property, which is crucial for the animation of matter, should coincide and combine with other processes occurring in a certain sequence. The presence of these factors may to a large degree be viewed as material evidence indicating that the natural environment found and the mechanisms operating therein have indeed direct bearing on the process of the emergence of living matter.

In the new plasma concept of the prehistory of life proposed in this book I am trying to demonstrate that during the expansion-away of the plasma torch processes can be realized and mechanisms may develop that are capable of ensuring the development of the key properties characteristic of the most primitive forms of living matter.

In particular, it was shown experimentally that in the plasma torch simple and complex organic compounds are synthesized, ordering of matter occurs, and complex molecular structures are assembled.

Experimental studies of the physical processes generated during the expansion-away of a plasma torch show that the emerging environment has all the necessary conditions for weak symmetry breaking. Thus the "true" factors of advantage arising as a result of magnetooptical processes produce local chiral physical fields, which could have determined the "sign" of the symmetry breaking of bioorganic world. Such fields could, by controlling spontaneous processes, have caused the symmetry breaking and, in rare cases, ensured the chiral purity of the environment from where polymer capture occurred that is required for the assembly of homochiralic molecular structures needed to ensure their replication. The conditions produced in the impact crater should have made possible the survival and evolutionary development of primitive forms of living matter over millions of years.

I believe that my analysis of the results of novel experiments carried out under laboratory conditions during the development of the concept must persuade the reader that an impact-produced plasma torch should be viewed as the most promising medium where conditions could have developed that were necessary for the formation of the primary forms of living matter. This environment is currently believed to have no alterative in terms of a number of key properties.

The results of numerous experimental and theoretical studies carried out and published by other authors and research teams and discussed in this book also confirm this conclusion.

The Russian edition of this book was published in Moscow in late 2009. Its translation took considerable time.

I used this time to make some changes and additional explanations. I defined some concepts more precisely and introduced an extra section after the conclusion. This section is dedicated to the plasma processes that occur during laboratory experiments or during natural experiments performed within the framework of the Deep Impact space mission.

I could include this material into the appropriate chapters of this book, but I deemed it to be more reasonable not to rearrange the Russian version of the book. I made this decision since (i) the information given in the supplementary section was consistent with the main conclusions listed in the concluding section of the book, and (ii) it provides information about possible directions for further research.

The proposed concept deserves to be viewed as credible because its fundamental conclusions are based on the results of laboratory experiments and on material evidence available in nature. It is also important that the realization of the individual hypothetical mechanisms proposed is highly probable because it is consistent with the basic laws of plasma physics, plasma chemistry, and other universally recognized laws of nature.

It is important that the proposed concept provides all that is needed to allow living matter to originate in accordance with known laws of nature without resorting to Divine Powers.

November, 2010

AUTHOR'S FOREWORD TO THE RUSSIAN EDITION

It was shown experimentally 10 years ago that organic compounds can be synthesized in a plasma torch produced by Q-switched laser radiation (Managadze, 2001, 2002b). This property of plasma torch has been previously unknown. A hypothesis has been suggested that the synthesis of organic compounds may also be possible in an impact plasma torch, because of the great similarity between laser and impact produced torches.

There are bona fide experimental data described in Chapters 4 and 6 indicating conclusively that the matter becomes ordered in the plasma environment generated in the process of the formation and expansion-away of the torch, and that emerging physical fields meet the main requirements for "true" local chiral fields. There are reasons to believe that these fields are capable of producing the initial weak breaking of the symmetry of enantiomers, which can then amplify substantially in the processes that occur in a highly superheated and nonequilibrium plasma that is far from the thermodynamic branch of equilibrium.

The new mechanism proposed for the synthesis of organic compounds may also be responsible for their formation in interstellar gas and dust clouds as a result of collisions of dust particles moving, e.g., with relative velocities exceeding 15—20 km/s, and ensures volume ionization of dust particles with a density of 2 g/cm^3 . Hence a hypervelocity impact, while being a tremendously destructive event, may also be potentially constructive because it spends part of its energy to create conditions necessary for the emergence of life.

Dedicated experiments for laboratory simulation of the impact and the study of the torch plasma properties showed that plasma expansion-away is accompanied by the ordering of the constituent structures of organic compounds in the torch and that the mass of these compounds increases proportionally to the characteristic size of the projectile. The discovered properties of torch plasma offer a new insight into the functional and structural complexification of organic compounds in the process of synthesis.

The experimentally found properties of torch plasma, where synthesis and ordering are combined with the possible breaking of the symmetry of organic compounds, suggest that this may be the most promising environment for the synthesis of primary forms of life from nonliving substance. The study of the main physical properties of plasma torches showed that this environment may, with certain restrictions, be considered what is referred to as the dissipative structure. Such highly non-equilibrium structures, which are far from the thermodynamic branch of equilibrium, may in the past have provided conditions necessary for the formation of complex molecular structures.

The most important features of a torch plasma described in this book provide conclusive evidence indicating that the proposed mechanism of plasma-chemical synthesis of organic compounds in a hypervelocity impact is a bona-fide established process of high efficiency. It is "infallible" in that the mechanism operates always and without fail every time the impact velocity exceeds a certain critical threshold.

A plasma torch is a short-lived and high-speed outburst of hot and dense plasma. In nature such an environment with a special expansion dynamics forms only in the process of a hypervelocity collision of fragments of matter, e.g., during meteorite impacts onto the Earth, if the impact velocity exceeds 11 km/s. The velocities of meteorite impacts onto the Earth are known to range from a minimum of 11 to a maximum of 72 km/s.

The extremely high catalytic activity of the torch plasma ensures very high rates of plasma-chemical reactions that involve the formation of new chemical compounds. The collision results in highly efficient synthesis of organic compounds provided that the constituent elements are available in the colliding bodies. Adiabatic plasma expansion is accompanied by fast cooling of the matter from 10^5—10^6 K down to the ambient temperature. Such expansion ensures the irreversibility of plasma-chemical reactions, fast removal of the products of synthesis from "hot" regions, and the participation of these products in subsequent lower-temperature processes, which result in the production of more complex final products.

The motions of dense and high-temperature plasma flows in the torch favor the generation of spontaneous magnetic fields with constant orientation, which, when combined with linearly polarized radiation, are generally recognized as «true» chiral factors. Such motions also excite plasma instabilities, which are very likely to generate circularly-polarized radiation in the plasma environment of the torch. The experimentally observed non-equilibrium unidirectional electromagnetic fields, in turn, are indicative of the «innate» asymmetry of the torch.

The enormous energy creates the conditions for the formation of an environment necessary for the survival of the emergent primary forms of living matter. Thus a major meteorite impact may heat rocks up to moderate temperatures and melt ice in a substantially large area adjacent to the impact site with the cooling time scale ranging from several hundred thousand to about ten million years, and fill this area with large amounts of organic compounds.

Under special conditions of a hypervelocity impact the complex organic compounds synthesized in the torch may be injected deep down into relatively warm layers of cosmic bodies with too low or too high surface temperatures.

Because of the great variety of their properties, the processes that occur during the impact can meet the conditions imposed by many models with their different scenarios of the emergence of primitive forms of life.

The main conclusions of the proposed concept rely on experimental results, and the mechanisms employed are based exclusively on physical, chemical, and plasma-chemical processes.

The work on the development of the concept and model experiments has been carried out mostly at the Laboratory of Active Diagnostics, Space Research Institute of the Russian Academy of Sciences (SRI RAS). The first results concerning the synthesis of organic compounds in a torch were obtained as early as 1992. However, the first results obtained within the framework of the new concept were published only in 2001 by the author of this book. Since 2003 the work has been carried out with the participation of Dr. W.Brinkerhoff in cooperation with John Hopkins University (USA), which is one of the leading American space research centers with which I collaborated from 1993 through 2007. Since 2008 the work has been transferred to the NASA Goddard Space Flight Center.

The results of the work on the development of the new concept were published in Russian and international research journals, reported at numerous conferences and symposia, and discussed at seminars held in Russia, Ukraine, USA, France, Greece, Bulgaria, and Serbia.

The evident realizability of the proposed scenario in nature subject to a minimum number of initial conditions helped the main ideas of the new concept to be rapidly adopted by the research community and made it highly popular among the audiences independently of their professional background level.

The factors described above played the crucial part in my decision to write this book.

The main ideas of the new concept are consistent with many earlier suggested scenarios of the origin of primary, primitive forms of living matter. The point is that meteorite impacts are processes with local, spatially limited effects. This circumstance allowed me to minimize the review of published work.

The primary aim of publishing this book was to promote the new ideas and the results obtained in order to further extend research in the field and draw the public attention and especially that of the young people to the new approach to the solution of the "Eternal Problem".

I am mostly indebted for the preparation and publication of this book to my best friend V.A.Gelovani, who by his attention and undiminishing interest since the emergence of the new idea has instilled confidence in the importance of the proposed concept. I am also truly grateful to W.Brinkerhoff for his great and incessant assistance.

My belief in the viability of the proposed concept grew stronger and became firmly established in the process of rather heated debate with many renowned experts in various fields of knowledge. I became friends with many of these researchers, whom I have not personally met before. This was an especially valuable result of my work. The advice and comments of these people often determined the direction of further research. I therefore feel it my duty to sincerely thank those who directly or indirectly contributed to the publication of this book and, first and foremost, my severe opponents, such as V.A.Avetisov, A.D.Altstein, E.A.Vorobyova, A.Yu.Rozanov, A.A.Sysoev, and A.S.Spirin. My consultants at different stages of this work were V.G.Babaev, N.G.Bochkarev, E.N.Brodskii, O.G.Chkhetiani V.A.Davankov, N.A.Inagamov, I.D.Kovalev, Yu.G.Malama, L.M.Mukhin, E.N.Nikolaev, D. Papadopulos A.A.Rukhadze, Yu. G. Shkuratov. G. V. Sholin, A.V.Vityazev, and R. Kh. Ziganshin.

I am grateful to L.M.Zelenyi, A.V.Zakharov, and V.M.Kuntsevich for providing the opportunity to deliver my first report about the new concept of the prehistory of life at a foreign conference and publish it in an international journal. I am grateful to R.R.Nazirov for his constant assistance, attention, and interest in the work, and also to N.A.Eismont and S.G.Bugrov for the cooperation that contributed to the development of the technique of "head-on collision" and identification of carbines.

It is with great pleasure that I express my special thanks to R.Z.Sagdeev for his assistance in the first full presentation of the new concept in the USA. This happened at the extended seminar of the research department of the University of Maryland, which headed and conducted the Deep Impact - first space experiment involving an artificial impact onto a cometary core. The seminar was held within the framework of the International meeting dedicated to the 50th anniversary of the launch of the Sputnik. This ensured an interesting discussion of the concept with the participation of many distinguished researchers including the members of the Russian Academy of Sciences.

I am grateful to my friend and colleague L. Kelner, the President of Fenixtec (USA), for his unfailing interest and support of works in Russia and USA whose results formed the basis for this book.

I am also grateful to the young researcher and my daughter N.G.Managadze for her invaluable hard work on the preparation of the manuscript, to my student A.E.Chumikov, with whom I have coauthored many publications, and to the young researcher D.A.Moiseenko for their assistance in the preparation of plots, and also to the artists V.M.Davydov and A.N.Zakharov for the preparation of some figures. I am also grateful to L.V.Romanova and T.A.Khalenkova for preparing the book for publication.

I am thankful to all of those who believed in the importance of the idea.

August, 2009

Chapter 1

EARLY SCENARIOS OF THE ORIGIN OF LIFE AND THE NEW CONCEPT

1.1. INTRODUCTION

The existence of life on Earth should be viewed as the greatest heritage of our planet. The well-known microbial kind of life in the form of living cell capable of self reproducing and transmitting this capability to succeeding generations is a system of extremely complex functionality and optimum structure - the truly highest form of organization of matter. According to the available material evidence (Schidlowski, 1998), full-blown cells existed on Earth as early as 3.8—4.1 billion years ago. Contributing factors included moderate temperature and the presence of water on the Earth's surface 200 million years after the formation of the planet (Wilde et al., 2001). Since its emergence on the Earth life has evolved into civilization - the highest known form of organization of matter, which combines Intelligence with High Technologies. However, of greatest importance are spiritual values of the human society: yearning to do good, be grateful, be able to forgive, love for one's neighbor, combined with unrestrained pursuit of knowledge and self-perfection.

The emergence of life and its evolution are among the most topical problems in modern science, which remain the most difficult to investigate. Mankind's interest in these problems is known to date back to antiquity and to have been ever increasing over the millennia. However, the factors that led to the formation of living substance from inorganic matter and determined its subsequent evolution into an inestimable variety of its material forms still remain beyond our understanding.

The deciphering of the cause-and-effect relations in the evolutionary process that resulted in the emergence of living substance becomes increasingly more complex as we move toward its origins. Barriers include poor and often unreliable data about the initial conditions that led to the emergence of primary forms of living matter, and the lack of reliable information about the contributing physical and chemical processes, natural mechanisms, and ambient properties.

The progress of the evolution of living matter during later stages of its development can be reconstructed by studying material evidence available in the form of fossils. However, the processes of the synthesis of organic compounds from inorganic matter during the so-called stage of "chemical evolution", which sets up the conditions for the emergence of the first, primitive living organisms, are extremely difficult to study, because such creatures, and maybe even their "lithic record", have completely disappeared. There is also no generally agreed scientifically grounded approach to the problem, because it is unclear what may have been the form of the first primitive living creature and whether the time when its first individual specimens appeared can be identified with the emergence of life.

We therefore do not know with certainty when and via which mechanisms a number of key extremely complex organic compounds and prebiotic structures formed in nature. According to the classical approach, these compounds with their new, unique properties must have ensured the successful completion of the important stage in the process of the emergence of life - that of chemical evolution - and the emergence of the first, primitive forms of living matter.

The main problems in the study of the origin of life have always been and remain: (1) finding a natural phenomenon capable of making possible the formation of the first primitive forms of living matter and (2) testing the capacity of this phenomenon by reproducing a laboratory analog of such a mechanism that makes possible abiogenous synthesis of the most important molecular structures associated with the emergence of life.

Considerable effort has been spent over the last 60 years to solve the first problem; however, to say that the required natural phenomenon has been found would still not be true. The second task is far from solved despite numerous hypothetical models of primary forms of living matter. The finding and experimental confirmation of a viable mechanism of the emergence of life under primeval Earth conditions will make it possible to choose the correct model of the first creature, whose appearance marked the beginning of the era of biological evolution.

According to the modern views, speculations about the first primary forms of living matter on the Earth must proceed from the fact that under natural conditions they should have formed abiogenically: i.e., via physical, chemical, and plasma-chemical rather than biological processes. Therefore the results of experiments involving laboratory simulations of natural processes of abiogenous synthesis of organic compounds have always been of key importance.

To describe in the most general form the emergence of the primary form of living matter, we must first clarify some complex issues using published research results to this end. The answers so obtained cannot be expected to be particularly reliable, because they are, as a rule, hypothetical in nature.

The most important of these questions are:

- What were the primary forms of life, their structural features, functional capabilities, and minimal characteristic sizes, and which of these forms can be viewed as the common ancestor of our contemporary biological world?
- Could the now known natural processes provide the preconditions for the formation of primary forms of living matter and create the conditions for its further survival and development at the biological stage of evolution?
- How close have the numerous laboratory experiments simulating the processes of the emergence of life come to the hypothetical primary form of living matter?

The answers to these questions may determine to a considerable degree where modern science is now on its way toward solving the "Eternal Problem" and how wide is the gap between laboratory prototypes and hypothetical primary forms of living matter.

Some of the above questions can be answered based on the criteria generally accepted in modern science. Thus, according to current views, for the matter to be considered "living" it must be capable of:

- reproducing similar structures and transmitting this capability to successive generations;
- participating in the process of natural Darwinian selection.

With minor reservations we can assume that life emerged after the primary form of living matter defined above had managed to produce not only its first, but many subsequent descendants, and started to reproduce via biological processes and subject to Darwinian selection.

The most unyielding problem was and still remains the determination of the structural features of the "primary form of living matter". This is due to the difficulty of identifying the time of the "vitalization" of non-living substance. It appears to be an easy task when addressed in the most general form without going into the details of hypothetical forms of life.

Thus structurally the primary form of living matter may have been something like a localized, but not necessarily spatially isolated, molecular system consisting of homochiralic oligonucleotides and peptide molecules linked via primitive genetic code. Its functional capabilities could have been determined by the specific nature of such a combination and make possible the reproduction of its structure including the synthesis of polypeptide, and the transfer of these capabilities to the successive generations subject to further selection of the Darwinian type.

However, the differences between the definitions of the "primary form of living matter" may become important when one considers particular hypotheses. Therefore, to correctly define this concept, we must use the formulations employed by the authors of these hypotheses.

I try to adhere to this principle in the book, especially when discussing the properties of the primary forms of living matter.

Our main models of the first organisms in Sections 1.3 and 1.4 are "single-polymer forms of life" of the ancient RNA world (Spirin, 2007) and the protoviroid, which, according to the author of the hypothesis, was the first living being on the Earth and the progenitor of the biosphere (Altstein, 1987).

In recent years, myriads of theoretical and experimental studies have been dedicated to the problem of the emergence of life. The authors of these works put forward new ideas, analyzed various natural mechanisms, and reported the results of often well-thought-out and well-performed laboratory experiments, which may indeed bear a direct relation to the problem. These are important publications of significant scientific value. We refer those who want to "dive deep" into the current state of the problem to the relatively recent reviews and original publications (De Duve, 1995; Fortey, 1998; Margulis, Sagan, 1997; Schorf, 2000; Jones 2004, Galimov, 2001). Of certain interest are also some older works (Orgel, 1973; Ponnamperuma, 1972; Dickerson, 1978; Folsome, 1979; Goldsmith and Owen, 1980;

Horowitz, 1986), because many problems addressed by the above authors still remain topical. We give a rather short review, because thorough analysis of early and modern works lies beyond the scope of this book, which is dedicated to the particular natural phenomenon and that is why in this review we mention only the works that are of potential interest in connection with the new concept.

We now return to the difficult questions posed above and point out that preliminary answers could have been found at least for some of them. However, we still have to answer the most difficult of these questions and, in particular, to find out how complex was the molecular structure of the primary form of living matter and whether it could have formed via known natural processes. To this end, we must familiarize ourselves with the earlier hypotheses about the first organisms and possible mechanisms of their synthesis.

1.2. Known Mechanisms of the Synthesis of Organic Compounds in Nature

As is well known, new origin-of-life concepts were proposed in the past century by A.Oparin (1924) and D.Haldane (see Bernal, 1967). These concepts were based on the idea of autogenesis, where chemical evolution played the crucial part. A.Oparin described his fundamental concept, whose main idea was to explain the emergence of life "from uncombined elements to organic compounds" and "from organic matter to a living creature", in his book "The origin of life" (1924).

Five years later, in his paper "The origin of life" D.Haldane independently analyzed the important aspects of this problem emphasizing the transmission of hereditary information as an important feature of living matter and proposed the idea that organic compounds are synthesized through the action of solar ultraviolet radiation on the atmosphere.

The works of A.Oparin and D.Haldane, which afterwards played the decisive part in the study of the origin of life, remained little known at the time and had no significant impact on the development of research in this field. However, the researchers continued to study the possibility of abiogenous synthesis of organic compounds and to attempt to recreate a living cell.

For example, during those years A.Oparin proposed the popular and later generally adopted proteinaceous coacervate theory of the origin of life. Its main idea was based on the possibility of spontaneous abiogenous chemical synthesis of protein monomers - amino acids - and their polymers - polypeptides, which were believed to play the key part in the process of the emergence of life.

Because of their high catalytic activity, some protein-like compounds of coacervates may have "simulated" assimilation, growth, and reproduction, which are typical of a living cell. Competently conceived and conducted laboratory experiments appeared to corroborate the proposed theory. However, this hypothesis lacked a mechanism for precise reproduction of accidentally developed efficient protein-like structures. Yet Oparin's works determined to a large extent the future choices in the origin-of-life studies.

We should consider the problem number one at present to be the search for new highly efficient natural factors, which, when acting on typical environments of the early Earth, would make possible abiogenous synthesis of polymeric organic compounds. This problem remains so far unsolved.

At present the possibility of abiogenous synthesis of simple organic compounds is beyond question. This primarily concerns intermediate reactive compounds, basic nitrogens, sugars, lipids, amino acids, and nucleotides. Such organic compounds, which have direct bearing to the process of the preparation for the emergence of life, were synthesized in experiments performed to study the effect of various laboratory analogs of natural factors on inorganic substances.

The central problem of abiogenous synthesis consisted in the lack of reliable and highly efficient natural processes that would allow the available monomers to combine into polymer chains, - the processes without which it is impossible to imagine the realization and development of the stage of chemical evolution. To overcome these difficulties, mechanisms have been proposed for the "concentration" of monomers in the process of the synthesis of polymer organic compounds at the surfaces of various minerals, e.g., clays (Bernal, 1967) or quartz.

One of the important sine qua non requirements to the mechanisms of the synthesis of both simple and polymer organic compounds is that it must be possible for that such a process to occur under the conditions that simulate natural conditions as close as possible. The point is that now in many laboratories organic polymers are synthesized without ferments, e.g., polypeptides. The conditions created in these experiments, which include the use of highly concentrated pure monomers, anhydrous solvents, protection of reactive groups using energy-releasing reagents, and other similar additional efforts and tricks, guarantee success (Horowitz, 1986). However, the starting conditions of these experiments, characteristics of the original substances, and conditions of synthesis are very far from real circumstances on the Earth, and therefore these studies cannot simulate natural processes.

The situation is more or less the same for the protection against phased or instantaneous destruction of synthesized organic compounds in the presence of solar or cosmic radiation, respectively (Shklovskii, 1965). In the former case organic compounds could not be protected against destruction without invoking complex and inefficient mechanisms, which are difficult to develop under natural conditions and which involve vertical circulation of organic compounds together with water in reservoirs. Thus every time the researchers found various hypothetical mechanisms associated with the chemical evolution stage to be difficult to realize under natural conditions, they usually invoked theoretical schemas that had little relation to the problem considered.

C.Folsome wrote about overcoming the problem of the "concentration gap" that "Hypothetically, there are ways to circumvent the concentration gap, but they all appear to be more wishful thinking than plausible facets of reality" (Folsome, 1979).

As far as this is concerned, the well-known experiments of S.Miller and H.Urey (Miller and Urey, 1965) simulating abiogenous synthesis of organic compounds under laboratory conditions were convincing, well prepared, and well conducted. In these experiments, laboratory analogs of natural phenomena – in particular, high-voltage electrical discharge and ultraviolet radiation, - reproduced to a considerable degree the effect of electrical discharges and solar radiation onto the early Earth atmosphere.

The source of problems and inconsistencies in these experiments is the composition of the primeval Earth atmosphere, which, according to the authors, was reducing. At the time many researchers supported this view, because oxidizing atmosphere appeared to prevent efficient synthesis of amino acids. The proposed model of oxygen-free atmosphere was a mix of hydrogen, methane, ammonia, and water vapor. The concentrations of the principal components were constrained to rather narrow intervals and the synthesis of organic compounds in model experiments became impossible should the concentration ratios differ even slightly from their optimum values.

The proposed atmospheric composition model has come under criticism from the very beginning, because the actual composition was unknown (Moroz and Mukhin, 1977; Galimov, 2001). Even now, it remains a subject of debate. Some researchers began to admit that the atmosphere may have been oxidizing and unsuitable for the synthesis of amino acids (Mukhin, 1980).

Early models yielded low volume density for the monomers produced in the process of the synthesis of organic compounds in the early Earth atmosphere. This prevented the subsequent conversion of monomers into polymer structures – the problem that later came to be called the "concentration gap". Moreover, amino acids synthesized in laboratory experiments simulating early Earth conditions showed no evidence for the breaking of isomer symmetry.

Now, about half a century after the first experiments of H.Urey and S.Miller, the researchers regard their results from a somewhat different standpoint. They are often viewed as experiments organized so as maximize the likelihood of obtaining the required final result by choosing the appropriate starting conditions. Furthermore, these experiments are unfairly criticized because of their failure, in spite of the assurances made, to reproduce even the most elementary forms of life.

However, if approached objectively, these experiments lead us to quite different conclusions. Because of the clear and easy-to-grasp nature of the proposed concept, the experiments of H.Urey and S.Miller have triggered the popularity of the origin-of-life problem among the researchers. This was of course due in part to the scientific authority and wide renown of H.Urey, but the most important consequence was that the interest in these experiments determined the direction of further work for more than one generation of researchers. These works demonstrated once again and more clearly that the plasma that forms in a high-voltage electric discharge is capable of synthesizing a number of key amino acids with no additional effort required.

It now becomes clear that electric discharge in a gaseous environment used to simulate the atmospheric composition in a laboratory can be regarded as a highly scaled-down model of the lightning discharge, and one should not expect more complex organic compounds to be synthesized in such small-sized environments. The mass of the compounds synthesized under such conditions may also be limited because of the one-dimensional (linear) nature of the discharge. The masses of the synthesized compounds could be increased several ten- and hundred-fold, resulting in a substantial complexification of the molecules, by increasing the size of the discharge gap. However, this was impossible at the time of Urey and Miller experiments - primarily for technical reasons, and it still remains an important task.

Later works on abiogenous synthesis of organic compounds addressed the problem of the influence of a wide range of various laboratory analogs of natural factors on different environments. Experiments of this kind abandon solar radiation in favor of other, more efficient factors thought to provide the optimum effect on the atmosphere.

Thus Matsu and Abe (1986) analyzed the possible consequences of hypervelocity meteorite impacts for impact degassing of planetesimals and generation of a high-temperature atmosphere. According to this hypothesis, such an impact-generated atmosphere could contain CO and CO_2 - compounds required for the synthesis of amino acids.

In their experiments Miyakawa et al. (1997, 1999) subjected a CO-N_2-H_2O gas mixture residing in a magnetic field to an arc discharge. The effect of the plasma arc was that amino acids had been found along with uracil and cytosine in the hydrolyzate of the final product.

Experiments were also performed (Bar-Nun et al., 1970) to simulate the effect of shocks produced by meteorites penetrating into the tenuous primeval Earth atmosphere. In these experiments several amino acids have been synthesized in laboratory conditions in a shock-treated gas mixture consisting of CH_4, C_2H_6, NH_3, and H_2O.

The results of these works did not differ significantly from those obtained by H.Urey and S.Miller. However, they marked the transition of the research toward more decisive actions, which consisted in replacing the target medium. Thus solid matter – minerals - replaced the model mixture used to simulate the early atmosphere as the target medium, and thermal heating, which resulted in the meltdown and evaporation of the samples, was used as agent instead of solar radiation.

Oro (1961) synthesized nucleic acids and adrenaline in particular by heating a mixture of hydrocyanic acid, ammonia, and water. Ponnamperuma (1963) synthesized adenine via electron-beam irradiation of a mixture of CH_4, H_2, and H_2O. Guanine was synthesized by subjecting a hydrocyanic acid solution to ultraviolet radiation. Nucleosides and nucleotides were synthesized in later experiments.

Considerable interest was sparked by the results of the works whose authors assumed that energy was provided by volcanic heat released during the eruption and the thermal heating of rocks in the vicinity of the meteorite impact crater.

The idea that led the specialists in abiogenous synthesis to think about volcanoes was very simple: organic compounds may form under the conditions of high-temperature volcanic lava made up of inorganic substances. This approach can be justified by the following reasoning: although volcanoes contribute several orders of magnitude less thermal energy than the thermal effect of solar irradiation, volcanic heat may have been more efficient because of its high concentration.

The experiments performed by Fox (Fox and Dose, 1977) to study the effect of volcanic heat on a methane-ammonia atmosphere showed that such reactions may produce a wide range of amino acids. However, the composition of the reaction mixture that Fox used in his laboratory facilities had nothing to do with that of the primeval atmosphere. In his experiments volcanoes served only as the source of thermal energy to power the synthesis of amino acids. However, they could also serve as sources of gas and molten minerals. Moreover, volcanic gases contained all the components required for the synthesis of organic compounds - methane, ammonia, hydrogen, and carbon monoxide. Could volcanic outbursts of gases and lava – including underwater outbursts – facilitate the synthesis and accumulation of organic compounds needed for the emergence of life?

Studies were carried out to simulate land and underwater volcanoes in laboratory conditions, and measurements were made in volcanic conduits (Mukhin, 1974). The results obtained led the authors to conclude that volcanic eruptions may produce cyanic acid, aldehydes, and amino acids, which, in turn, may serve as the building blocks for the required macromolecules.

Mukhin et al. (1989) and Gerasimov et al. (1991) were the first to demonstrate the possibility of the formation of organic compounds in the case of an impact heating with the evaporation of rocks by the heat released in meteorite impacts. The above authors believe that heating of the matter up to 3000—4000 K should result in the formation of a gas-vapor cloud. A 1.06-μm neodymium laser with pulse duration of ~1ms is used to model this process in the laboratory (Gerasimov et al. 1991). Such duration of the laser pulse is ~10^5 times longer than needed to produce a bona fide vapor cloud and for thermodynamic equilibrium to be reliably established in the process of evaporation.

According to Gerasimov et al. (1991), in low-velocity impacts only up to 30% of the matter of impacting bodies are converted into the gaseous state creating a mixture of oxidizing and reducing components and carbohydrates with 0.1% admixture of hydrogen cyanide and acetaldehyde. In the experiments mentioned above, which involved thermal heating, a great variety of reactive organic compounds combined with high temperature make the formation of the compounds reported by the authors quite likely. However, the above authors did not consider plasma processes occurring in low-velocity impacts, whereas no plasma was generated in laboratory experiments performed to simulate impacts. The results of these studies are described and discussed in detail in Section 1 of the Afterword.

The idea of the new scenario of the emergence of life was inspired by the discovery of anaerobic microorganisms in the depths of the ocean. Colonies of microorganisms whose metabolism requires no solar light have been found near underwater volcanoes – the so-called "black smokers". Unlike the authors of the early scenarios based on the classical experiments of S.Miller and involving an external energy source, G.Wächtershauser (1992) suggested that the energy required for the creation of primitive organisms could be released via reducing reactions involving sulphides of metals – iron and nickel in the first place. He found that in the presence of a sufficiently high concentration of the constituent elements of organic compounds such energy is sufficient not only for the synthesis of simple molecules, but also for the formation of oligomers and polymers. A hypothesis has been put forward suggesting that metabolically active structures capable of self-replication can form in autocatalytic mode in the "Wächtershauser system". However, actual experiments showed the efficiency of the synthesis of dimers and trimers of amino acids to be extremely low under anaerobic conditions. These results so far prevent a definitive conclusion concerning both the viability of the main idea that life may have originated under water and its prospects based on the mechanisms mentioned above. However, we cannot rule out completely such a possibility.

Among the relatively new scenarios of the emergence of life the "metabolism-first" scenario proposed by R.Shapiro (1984, 1995) has become especially popular in recent years. It assumes that in nature the synthesis of a big replicating RNA molecule is unlikely and therefore reactions between small molecules must have played the crucial part in the emergence of life. According to this model, in the beginning was metabolism and the emergence of life started with spontaneous compartimentation. Some of the compartments formed contained groups of molecules, which entered into chemical reactions and formed cycles. With time, the latter became increasingly complex and made possible the synthesis of polymer molecules – stores of information. Despite the relative popularity of this concept, experimental proofs of its realizability are rather slow to obtain. Its main idea is similar to that of the Wächtershäuser hypothesis (Wächtershäuser, 1992). In both cases experimental tests of the concepts are scanty and the available experiments are limited to the demonstration of the realizability of a certain part of the closed cycle. These experiments remain fragmentary and far from completion, although the main idea of the "small molecule world" concept – an alternative to the RNA world – appears to be quite realistic. However, the results of laboratory modeling show that both concepts are actually at the limit of their capabilities and have no sufficient "safety margin" to allow the synthesis of substantially more complex molecules.

S.Fox performed experiments on abiogenous synthesis of polymer organic compounds (Fox and Dose, 1977; Fox and Nakashima, 1980) using pure and dry amino acids as primary materials instead of inorganic compounds. Heating these amino acids under various conditions to temperatures in the interval from 100 to 170 C resulted in the formation of a polymer substance called protenoid with a molecular mass ranging from 1000 to 30000 atomic mass units. The importance of the results of such a limited or intermediate synthesis is that the organic compounds obtained exhibited experimentally detectable zymogenic activity.

According to the current ideas, the most important processes are those that lead to the synthesis of oligonucleotides. Some of the results demonstrate the synthesis of polymer chains of oligoribonucleotides consisting of up to 40 monomers (Ferris and Ertem, 1993; Ferris et al., 1996, 2004; Huang and Ferris, 2003). In these experiments the synthesis was performed in an aqueous medium using clay-type surfactant mineral catalysts (monymorillonite) made of nucleoside-phosphoramidates, and resulted in the formation of 3'-5' inter-nucleoitide bonds.

The possibility of the synthesis of organic compounds in the protoplanetary cloud was theoretically confirmed by B.Parmon (Parmon, 2002, 2007) and his colleagues. The above author used computations and numerical simulations to show that gas-dust protoplanetary clouds provide all the conditions, organic building-block elements, and simple chemical compounds that are necessary to make possible the synthesis of various organic compounds. Parmon named the presence in the medium of catalyst particles containing iron, nickel, and silicon as the must-have condition for the synthesis.

This approach allowed the researchers to develop the so-called "catalytic" hypothesis according to which the synthesis of primeval organic compounds and the formation of planets are two sides of the same coin. Hence for the Earth the problems of the origin of life and formation of planets converge to within the same time interval.

As far as astrocatalysis is concerned, of considerable interest is also the hypothesis proposed by V.Snytikov (Snytikov, 2007a, 2007b), who suggested that organic compounds could be synthesized both in the pre-planetary circumstellar disk and in the process of the formation of the Earth, and could proceed through the development of collective instability, i.e., during a simultaneous merger of a very large number of small bodies.

These organic compounds may have arrived to the Earth's surface undestroyed by high temperatures during the passage of the cosmic body through the atmosphere provided that these compounds were located in larger-than-dust particles or the meteorite fell into a deep water reservoir (Shuvalov et al., 2008).

Once the main mechanism of synthesis is identified, the possibility of the formation of complex polymer organic compounds in the circumstellar disk needs to be confirmed by experiments, which are very difficult to perform. This is of further importance because the pre-planetary circumstellar environment remains, like plasma, a poorly studied substance, which may have many other unknown and unpredictable properties. Therefore such environment with so far uncovered properties merits the most detailed investigation.

Ostrovskii and Kadyshevich (2007) propose a "hydrate" hypothesis of the origin of the living-matter simplest elements. According to this hypothesis, the so-called living-matter simplest elements originated repeatedly and, maybe, even now continue to form from methane, niter, and phosphate within boundary layers of the solid phases of gaseous hydrates of the hydrates of the simplest hydrocarbons. Despite the purely theoretical nature of their work, the above authors also discuss the possibility of testing the hypothesis by conducting a rather hard-to-perform experiment. The mechanism of the origination of homochiralic macromolecules is not yet entirely understood and it is not clear what factor determined the polarity "sign" of the biological world.

It follows from the above that natural processes that occurred on the primeval Earth and that are now reproduced in the laboratory involved abiogenous synthesis of gases, reactive intermediate products, small organic compounds and amino acids in particular, fatty acids, carbohydrates, nitrogen bases, and short polymer chains of some monomers. Thus highly efficient synthesis of amino acid monomers occurred during a discharge in gas, and they polymerized in the processes of further heating of monomers. Organic compounds so obtained, which S.Fox called protenoids, exhibited certain enzymatic activity despite the fact that they differed substantially from proteins. Such results inspire hope.

To conclude this section, let us briefly discuss the specific features of the results of plasma experiments involving the synthesis of organic compounds.

Note that because of their high catalytic activity plasma formations proved to be among the most efficient environments for the synthesis of a wide range of organic compounds. This has been repeatedly confirmed by various experiments beginning with 1906, when Löb (1906) synthesized aldehydes and glycine in a glow discharge for the first time.

The possibility of the synthesis of organic compounds in plasma formations has later been convincingly confirmed. The most reliable results are those obtained in the experiments of S.Miller (Miller, 1953). As we mentioned above, up to five amino acids were synthesized in these experiments with simulated lightning, which is a kind of a natural plasma discharge.

Recently, J.Bada (Bada, 2003) reenacted Miller's experiment performed in 1953 and analyzed the resulting substances using modern analytical instruments. The results of the new measurements revealed much more – at least 22 – amino acids and, in addition, amides (Johnson, 2008).

C.Simionescu and F.Denes (Simionescu and Denes, 1986) made an attempt to synthesize organic compounds in experiments simulating the conditions in the cold ionospheric plasma. They managed to synthesize and identify a number of key origin-of-life related organic compounds. In particular, besides amino acids they identified nucleic acid bases, porphyrins, polysaccharides, and lipids.

Of interest from this viewpoint are the results of Miyakawa et al. (1997, 1999) mentioned above. These authors synthesized amino acids along with uracil and cytosine by subjecting a gas mixture consisting of CO, N_2, and H_2O to an arc discharge.

These and similar works show that steady-state and pulsed natural plasma processes can be expected to result at least in the synthesis of monomers of organic compounds required for the formation of more complex polymer molecular structures.

The predominant synthesis of monomers in the above experiments must be due to the fact that the corresponding plasma formations were created in a gaseous environment and therefore the concentration of synthesized monomers was insufficient for the subsequent formation of extended polymer chains in continuous processes in a tenuous plasma environment.

However, the importance of the results obtained is beyond question, because they have demonstrated the possibility in principle for polymer organic compounds to be synthesized in the case of higher density of the primary matter, which, in turn, could ensure high concentration of monomers in plasma formations that form when they are subject to natural or artificial external factors. However, the major problem is that all synthesized organic compounds had very little in common with the building blocks of complex polymer macromolecules, whereas the earlier proposed mechanisms were incapable of providing the required high density of their localization.

1.3. Living Systems from the Ancient RNA World

From time to time, descriptions of a hypothetical model of the most primitive living organism appear in the scientific literature. Thus in his book published in 1979, C.Folsome (Folsome, 1979) tried to describe a model of the living "protocell" matter in its most primitive version. According to Folsome's hypothesis, first proteins of such a creature may have contained five to seven amino acids, and primary amino acids, 10 to 12 bases. Folsome also analyzed a possible process of the transition of inorganic matter to life. Such a hypothetical creature would fit into a sphere of about 5~nm in diameter (without walls). Its molecular mass must have been no greater than 6 kDa. The idea to describe the most primitive form of living matter may have been inspired by the interest in the RNA world idea that was becoming increasingly popular at the time.

The main idea of the RNA world hypothesis is that RNA molecule is, by the combination of its most important properties and characteristics, the best candidate compound capable of ensuring the development of living matter. The accumulated knowledge about the properties of nucleic acids led the well-known researchers – originators of molecular biology F.Crick, C.Woese, L.Orgel, and W.Gilbert (Crick, 1968; Woese, 1967; Orgel, 1968; Gilbert, 1986) – to this fundamentally new idea.

The discovery of catalytic properties of RNA in biological systems (Kruger et al., 1982; Guerrier-Takada et al., 1983,) and a number of other works provided a powerful stimulus for the development of the RNA world idea. The concept of the RNA world began to be viewed as the idea of a self-sufficient world – precursor of life (Gilbert, 1986; Joyce, Orgel, 1993). Over time, the results obtained by Cech and Bass (1986), Woese (1998), and Orgel (1998) allowed the concept of the RNA world to acquire a number of exceptional advantages and, after having resolved many controversies, to become the dominant origin-of-life hypothesis.

A.Spirin was another important contributor to the study of the RNA world (Spirin, 2001, 2003, 2005). He suggested, from the standpoint of the self sufficiency of the RNA world, an original version of the process of the RNA evolution that leads to the development of the protein biosynthesis mechanism. Spirin showed in the above works that RNA is capable of performing all or almost all functions required for the emergence and existence of living matter. He also proved that abiogenous synthesis of ribonucleotides and their covalent grouping into oligomers and RNA-type polymers could occur under the same conditions and in the same chemical environment as those postulated for the formation of amino acids and polypeptides. The main ideas were corroborated by the results of the experiments performed by Chetverina et al. (1999), which demonstrated the possibility of spontaneous rearrangements of at least some polyribonucleotides (RNA) in common aqueous media. Such a recombination could have resulted in longer polyribonucleotides and possible development of catalytic activity of these molecules.

A.Spirin (Spirin, 2007) analyzed the hypothesis about the ancient RNA world as the primary form of life that was the predecessor of the modern RNA world based on three polymers: DNA, RNA, and proteins (Woese, 1967; Crick, 1968; Orgel, 1986; Gilbert, 1986). The evolution of the ancient RNA world could have led to the development of the protein biosynthesis mechanism, DNA-based specialized genetic apparatus and, ultimately, cell organization of the living matter (see Joyce, 2002; Orgel, 2004; Joyce, Orgel, 1999, 2006; Spirin, 2001, 2003).

Ensembles of molecules of the ancient RNA world may have been self-sufficient formations maintaining their proper existence, growth, and reproduction, i.e., analogs of living creatures, but consisting of single kind of polymers exclusively. Such formations could be considered the primitive form of life of the ancient RNA world, but it remained to determine their characteristic size, structure, and functional capabilities.

Spirin (2007) also analyzed the initial conditions of the environment required for the existence, amplification, and evolution of the ancient RNA world, the problems facing abiogenous synthesis of a single-polymer life form, and paradoxical situations arising in relation to its stability, functioning, and the place of the RNA world in the geological history of the Earth.

In his paper Spirin discussed:

- the water paradox, i.e., the incompatibility between the chemical instability of the covalent RNA in the aqueous environment and the need for water for the formation of the functionally active conformations of this structure;
- the conformation paradox, which consists in the incompatibility between the stable double-stranded RNA structure required for its replication and stable compact conformations of single-chain RNAs needed for catalytic functions;
- the geological paradox, which consists in the too short duration or total lack of the time interval in the geological history of the Earth between heavy meteorite bombardment and the emergence of the first evidence of life found in terrestrial rocks.

The concept faced other problems as well. Spirin (2007) writes, citing the results of Orgel (2004) and Joyce and Orgel (2006), that "The conclusion that can be reached based on all available data is not encouraging; despite many attempts to simulate various conditions of the primeval Earth, the complete abiogenous synthesis of nucleotides, which are the components (monomers) of RNA, have not been successful"

In Spirin's opinion, no solution whatsoever has been found for the problem of chirality: products of abiogenous synthesis of nucleotides were always a racemic mixture, whereas true RNAs can form only from homochiralic mixtures.

The situation is not better for nonenzymic polymerization of nucleotides. Certain achievements in this direction (Ferris and Ertem, 1993; Ferris et al., 1996; Huang, 2004; Ferris, 2003), where phosphoramidate substrates were used for the synthesis of oligoribonucleotides consisting of 40 monomers, proved to be illusive, because the latter appears to have been hardly present on the primeval Earth.

The above problems and other difficulties arising in the attempts to overcome the complexities of abiogenous sythesis of RNA led many researchers including L.Orgel to a new idea. It consists in the assumption that RNA itself did not form abiogenically, but was "invented" by a more simple system that predated the RNA world, and this early world could have existed under more extreme conditions (Orgel, 2004; Joyce and Orgel, 2006).

The researchers suggested that possible predecessors of RNA could have been their artificial analogs - peptide nucleic acids (PNA), which have a polypeptide rather than a sugar-phosphate backbone (Egholm et al., 1992, 1993). However, neither of these structures proved to be less complex than RNA, and the replacement of a genetic structure by another genetic structure appears to be unlikely.

Other extra conditions for the existence and evolution of the RNA world considered by some researchers include the presence of RNA-adsorbing surfaces with drying and heating-and-cooling cycles and, finally, the invariable presence of some mechanisms to protect RNA against cosmic radiation or ensure its temporary conservation.

All these "paradoxical" controversies led A.Spirin to conclude that the emergence and existence of the RNA world, as well as its evolution into cellular forms of life on the Earth was highly unlikely and made him turn to the idea of the cosmic origin of terrestrial life.

One of the versions of such a model assumes that life originated in cometary nuclei (Hoover and Rozanov, 2002; Hoover, 2006).

According to A.Spirin, such a transfer of the RNA world from the Earth into space is due to problems of a different kind. The conditions in various parts of the Universe are practically unknown and this forces us to adopt the first cellular forms of life that originated from the RNA world as "as the creation of some unimaginable conditions and forces, the products that were supplied ready-made to the Earth, and probably to some other planets and bodies in the Solar System."

It follows from the above that the first primitive form of life of the ancient RNA world was a single-polymer structure consisting of RNA molecules exclusively. Its main molecule contained nucleotides with an average total number and mass of up to 200 and 70000 atomic mass units, respectively. The ensemble of such RNA molecules was possibly an autonomous formation sustaining its own existence, growth, and reproduction, and it may have been an analog of a living system. The time of their appearance can be, with certain reservations, identified with the time of the emergence of life. Such formations can, with certain reservations, be viewed as the first form of living matter of the ancient RNA world.

It is important to note that these formations could neither originate nor exist if the conditions on the Earth allowed the realization of the paradoxes mentioned above. However, the origination of the first forms of life of the ancient RNA world cannot be ruled out under other conditions, e.g., in the absence of the "water paradox".

1.4. Protoviroid – The Progenitor of the Biosphere

Altstein in his paper (Altstein, 1987), which A.Spirin (Spirin, 2001) characterizes as "elaborate and attractive" (Spirin, 2001), proposed a model of the first living being on the Earth – the progenitor of the biosphere. According to this hypothesis, the first living system consisted of two molecules – a polynucleotide and its coded protein (processive polymerase).

The system reproduced via the unique process of replicative translation based on template principle similar to that of the modern biological world, and evolved in accordance with the Darwinian principle “heredity – mutability – natural selection”.

To explain the emergence of such a rather complex system, the hypothesis postulates the principle of “progenes” – mixed anhydrides of a nonrandom amino acid and 3’-gamma-phosphade trinucleoitides. According to the main postulate of the hypothesis, progene forms via bonding of dinucleotide (DN) and aminoacyl-nucleotide (AAN) so that the amino acid of the latter specifically interacts with DN facilitating the formation of unstable “triplet” (without the covalent bond between DN and AAN). Two such “triplets“ overlap complimentarily. Such an interaction improves the stability of the complex and facilitates the formation of the phosphorus-ester bond between the second and third nucleotides. The result is the formation of a progene – trinucleotide with nonrandom amino acid bound to the gamma-phosphate of the third nucleotide, and a set of nucleotides capable of complementary interaction.

The chemical structure of progenes is such that they can serve as a substrate for the simultaneous synthesis of polypeptide and polynucleotide. The first genetic system (single gene, single polypeptide, processive polymerase, progenligase) forms from progenes as a result of a very rare event. Association of molecules (always involving the template principle) results in the simultaneous formation of oligonucleotide and oligopeptide, where the latter facilitates the growth of the system. If after the addition of a progene the structure of the system is favorable for the addition of the next gene, the system continues to grow, otherwise the growth halts and the system breaks up.

Thus an enzyme (a processive polymerase using progenes, i.e., a progenligase) forms in close association with its substrate. The growth of a substantially long nucleic-acid molecule is ensured by the enzyme that grows at the 3’-end of the chain and is coded by the extending nucleotide sequence. The resulting genetic system then undergoes replication based on progenes, which serve as protein synthesis adaptors similar to the transport ribonucleic acid and, at the same time, as the substrate for the synthesis of a complimentary nucleotide chain and the polypeptide coded in the template.

This hypothesis offers a unified solution for a number of hitherto unsolved problems. In particular, the mechanism of the formation and association of progenes allows us to understand:

- natural selection of monomers and the origin of the homochiralic nucleotide and almost homochiralic amino-acid sequences; the fact that nucleotide has to be activated at its 3’-end leads us to conclude that nucleotide components of the first genetic system must have been mostly of the DNA type;
- the emergence of the physicochemical group genetic code; stereochemical proofs have been obtained that indicate a certain correspondence between the code based on the progene hypothesis and modern genetic code (Altstein and Efimov, 1988);
- origination of the “genotype-phenotype” association in the process of the development of the genetic system;
- description of the nature of the first self-reproducing genetic system and the first molecular-genetic process (replication – transcription – translation, or replicative translation).

In other words, Altstein analyzed a model of the first living being and the process of its appearance on abiogenous Earth. This being, as shown in Figure 1, was a nucleoproteid whose protein part consisted of ~100 amino-acid residuals (processive polymerase), and was coded by a single-stranded deoxyribopolynucleotide consisting of ~300 nucleotide residuals, each with a 3'-phosphate end group. The system has the main characteristics of a living being – the ability to replicate and evolve based on the template principle. At this stage of evolution the two most important components of living matter (compartimentation and metabolism) are provided by abiotic nature. To appear and replicate, such a living system needs nothing but progenes and conditions where they can arise. The system has a virus-like (protoviroid) nature, it migrates from one unstable abiogenous source of progenes to another (we are rather dealing with liposomal structures at the time of their origin). Such a system is capable of biological evolution with natural Darwinian selection and can be considered to be the first living being on Earth – the progenitor of the biosphere. An analysis of the properties of the protoviroid leads us to conclude that it is capable of stable replication of nucleotide and amino-acid sequences based on the template principle. According to A.Altstein, the "error catastrophe" principle (Eigen and Schuster, 1979) can be circumvented in this case by using triplets rather than monomers for replication, and because of the presence of primitive processive polymerase. The same factors make it possible to achieve polynucleotide homochirality.

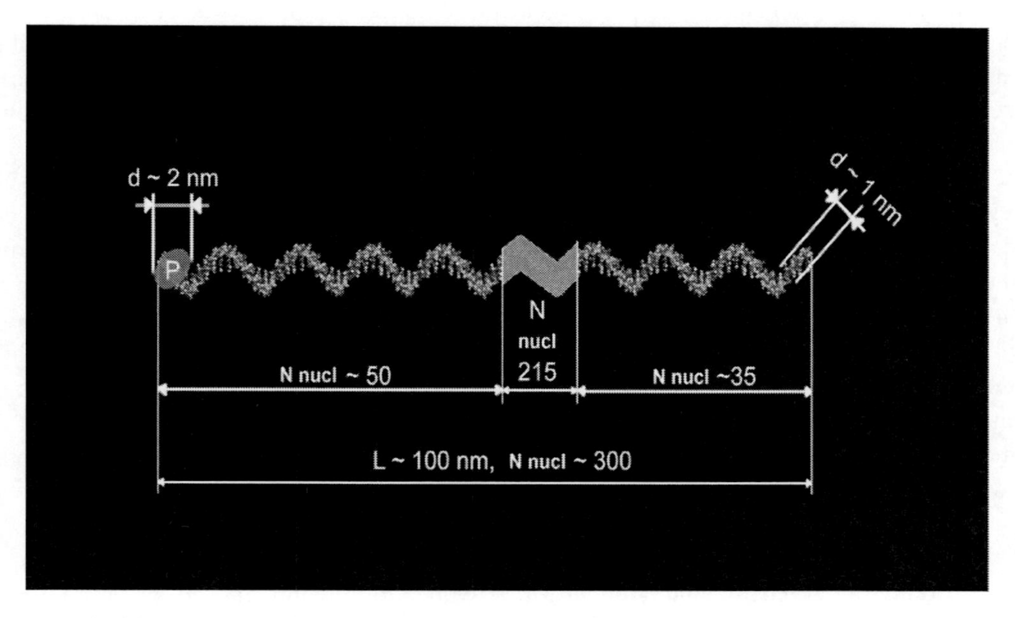

Figure 1. The block diagram of Altstein's protoviroid consisting of a single-stranded polynucleotide including ~300 monomers and procesive polymerase (it is designated by the letter P), which, in turn includes ~100 amino acids. The molecular mass of such a formation is ~100 000-120 000 a.m.u. Color spheres in the polynucleotide correspond to the following elements: carbon (blue), oxygen (red), nitrogen (dark blue), hydrogen (white can be seen if zoomed in), and phosphor (yellow), and their arrangement reproduces the structure of the polynucleotide. The progress of the polymerase along the strand from left to right (from the 3' to the 5' end) results in the replication of complimentary polynucleotides and formation of polymerase with the order of amino acids determined by the order of triplets in the polynucleotide, i.e., in the emergence of the process replicative translations.

Thus A.Altstein's hypothesis indicates a way that differs fundamentally from the idea of the RNA world. According to Altstein, chemical evolution at its final stage was capable of producing the most primitive living substance on Earth based on understandable mechanisms that are quite realizable in nature, and thereby mark the start of biological evolution.

It is also of great importance that such substance could have had minimal molecular mass of ~100000 a.m.u. and consisted of ~100 amino-acid residuals and ~300 nucleotides. A macromolecule of such a characteristic size would fit into a cylindrical volume with a length and diameter of ~100 nm and ~10 nm, respectively.

A serious disadvantage of this hypothesis is that so far only stereochemical and no experimental evidence is available to prove its realizability in nature.

1.5. Main Difficulties

It follows from the previous section that the constituent molecular structures of the simplest forms of living matter, and oligonucleotides and polypeptides in particular, must contain a certain sequence of monomers with established correspondence between their alternation. Such a correspondence conforms to a genetic code. These molecules must be homochiralic and have exceptional properties characteristic of biological level of complexity. However, according to generally agreed statistical estimates, such molecular structures, which arise abiogenically in nature, could not form as a result of random trials of possible combinations. The existence of life on Earth indicates that the simplest form of life must have arisen abiogenically, but we are so far unable to identify the prerequisite conditions, processes, and mechanisms.

Avetisov and Goldanskii (1996b) provide important information concerning this issue. They consider two possible ways of the formation of homochiralic structures – of chemical and biochemical levels of complexity. According to the estimates reported by the above authors, whenever the length (N) of the polymer does not exceed a couple of dozens (up to N=40), conditions of the chain formation can be so chosen that the resulting polymer chains will involve the full spectrum of possible sequences of links, including the "right", homochiralic sequences of monomers. Avetisov and Goldanskii refer to macromolecules of such length as the structures of the chemical level of complexity. Hence the formation of homochiralic structures of the chemical level of complexity requires no specific functions. However, such short molecular structures are incapable of ensuring the formation of homochiralic macromolecules of the biochemical level of complexity with the specificity required for the production of the simplest forms of living matter.

According to Avetisov and Goldanskii (1996b), for chains of 150 or more units, statistical constraints are rather severe. In the case of a chain of ~150 monomer units 10^{40} molecules are needed to obtain the "right" molecule (which must also be homochiralic), and this is comparable with the number of bioorganic molecules on the Earth. This implies that any existing sequence involving 150 links or more is unique, because most of them could not have formed at all. Such a level of molecular complexity, which the above authors call biological, is typical of enzymes, RNA, and DNA. The fraction of realizable sequences is always extremely small, whatever the physical or chemical conditions may be. In other words, homochiralic macromolecules consisting of N=150 or more units cannot arise as a result of random trials of all possible combinations. Any hypothesis claiming to explain the mechanism of the origin of life must respect this constraint and explain how it can be circumvented under natural conditions.

Thus the constraints considered above prevent the realization of abiogenic synthesis of molecular structures with the properties required for the formation of the simplest organism. Moreover, despite a great number of hypothetical processes of the synthesis of such structures suggested by the researchers, no natural mechanisms have been found so far that could be capable of overcoming these constraints to ensure the assembly of homochiralic molecules with the level of complexity and the minimum set of chiral defects matching the information and functional level of real biological objects.

In one of the sections of their paper Eigen and Schuster (1979) consider the constraints due to the "errors catastrophe". They describe the problems arising when the number of errors and defects in the process of the assembly of a homochiralic structure exceeds a certain threshold value. If this threshold is exceeded, the equivalent amount of useful information gradually decreases in successive replications, and the ultimate result is the complete loss of genetic information.

Eigen (1973) tried to overcome the above constraints and difficulties. He was the first to formulate the concept of the formation of ordered molecules from unordered matter based on template replication and subsequent selection. In this paper Eigen for the first time applied an analog of Darwinian selection to the chemical stage of evolution and to the process of abiogenic synthesis of molecules that make possible the emergence of the simplest organism.

The new concept was based on the hypothesis of catalytic cycles, which, in turn, was based on the idea of "cross-catalysis" (Eigen and Winkler, 1975). In this model nucleotides can produce proteins, which, in turn, produce nucleotides. The resulting cyclic reaction pattern, which was later called the hypercycle (Eigen and Schuster, 1979), demonstrates the principle of natural self-organization of molecular structures. Hypercycles are believed to be capable to become more structurally complex in the process of competition while undergoing mutations and replication. The above authors linked the proposed model of "realistic hypercycles" to the origin of molecular organization of the apparatus that ensures primitive replication and translation. Note that Eigen and Winkler were the first to use the apparatus of the qualitative theory of dynamical systems.

The concept of M.Eigen provided a general principle of selection and evolution on the molecular level based on the stability criterion of the thermodynamic theory of steady states. This concept established the qualitative base for laboratory experiments on evolution. It could also show the way how to construct simple molecular models of possible precursors of living matter.

M.Eigen believed (Eigen, 1973) that evolution began with random events in the molecular chaos of primeval Earth. "Realistic hypercycles" were to make it possible to overcome the difficulties and constraints mentioned above.

The model proposed by Eigen proved to be vulnerable despite the mathematical and thermodynamic validity of hypercycle and the ingenious scheme of its operation. Hypercycles could not be reproduced in laboratory. Moreover, they fundamentally differ from a living cell and this restricted substantially the possibility of the transition from hypercycle to cellular structure.

D.Chernavskii (Chernavskii, 2000) suggested an ingenious solution to circumvent the difficulties associated with the origin of the simplest living beings from nonliving matter.

According to Chernavskii, these difficulties arise from the misconstruction of the term 'coding'. They are overcome once we abandon the literal understanding that *in primary organisms (hypercycles) "polynucleotide coded protein-replicase*" in favor of a broader interpretation: *"polynucleotide catalyzed the formation of protein-replicase"*.

The proposed stage-by-stage scenario of such a process involves (1) the formation of a protein, which is a replica of a DNA molecule and which is, like modern replicase, capable of catalyzing DNA replication, and (2) the emergence of adaptor proteins. Note that in this case the maximum length of a random DNA may be as small as 30 nucleotides or even less (Chernavskaya and Chernavskii, 1975; Romanovskii et al., 1984). Hypercycle, or the simplest version of modern biosynthesis, which contained the above ingredients, may have combined the processes of codeless protein synthesis and coding, which are similar to present-day biosynthesis. This opened up a wide range of possibilities for further biological evolution. According to D.Chernavskii, the proposed scenario is free from the problem of the low probability (due to the "errors catastrophe") of the formation of the primary hypercycle from nonliving substance. Unlike Eigen, Chernavskii explains the existence of only one version of the genetic code by assuming that it was chosen (not selected) as a result of interaction between different populations. Chernavskii supports this hypothesis by a detailed discussion of the validity of such a choice.

In his analysis of the problem of the symmetry breaking D.Chernavskii (Chernavskii, 2000) concludes that the "sign" of the asymmetry of the biological world is the result of a random choice. He also believes that "racemic organisms" may have existed, which did not survive, because they may have been too unwieldy and unadapted for the struggle for survival. Symmetry was broken during, and not before or after, the formation of the primary hypercycles. Chernavskii introduced important corrections, which reduced the number of nucleotides in DNA to 30. This made some of the constraints much easier to overcome. However, this original hypothesis lacks experimental support.

According to the hypothesis of A.Altstein (Altstein, 1987), a ~300 nucleotide single-stranded DNA was needed for a protoviroid to form from progenes. Given the above constraint, it is difficult to imagine how such a long macromolecule with the "right" sequence of nucleotides could be synthesized without biological mechanisms.

Abiogenous synthesis of such a complex macromolecule could be made possible by the scheme proposed by A.Altstein, where the origination of homochirality in the process of the formation of progenes was facilitated by steric interactions between amino acid and dinucleotide, and by prebiotic selection of the "right" monomer sequences with "wrong" sequences denied the possibility of further growth. A.Altstein, who is of the same opinion as D.Chernavskii, believes that the "sign" of the asymmetry of the protoviroid is a result of a random choice.

Galimov (2001) focuses on searching a solution for the problem of evolutionary ordering. His analysis is based on the principle of maximum production of entropy, which is valid only in linear nonequilibrium thermodynamics and applies to processes that are not too far from equilibrium. Closeness to equilibrium is understood as significant reversibility of the processes that approach equilibrium. E.Galimov believes that chemical equilibrium means the death of a living system and therefore life is continuous struggle with the tendency for the transition to the equilibrium state.

Despite the apparent incompatibility of the closeness to equilibrium with the highly ordered and highly nonequilibrium nature of biosystems, close-to-equilibrium processes may dominate in many events during the chemical and biological stage of evolution. That is why the theoretical models of the origin of living matter discussed above chose in favor of steady-state thermodynamic systems of irreversible processes. Such processes are characteristic for a considerable number of natural systems of certain type. This approach is valid only for linear nonequilibrium thermodynamics, which describes processes that are not very far from equilibrium.

There is, however, another point of view on this problem. The fundamentals of this approach were developed by I.Prigogine and other representatives of the Brussels school. According to their concept, quite different processes may also occur in nature, and nonequilibrium systems or dissipative structures may develop that are far from thermodynamic equilibrium and that are capable of establishing "order via fluctuation" and break symmetry via loss of stability and bifurcation (Prigogine, 1980). Many renowned researchers associated macromolecular self-organization and breaking of mirror symmetry with the processes of this kind and the theory still remains popular and topical.

Thus, according to E.Galimov, many events that occur during evolution can be viewed as bifurcations, as structural orderings caused by nonlinear processes far from equilibrium. "However, these events are only singular points on the path of evolution" (Galimov, 2001).

We are now short of ideas on the possible mechanisms of the formation of the simplest form of life, and the available information about the properties and capabilities of the processes occurring at such "singular points" remains very scarce. We can hope that at a "singular point" the potential capabilities in energy-intensive nonequilibrium processes would be much ampler than those of equilibrium processes. Therefore having studied the main characteristics of a "singular point" we should not miss the chance to use it to bridge the gap between nonliving and living nature. The attempts to achieve it within the framework of linear nonequilibrium thermodynamics in the processes not very far from equilibrium have so far been unsuccessful.

To perform such studies, one has to: (1) find a natural phenomenon with the properties typical of dissipative structures and (2) correctly reproduce such a phenomenon under laboratory conditions in order to thoroughly study its properties. Note that the processes of ordering, symmetry breaking, and synthesis of homochiralic macromolecules in dissipative structures needed for the origination of the most primitive organisms (Prigogine, 1980) have not been experimentally studied earlier.

Multivariate analysis performed with the results of earlier published works taken into account makes it possible to reconstruct to a first approximation the model of events that may have involved the emergence of the primary form of living matter. Figure 2 shows schematically the interrelations between the principal processes and their main properties. Below we describe it in more detail.

The existence of life on the Earth implies that there must be a natural phenomenon that shapes the environment where the energy of this phenomenon may have powered the synthesis of complex polymer organic compounds in nonbiological processes. The synthesis of these compounds must have been accompanied by the ordering of matter and prebiotic selection of the molecular structures produced. Furthermore, the properties of the environment must have ensured a weak breaking of symmetry, supposedly, due to local chiral physical fields, which may have later determined the "signs" of the asymmetry of the bioorganic world and spontaneous symmetry breaking needed for the formation of homochiralic macromolecular structures. It is only after the realization of these processes that one may expect the emergence of the first primary forms of living matter whose origination hypotheses we discussed in the previous sections. The emergence of the first primary forms of living matter can be considered to be the time of the origin of life. However, for the emerging forms of living matter to survive, some natural phenomenon must provide an environment with a moderate temperature with high concentration of water and organic compounds. Hence at the final stage of the process the emergent living matter must be located in an environment that would ensure its "feeding" and, consequently, "survival".

An analysis of published origin-of-life studies shows that until recently, no natural phenomenon with the properties mentioned above has been known to generate environments with the required properties.

1.6. Meteorite Impact and the New Concept

We suggest that meteorite impact may be a natural phenomenon that would meet the conditions formulated at the end of the previous section. The plasma torch produced by the impact of a meteorite creates the conditions for the synthesis of complex organic compounds and for the formation of a highly nonequilibrium environment, which makes possible the development of local chiral physical fields. Moreover, we show below that torch plasma qualifies as a dissipative structure, where intense fluctuations may result in a spontaneous symmetry breaking.

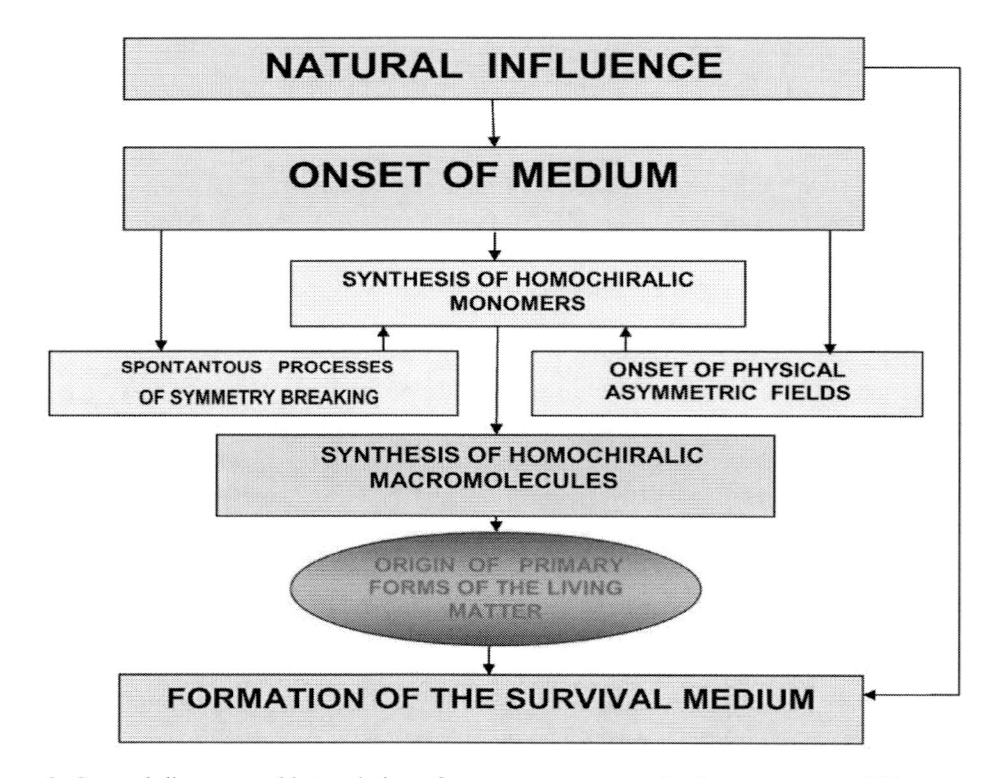

Figure 2. General diagram and interrelation of processes necessary for the emergence of life.

However, it will be useful, before we start a detailed analysis of complex physical processes that occur in a plasma torch, to have a brief glance over the main concept of this book. We assign a crucial part to the hitherto unknown properties of torch plasma and its possible relation with the emergence of living matter – the relation whose discovery was made possible by a special approach to the problem.

Thus the dominating approach at the early stages of the origin-of-life studies, which have about two centuries-long history, can be viewed as empirical.

Even the first such studies can be considered to be scientifically grounded. However, the researchers at that time believed that if all the key organic compounds needed for the creation of living matter could be reproduced in laboratory analogs of natural effects, the combined presence of such compounds, e.g., in warm lakes or in the ocean, should be sufficient for life to emerge.

The lifetime of the Universe was then believed to be infinite implying that the event - formation of molecular structures satisfying the criteria of living matter – may have occurred "sometimes and somewhere" in such a substrate.

That is why at that period the researchers focused on demonstrating that various organic compounds of greatest importance for life could be synthesized from nonliving substance. Note that these works continued to be relevant until mid 20th century and they achieved considerable progress. We already showed in the previous sections that many laboratory simulations of natural factors including those involving the generation of various plasma formations have demonstrated the possibility of the synthesis of practically all organic compounds or monomers whose combination would make it possible to create the molecular structures required for the emergence of life.

Hence within the empiric approach of the time no one doubted that abiogenically synthesized key monomers could combine into macromolecules.

By the second half of the 20th century it became clear, owing to the results obtained by M.Eigen, I.Prigogine, D.Kondepudi, V.Goldanskii, V.Avetisov, D.Chernyavskii, and other researchers, that mirror symmetry breaking is the crucial factor that makes possible the emergence of living matter. It was shown that the emergence of living matter is impossible to implement without homochiralic macromolecular structures being produced in abiotic processes. Moreover, the emergence of life was found to require global breaking of mirror symmetry, supposedly in spontaneous processes. Processes that ensure such a symmetry breaking are difficult to find in laboratory analogs of natural factors.

This circumstance motivated the adoption of the systems approach in the studies of the emergence of primary forms of living matter – the problem that we address in this book. In accordance with this approach, the requirements to the processes needed for the origination of the primary forms of living matter are analyzed jointly and compared to the main properties of various natural phenomena that may have produced the environments meeting these requirements.

Systems approach made it possible to reject many natural processes as unsuitable for the task. It permitted identifying the currently the only natural phenomenon that, by the main properties of the environment it creates, meets the conditions necessary for the emergence of primary forms of living matter.

The underlying idea of the new concept is based on the hypothesis that the processes accompanying the well-known natural phenomenon of a hypervelocity meteorite impact are capable of creating the initial conditions that make possible the emergence and survival of primary forms of living matter.

In the case considered the crucial processes among those that accompany a hypervelocity impact are the generation of the plasma torch (Zel'dovich and Raizer, 2002) and the formation of the impact crater (Ivanov, 2008a, 2008b; Alvarez, 1997; Bronshten, 1987). The former may have been responsible for the synthesis of complex homochiralic organic polymers in the torch plasma and the latter, for the formation of the survival environment during the formation of primary forms of living matter.

This book focuses on the discovery and study of hitherto unknown properties of the torch plasma. A plasma torch, which was completely identical to an impact torch, was simulated in laboratory experiments, which involved modeling a hypervelocity micrometeorite impact via Q-switched laser radiation (i.e., with a pulse duration ranging from 0.5 to 10 ns).

In these simulations new properties of the plasma torch have been experimentally found and confirmed. These properties consist in the possible synthesis of complex organic compounds in the process of adiabatic expansion-away of the plasma torch produced by a laser pulse and identical to an impact-generated torch (Managadze, 2001, 2002a, 2003; Managadze et al., 2003a, 2003b, 2003c).

The newly discovered properties of the laser-generated plasma torch proved to be similar or identical to those of the impact-generated plasma torch and, what is very important, this is especially true for the synthesis of new compounds. These results were obtained experimentally by comparing the mass spectra. Thus it became possible to perform scientifically grounded laboratory simulations of impact-driven synthesis of organic compounds using laser radiation in order to study the structural and functional complexifications of the resulting products including high-molecular polymer structures (Managadze, 2005b; Managadze et al., 2006; Managadze and Managadze, 2007a).

Later a hypothesis was proposed (Managadze, 2005a) that unipolar asymmetric and nonequilibrium electric and magnetic fields with complex configuration that are generated during the expansion-away of the torch plasma may ensure the generation of local chiral physical fields (Barron, 1986, 1994; Goldanskii and Kuz'min, 1989). This hypothesis is based in the results of earlier performed experimental studies (Korobkin et al., 1977; Bychenkov et al., 1993; Stamper, 1991), which pointed to the innate natural asymmetry of the plasma torch of arbitrary nature. Thus in accordance with the new idea proposed by Managadze (2007b), some physical factors that had earlier been found experimentally in a laser-generated plasma torch were also applied to the case of impact-generated torch. In the above paper I assumed that this idea may explain not only the synthesis of organic compounds, but also the breaking of mirror symmetry during the formation of enantiomers in a hypervelocity impact. The same effect could also result from electromagnetic radiation generated in the torch plasma – radiation that was supposedly circularly polarized and, according to Goldanskii and Kuz'min (1989), was also capable of breaking the symmetry.

The hypothesis that the synthesis of organic compounds in the impact-generated plasma torch may have involved the breaking of mirror symmetry of the resulting isomers (Managadze, 2005a, 2007a, 2009, 2010a) was not purely theoretical, but had a sound, albeit a tentative, experimental basis (Korobkin et al., 1977; Bychenkov et al., 1993; Stamper, 1991). The hypothesis was based on the fact that electromagnetic fields generated in the torch did not change their orientation from one act of influence to another because their polarity was determined by plasma processes exclusively.

The post-impact processes in the crater, which provide "comfortable" conditions for further evolution of the end products of the synthesis over several hundred thousand or even several million years (Ivanov, 2008a, 2008b), could also have played an important part. Thus well-ground, fine-dispersed wet rock matter, which has moderate temperature inside the impact crater and which is full of organic compounds produced via impact synthesis, may have served as a "survival region" for the simplest organisms at the initial stage of their "enzymeless" existence.

These phenomena and processes, which accompany the impact, and many of which have been hitherto unknown, have never been considered as having anything to do with the emergence of life.

As a highly catalytic environment, plasma torch makes possible the synthesis of high-molecular organic compounds within the plasma expansion-away time scale. Hence one of the most intriguing problems is the study of maximum achievable structural and functional properties of the end products of the synthesis of organic compounds in plasma torches produced in laboratory conditions and in space (Managadze, 2003; Managadze et al., 2003a).

We also assigned high priority to the theoretical studies of similar problems involving a major meteorite impact and, in particular, to the study of how the basic properties of synthesized compounds may depend on the initial velocity and characteristic size of the meteorite and on the configuration of the impact.

As the investigation of this phenomenon progressed, it became clear that the processes that accompany a hypervelocity impact and that may be directly related to the emergence of life on the Earth are not limited only to the synthesis of organic compounds in the torch plasma. It turned out that during and after the meteorite impact an entire sequence of physical processes develops, which makes maximum use of the impact energy to maintain favorable conditions for the emergence of life.

Obviously, if the proposed new mechanism could operate on the primeval Earth, its energy and spatial scales, which were characteristic of a meteorite impact, should have exceeded by many million fold those achieved in laboratory simulations. It was therefore important to determine the correct procedures for extrapolating the results obtained on small energy and spatial scales of artificial microparticles in a laboratory to the characteristic scales of major meteorite impacts.

During the period of the formation of the Earth meteorite impacts delivered matter from space to feed the growth of the planet (Pechernikova and Vityazev, 2008). The meteorite influx consisted of 90% ordinary chondrites and 10% carbonaceous chondrites, whereas icy cometary nuclei accounted only for 1% of all meteorites. Some of these bodies were truly enormous with sizes amounting to 3000 km. The meteorite influx was also enormous. Thus during the first 200 million years about 10^7 ~10-km large meteorite bodies fell on the Earth. With such a high meteorite influx global-scale impact catastrophes occurred, on the average, every 20 years.

An impact of a meteorite of such a size moving at a velocity of 20—30 km/s must have produced a plasma torch as high as 100 km. The diameter and depth of the impact crater could have amounted to 100 and 30 km, respectively. Thermal heating from such an impact lasted for up to 10 million years (Ivanov, 2004). The meteorite bombardment of the Earth, which lasted about 600 million years, must have been the most energetic natural influence in the geologic history of the planet. This influence must have affected all the ensuing history of the Earth.

We also showed (Managadze, 2003) that organic compounds observed in interplanetary gas and dust clouds are most likely synthesized in the process of hypervelocity collisions between dust particles in plasma processes identical to those described above.

Further development of these studies led us to some new ideas that are of separate interest. The most important of these ideas proved to be:

- the possibility of the development of conditions needed for the emergence of life in deep layers of planets and moons with low surface temperatures. Such conditions can be created by a penetrating impact of a large meteorite
- hypothesis about the possible mechanism of the formation of water in the Universe in condensed-state processes during the interaction of hydrogen ions of the stellar wind with the surfaces of oxide-containing microparticles and atmosphereless planets or small bodies;
- the possibility of the formation inside an impact crater of an environment providing the conditions necessary for the survival of a primitive organism, for the preservation and development of life over several hundred thousand or even several million years;
- development of a new technique for conducting unique, controlled impact experiments that are impossible to perform in ground-based laboratories, in the near magnetosphere, where supercritical impact velocity can be achieved in a collision between the projectile and target launched from two satellites moving in opposite directions.

Remote mass-spectroscopic measurements of the products of synthesis in these experiments could be made from distances ranging from 100 m to 4 km (Managadze and Eismont, 2009b).

The results of these studies led us to conclude that highly efficient synthesis of organic compounds in nature is made possible in plasma-chemical reactions owing to the mechanism that operates in the process of the expansion-away of torch plasma during a hypervelocity impact. This natural phenomenon could have contributed to the emergence of living matter on the Earth and origination of its possible extraterrestrial analogs on other Solar-system bodies subject to meteorite bombardment. The synthesis of organic compounds found in interstellar gas-and-dust clouds can be explained by similar physical processes associated with the generation of torch plasma, which forms in hypervelocity collisions of interstellar dust particles (Managadze, 2003; Managadze et al., 2003c).

1.7. Plasma Torch as a Dissipative Structure

A comparison and analysis of the results obtained in the papers mentioned above raises questions of extreme importance, which come to this: can plasma torch be considered to be a dissipative structure? If so, we can hope that in the process of the expansion-away of a highly catalytically active torch plasma the processes characteristic of dissipative structures should occur, such as ordering and self-organization of matter; development of the instability of the system, which results in a spontaneous symmetry breaking, and the symmetry breaking with constant polarity "sign" caused by local chiral physical fields of the torch. Other processes may also occur that are possibly caused by strong fluctuations and bifurcation.

According to the definition given by I.Prigogine, a dissipative structure is a nonequilibrium and irreversible physical or chemical process far from the thermodynamic branch of equilibrium, which makes possible the development of a bifurcation. It is especially important to achieve this state because only bifurcation allows the system to attain the required freedom of choice and unpredictability.

Among the various dissipative systems analyzed by G. Nicolis and I. Prigogine (Nicolis and Prigogine, 1989) the closest one to the torch proved to be the system with explosion-like behavior. Autocatalytic chemical burning is a typical example of a phenomenon where these conditions are satisfied. Nicolis and Prigogine (1989) illustrated such type of dynamics for the case of adiabatic explosion. The simplest example of such an explosion is irreversible exothermal reaction. It was shown that a transient process in such a system consists in the development of a bifurcation with time.

The above comparison of the processes that occur during fast chemical burning and in the case of explosive expansion-away of a plasma torch shows that these phenomena can be viewed as similar. First, they have similar development dynamics in that they result in fluctuations, instability, and bifurcation – which eventually lead to self-assembly and ordering. However, plasma torch is vastly superior to burning process in many other respects. The point is that processes in a torch occur in a different – plasma – environment. Reactions that occur in such an environment are of plasma-chemical nature because they may produce compounds that are impossible to synthesize in chemical reactions. Not least important is plasma's superhigh catalytic activity due to its high temperature and state of matter, which is in active phase. It is therefore difficult to find in nature another phenomenon that would make possible the formation of more developed and energy-intensive dissipative structures capable of "creating order through chaos".

Thus by its properties a hypervelocity impact torch can be viewed as a dissipative structure where matter is in the forth aggregate state, i.e., in the state of plasma. According to the above data, this is a positive fact, which, however, raises certain difficulties. The point is that until recently no theoretical studies have been performed to analyze plasma-chemical processes in the torch plasma that result in the synthesis of new compounds. One should therefore expect that in the near future while the theoretical aspects of this process remain undeveloped most of the information about the properties of torch plasma will be obtained experimentally.

In Chapters 4, 5, and 6 we describe the most important experimental data about the properties of the plasma torch obtained in experimental simulations of a hypervelocity impact. These data raised no doubts or questions because the phenomenon was reproduced exactly in accordance with its natural analog. The results were adopted unchallenged in the discussions involving experts in the appropriate fields. Our data were often corroborated by the results obtained by other researchers who studied similar processes. The initial conditions in the experiments on impact modeling are based on material evidence exclusively.

We showed that during its expansion-away plasma torch has the following important properties:

(1) Capability of highly efficient synthesis of simple and complex (including polymer) organic compounds provided that the torch contains the constituent elements needed.

In the case of a "penetrating impact" configuration two equivalent torches form on both sides of the "thin" target. These torches expand in opposite directions and have approximately the same amount of compounds synthesized in them.

(2) Self-organization of organic compounds produced in the torch that occurs during their synthesis. Thus linear-chain hydrocarbons, cumulenes and polyenes, fullerenes and their fragments, and hyperfullerenes have been found and identified in the experiment involving a carbon target with a small admixture of hydrogen. With nitrogen and oxygen added to the target, the observed periodicity of the mass peaks in the spectrum was indicative of the formation of organic compounds that can be interpreted as hyperbranched acetylenic hydrocarbons with molecular masses amounting to 600 a.m.u., dendrimers of fourth-generation amino acids with molecular masses up to 4000 a.m.u., and polypeptides with molecular masses up to 5000 a.m.u.

(3) The earlier found and experimentally studied unipolar nonequilibrium electric and magnetic fields, which have all the properties of local chiral fields, produced a small, but detectable symmetry breaking in direct measurements of the products of synthesis. The symmetry breaking was too weak to make the environment homochiralic. However, in a strongly nonequilibrium environment, which ensures spontaneous symmetry breaking, such external factors could be sufficient to maintain the environment in the state of asymmetry. However, under the conditions of a strongly nonequilibrium environment that ensures spontaneous symmetry breaking, acts of external influence could (Prigorige and Stengers, 1993; Avetisov and Goldanskii, 1996b) be sufficient to maintain the environment in the asymmetric state by determining the "sign' of its polarity.

(4) It was shown that an impact of a ~100-nm diameter micrometeorite containing C, O, N, and H involves the synthesis of high-molecular organic compounds. An extrapolation of these results shows that the impact of a 1-cm diameter meteorite could result in the synthesis of organic compounds with two orders of magnitude higher masses amounting to 5×10^5 a.m.u.

(5) A proper comparison of the time scales of the synthesis of complex organic compounds for the conditions of equilibrium chemical reactions and for nonequilibrium plasma-chemical processes in a torch shows that the rates of plasma-chemical reactions are considerably higher than those of chemical reactions, by factors ranging from 10^7 to 10^9 with a high degree of confidence.

It is immediately apparent from the above-mentioned plasma-torch properties concerning the synthesis of complex molecular structures of organic compounds that all the crucial torch properties that are necessary for the formation of extended molecular structures within the plasma expansion-away time are present, and that it only remains to experimentally confirm the possibility of spontaneous symmetry breaking.

Currently no bona fide recommendations can be produced how to experimentally detect traces of spontaneous symmetry breaking in a plasma torch or in the products of plasma-driven synthesis. If, for example, spontaneous symmetry breaking has the same occurrence rate as self-organization of the products of synthesis – once per thousand acts of influence – the effect will be impossible to detect because of the small amount of the sample. There is no point in collecting such a sample in order to increase its mass, but there is another way.

Avetisov and Goldanskii (1996b) paid special attention to spontaneous symmetry breaking in chemical systems. The above authors concluded, based on the results of the experiments performed by Soai et al. (1995) who constructed a chemical analog of the model of F.Frank (Frank, 1953), that spontaneous symmetry breaking is undoubtedly possible not only on the biological, but also on the chemical level and, moreover, it can occur within the class of autocatalytic functions. However, Avetisov and Goldanskii continue that to achieve homochirality, a set of enantiomers is needed that can be produced only with enzymes and biological processes. The above authors believe that this conflict does not refute the hypothesis of asymmetric emergence of life, but rather emphasizes the serious difficulties it faces.

When discussing the problem we have to take into account the fact that the above difficulties apply to chemical systems. It is therefore unclear whether these difficulties also apply to a different, plasma environment, where plasma-chemical reactions occur at rates that exceed significantly those of chemical reactions. This is especially true given that the primary structures of functional carriers may have been non-universal and varied with varying environment.

Other features of the torch plasma may also have influenced the symmetry breakings and, possibly, the processes of polymer capture. Thus from the very beginning the synthesis of organic compounds in a plasma torch occurs in a nonequilibrium unidirectional asymmetric, local chiral field of the torch. These and the above properties of the torch plasma are very complex and poorly studied, and that is why we can now determine them only experimentally.

Hence the experimental discovery of self-organization and high catalytic activity of the plasma environment should be viewed as a new, very promising, and inspiring result. The preliminary results of direct measurements, which are indicative of the symmetry breaking in local chiral fields of the torch, acquire the needed reliability when analyzed jointly with the electromagnetic field properties determined by independent measurements.

Hence the combination of the newly found properties of the torch plasma may potentially provide the conditions needed to further the emergence of primary forms of life. These properties could have been sufficient and instrumental. We report the experimental results that confirm the properties of the torch plasma in Chapters 4, 5, and 6, and in Section 1 of the Afterword.

The question that often arises in the studies of the synthesis of organic compounds in a plasma torch originates from a misunderstanding of the process of the formation of such “fragile” compounds as amino acids and nucleotides and their oligomers in such a high-temperature environment as the plasma torch. This is usually due to the wrong perception of the processes that develop in a plasma torch. In particular, the high temperature in the torch is achieved only at the initial stage of plasma formation and it is necessary to ensure: (1) explosive expansion-away of the torch powered by high pressure due to fast heating and high temperature and (2) complete atomization of matter and 100% ionization of the elements produced. This process is necessary, among other things, to “release” carbon from the atoms of other elements and convert it into atomic state. The subsequent redistribution of elements must ensure highly efficient synthesis of organic compounds.

As the torch plasma freely (adiabatically) expands, its temperature begins to decrease abruptly. However, the matter inside the torch remains in the plasma state. As the expanding plasma cools down to a temperature that is equivalent to the energy of interatomic bond, atoms begin to combine into diatomic and triatomic molecular ions. The expansion-away allows these molecular ions to rapidly escape from the high-temperature environment where they could have again and again been disintegrated into atoms, and carries them into cooler regions of the plasma torch. This makes plasma-chemical synthesis reactions irreversible. Under the new conditions, reactions become even more complex at a lower temperature and produce polyatomic molecular ions of organic compounds. The "zone of synthesis", where the process of chemical complexification takes place, has a rather long lifetime depending on the characteristic size of the plasma formation region, which, in turn, must depend on the size of the meteorite.

Hence a plasma torch, which is a highly catalytic dynamic environment, ensures high rate of plasma-chemical synthesis reactions during the plasma expansion-away time. In this environment, other reactions can occur, which are impossible to perform under the best conditions available in chemical laboratories. The production rate of these reactions is rather high because the matter in the "synthesis zone" may be under a pressure of 10^5 to 10^6 atm. As is well known, compounds that are beyond the reach of classical chemistry, e.g., halogenides of inert gases, can be synthesized only in industrial plasma-chemical facilities.

The exceptional properties of the torch plasma discussed above make it possible to approximately estimate the characteristic size of a meteorite whose impact makes possible the synthesis of a macromolecular structure with the mass comparable to that of a protoviroid.

An extrapolation of the results presented in Figure 27 in Chapter 5 shows that this can be achieved by an impact of a 1-cm diameter meteorite. According to the data presented in Sections 3.3 and 3.5 of Chapter 3, the diameter of an intermediate-sized crater and the characteristic longitudinal size of the synthesis zone in the torch should in this case be equal to 5 cm and at least 40 cm, respectively. Given that the plasma expansion-away velocity does not exceed 3 10^6 cm/s, the matter should spend at least 10 μs inside the "synthesis zone", which is enough to synthesize a macromolecule comparable to a protoviroid. Obviously, this estimate does not imply that a protoviroid-like macromolecular structure ought to have formed in the torch plasma of every single meteorite impact. However, the newly discovered properties of the plasma torch open up such a possibility for extremely rare, exceptional cases.

Long and painstaking experimental and theoretical studies are needed to make sure that the torch plasma is the environment that can bridge the gap between nonliving and living matter. The success of these studies will depend crucially on the highly sensitive and highly selective next-generation analytical equipment. Hence this task cannot be accomplished in the nearest future.

However, there are tasks to be addressed without delay, e.g., those associated with the survival of the simplest forms of life in a meteorite impact crater. This process was studied assuming that the first primary forms of living matter formed in a torch plasma. However, we cannot rule out the possibility that it may have formed in a crater. In this case, individual macromolecular structures that are required for the formation of the most primitive forms of living matter and that are synthesized in a torch plasma may have come to the impact crater and then contributed to the emergence of life.

The emergence of the simplest forms of living matter was the event of greatest importance in the Earth's history. They could have then evolved – via biological processes - into a full-blown living cell. Further development of the biosphere destroyed the traces of the first forms of living matter. It is safe to suggest that the only remaining way to study these forms of living matter is to reconstruct them under laboratory conditions. Let us now see how feasible this possibility is.

The results of the laboratory experiments described in Chapters 4 and 5 show that the probability of the formation of an organized structure in a torch plasma is extremely low. Thus ordered structures in the mass interval from 1000 to 4000 a.m.u. are observed to assemble once in 10^3—10^4 acts of influence. However, the necessary conditions for the synthesis of homochiralic molecular structures include not only ordering of matter, but also, at least, the emergence of spontaneous asymmetry combined with the presence of unidirectional local chiral physical fields of the torch. It is currently impossible to estimate the probability of these processes occurring simultaneously with the processes of ordering. It only remains to assume that for molecular structures combining the above properties to form, the total time of the laboratory experiments will have to exceed several dozen years.

Hence a new approach to the experiment and more sophisticated equipment are needed to reconstruct living matter under laboratory conditions.

At the same time, if the proposed scenario of the emergence of living matter in the impact-produced torch plasma is realizable, it will be possible to find living substances in small icy impact craters on celestial bodies including cometary nuclei. It is entirely possible for living systems to be eventually found in young impact craters in the surface layers of cometary nuclei will prove to be viable.

When studying the problem of the emergence of life we need a proper definition of the concept of “emergence”. Can the time when a spatially isolated protoviroid-like primary structure capable of replication and translation be adopted as the “emergence day”, or “birthday” of terrestrial life? A.Altstein answers “Yes!” because he views his protoviroid as the first living being on the Earth and the progenitor of the biosphere. We should agree with that because it would be more convenient to associate the concept of “life” with a local molecular structure rather than with the emergence of the biosphere.

It goes without saying that the problem is not so simple. According to the concept proposed by G.Zavarzin (Zavarzin, 2006a, 2006b), a stable state of the biosphere is possible only in the case of relatively closed biogeochemical cycles. If these conditions are not met, living matter would very soon exhaust all the available resources and get intoxicated with the products of its own metabolism. For the cycles to be closed, several different microorganism species must exist with segregation of biogeochemical functions. However, this subject falls within the domain of biology, which we would like to avoid in this book.

The task is more obvious for the stage of chemical evolution because a single organic compound cannot persistently self-replicate and maintain homeostasis without a polymer community with active exchange of hereditary material between the organisms. Therefore the primordial properties of viable organisms must include variety, symbiosis, segregation of functions, and exchange of information. Does this imply that the temporal criteria of the emergence of life on the Earth should be linked not to the “birthday” of the protoviroid, but to the time when its many different “cousins” come into being? The answer should most likely be positive and it would be correct from the viewpoint of a biologist.

However, the main subject of this book is the experimental study of the possibility of the emergence of primary forms of living matter in the processes that accompany hypervelocity meteorite impacts, and it would therefore be reasonable to confine our analysis to this problem. Hence in the case considered, the formation of at least one organism capable of ensuring the generation of at least several succeeding generations can be viewed as the "birthday" of life on the Earth with no need to wait for the emergence of other communities. This compromise decision should be adequately received by the advocates of polymorphous communities.

Note that the new concept involving plasma mechanism of abiogeneous synthesis cannot be viewed as yet another scenario like the many other ones already proposed. The point is that the main concept and the physical basis of this mechanism differ fundamentally from those hitherto proposed, which involve the effect produced on the planet by solar radiation, electric discharges in the cool ionospheric plasma, and atmospheric shocks generated during meteorite impacts (Kobayashi and Saito, 2000; Miller and Urey, 1959; Matsu and Abe, 1986; Bar-Nun et al., 1970; Simionescu and Denes, 1986; Miyakawa et al., 1997, 1999). The physics of plasma processes in hypervelocity impacts also differs fundamentally from that of mechanisms involving the effect of high temperature on geological rocks by the heat released during meteorite impacts or during eruptions of volcanoes including underwater ones (Gerasimov et al., 1991; Mukhin et al., 1989; Mukhin, 1974).

This difference is due to the fact that in the above papers the rates and energy contributions of the processes are insufficient to ensure the production of a three-dimensional plasma outburst that would shape a feature meeting the criteria of a plasma torch (Delone, 1989; Anisimov et al., 1970). Note also that an impact produces, with no additional efforts, all the conditions required for the synthesis, assembly, and ordering of polymer organic compounds, and for the breaking the symmetry of enantiomers and creation of the "survival zone" - and all this in a single and continuous act of natural influence with precisely known initial parameters and characteristics.

The spatial and energy scales of a meteorite impact are practically unlimited. Thus in a laboratory experiment the ratio of the diameter of a large meteorite to the maximum diameter of the projectile is equal to 10^8 to 10^9. No other processes used in the origin-of-life scenarios have such a great scale-up reserve in terms of the characteristic size of the interaction. This also means that the current work is carried out at the lower end of the characteristic sizes of meteors, and hence major discoveries await us in the future. Neither do we know the capabilities of large-scale plasma torches in the synthesis of various organic compounds, or their functional and structural properties. It is therefore not improbable that a comparison of the prerequisites for the emergence of the most primitive forms of living matter of the ancient RNA world with the actual capabilities of a hypervelocity impact might allow us to find acceptable conditions for this process on the warm Earth rather than in the cold space with its severe conditions.

***

Thus to ensure the formation of the most primitive forms of living matter under the conditions of the early Earth we had to circumvent a number of very difficult barriers. These included ruling out the synthesis of complex organic compounds with the required functional and structural properties of the biochemical level of complexity via simple enumeration. The various scenarios of natural processes considered so far and the mechanisms of their realization are incapable, for a number of objective reasons, to satisfy the most important requirements – i.e., to ensure, along with the synthesis of organic compounds, their ordering, homochirality, and extremely high catalytic activity of the medium. It is also important that the energy characteristics of the synthesis mechanisms proposed in these scenarios are from the very beginning considerably lower than required. They are lower, in particular, than the level required for the formation of a nonequilibrium medium or of dissipative structures. Moreover, the earlier proposed synthesis mechanisms are practically at the upper limit of their capabilities. Therefore considerable advances are hardly to be expected on the basis of these scenarios and mechanisms. New input is needed in the form of new concepts, mechanisms, and ambient properties.

That is why the discovery of new properties of impact actions and torch plasma can be viewed as timely, important, and of current concern. The new concept, which has a sound experimental basis, is now viewed as quite realistic. It is currently at the starting point of its development and not yet capable of solving all the problems associated with the extremely difficult nature's mystery. However, we have to admit at the same time that today it is hard or even impossible to propose a natural phenomenon possessing a combination of required properties on a par with those that characterize the torch plasma and impact processes. Thus, the exceptional properties of the plasma torch ensure the realization of such important processes required for the emergence of life as the synthesis, assembly, and ordering of the structures of organic compounds, and, supposedly, the symmetry breaking of enantiomers. These processes are also characterized by high catalytic activity, impressive characteristic sizes and energy reserve for their realization. Of great importance is also the fact that a meteorite impact produces a crater, which serves as an environment with the conditions that make possible not only the survival of the first organisms, but also, possibly facilitates their origination. The above properties of the torch plasma and impact crater lead us to a reasonable conclusion that the processes that occur during a meteorite impact are capable of creating all the conditions needed for the origination and survival of primary forms of living matter. What is very important is that this scenario does not rule out the emergence of the first forms of life in the torch during its adiabatic expansion-away, which ensures intense cooling of the initially hot plasma. And, what is equally important is that these processes may occur not only on the Earth, but also elsewhere.

In the next chapter we discuss the initial conditions for the formation of planets and show that if a star has a planetary system then all the preconditions for the emergence of life are bound to be provided by nature. This conclusion applies not only to the "well-to-do" Earth, Solar System planets and their moons, but also to the planetary systems of other stars.

In Chapter 3 we discuss the physical principles of the process that makes possible abiogenous synthesis of organic compounds in the case of a hypervelocity collision of fragments of matter. We show, based on the reported experimental results, that the hitherto unknown properties of the plasma torch may show up independently of the presence of solar radiation in the atmosphere of any density and composition, irrespectively of the ambient temperature, and that they make possible the synthesis of organic compounds on the Earth and in interstellar gas-and-dust clouds.

Chapters 4 and 5 describe the experimental results indicating that in a plasma torch processes of organization of matter may occur alongside the synthesis of organic compounds. Such processes can be viewed as natural selection on a molecular level at the prebiotic stage of evolution.

In Chapter 6 we discuss the processes of mirror symmetry breaking in a plasma torch during the synthesis of enantiomers. In this chapter we present the results of measurements that are indicative of the possible symmetry breaking in the product of synthesis in a plasma torch.

Chapters 7 and 8 are dedicated to the processes that contribute to the origination of extraterrestrial life in the case of a penetrating meteorite impact involving the injection of organic compounds synthesized in the torch into the subsurface layers of cosmic objects. We describe the original techniques and new onboard mass-spectrometric instruments that allow signs of life on Solar-System planets and their moons to be detected from interplanetary stations and landers. We propose a new concept of experimental procedures involving the realization of hypervelocity impact interactions that are impossible to perform under ground-based conditions. These experiments should involve the collision of a projectile and a target installed onboard two satellites moving in head-on orbits in the Earth's magnetosphere, and mass-spectrometric measurements of the products of impact synthesis.

Chapter 2

Initial Conditions Provided by Nature

2.1. Event – Phenomenon – Environment

To familiarize ourselves with the basic physical processes responsible for the formation of the plasma torch, which we consider in Chapter 3, let us determine the conditions that make possible the phenomenon of hypervelocity meteorite impact. For this process to be realized in accordance with the terrestrial scenario, we need a system of planets orbiting a star. Hence the formation of planets near a star can be called an "event", and such events occur rarely in nature.

As is well known (Safronov, 1972; Safronov and Vityazev, 1983; Jones, 2004), a sun-like star with a planetary system represents a class of objects intermediate between single stars and binaries. One may expect that the laws of planet formation and the structure of planets should differ for different stars. However, these differences cannot be so radical as to result in the formation of protoplanets without such processes as accretion or coalescence of interstellar dust (Figure 3) and without the formation of planetesimals, i.e., primary celestial bodies, which later served as building blocks for the planets.

A germ of a planet – a protoplanet – forms via mergers of planetesimals moving in close orbits. A protoplanet is capable of capturing, by its gravitational field, other planetesimals located along its orbit, and it grows rapidly during the stage of its formation. As the mass of the protoplanet increases, its gravitational field becomes stronger and with time, the sum of the geocentric velocity of a meteor body and the velocity acquired as a result of gravitational attraction may exceed $V_{CR} \sim 15—20$ km/s, which is enough to ensure volume ionization of the target and projectile if the latter has a density of 2 g/cm^3

Under these conditions, a meteorite impact onto the surface of a planet produces an "event" resulting in the generation of a full-blown torch, i.e., in the formation of the "environment".

One may expect that the formation of planets around newborn stars, like the birth of the Solar System, should involve the formation of cometary-body reservoirs similar to the Oort cloud or the "Kuiper belt" (Busarev and Surdin, 2008). In this case, the geocentric velocities of comets should be substantially higher than those of planetesimals, resulting in a more efficient generation of the torch plasma in the impact process.

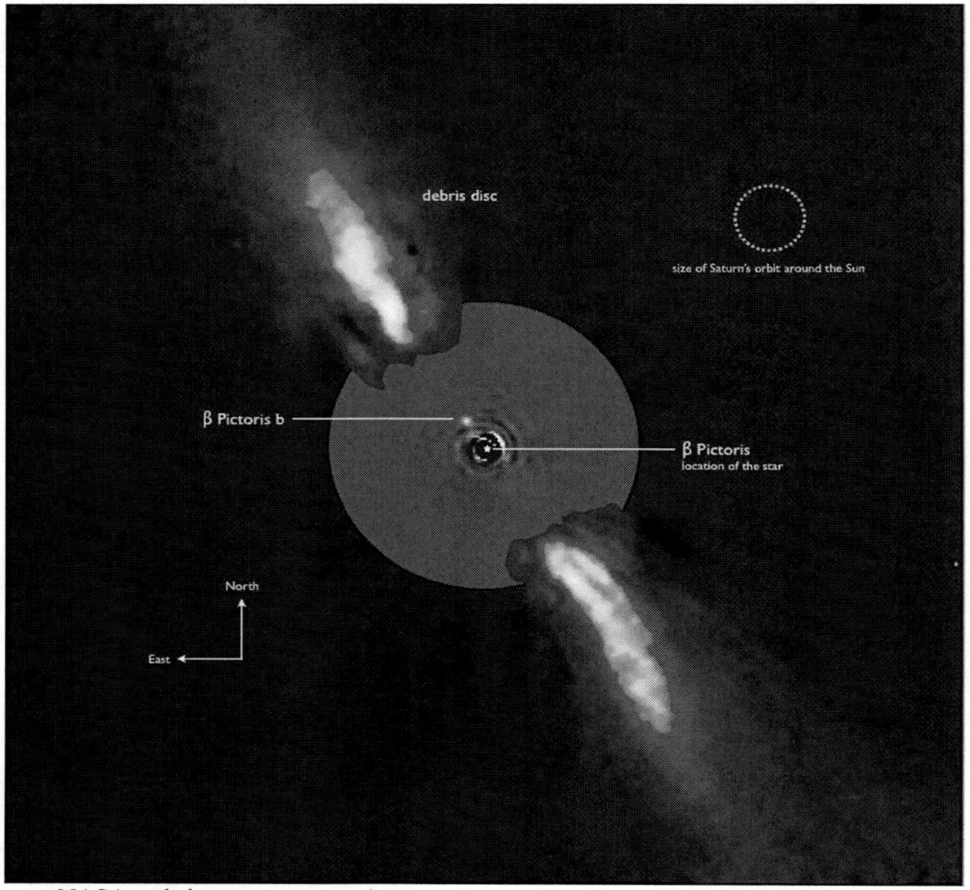

(Source – NASA website www.nasa.gov).

Figure 3. At the time of its discovery by B.Smith and R.Terril (USA) the dust disk surrounding Beta Pictoris was interpreted as evidence for the possible future formation of planets around this star. However, the situation proved to be much more interesting. In particular, 25 years later, the efforts of the researchers of Observatoire de Grenoble headed by Anne-Marie Lagrange culminated in the discovery of an exoplanet – a gas giant with eight times the Jovian mass – which proved to be the closest to its parent star among all known planets of this type.

We refer the reader to B.Jones's book "Life in the Solar System and Beyond" (Jones, 2004) for an overview of the current state of the problem of the origin of planetary systems including the recent advances in this field. This book also addresses the issues of the search for "exoplanets" around other stars and of the studies aiming to determine to what extent these relatively new cosmic objects can provide the conditions needed for the emergence of life on them. Jones also analyzes the possibility that these objects may develop an environment making possible the survival of living matter once it has formed. The book also attempts to demonstrate that life may originate not only on planets where the conditions "fall" within the so-called "life domain", but also in many "corners" of the Universe and, in particular, in the interiors of planets with either too low or too high surface temperatures. It must be emphasized that such conclusions about the possible existence of life on, e.g., Enceladus or Titan, have been appearing increasingly more often in scientific publications in recent years. However, such hypotheses are proposed without explaining how this could happen and without indicating any particular viable mechanism. The explanation is very simple: impact remains the only scenario and realization mechanism that can ensure both the synthesis of various organic compounds and their introduction into the deep layers of cosmic objects.

In accordance with the new concept we can assume that the formation of planetary systems anywhere in the Universe should automatically enable the realization of the "phenomenon – environment" association, resulting in the formation of a plasma torch, and that for this to happen it is sufficient for the meteorite to move at an overcritical velocity. This means that the formation of the environment could have proceeded in accordance with the scenario similar to that of the birth of the Solar System. Hence organic compounds could nevertheless be preserved and accumulate on these planets provided that the temperature was lower than the organic destruction temperature. At the same time, life can originate in accordance with the terrestrial scenario should conditions develop at the surface of a planet that allow water to form and be retained.

In other words, in the processes of hypervelocity impacts - a phenomenon that is very common in the Universe – the chemical composition of the environment or that of the torch plasma could be determined primarily by the composition of planetesimals and cometary nuclei. The maximum carbon content in these bodies could amount to 7 and 15%, respectively. The intense environment-forming processes that occurred during the impact resulted in the formation of complex organic compounds and, if water or moisture is present in the impact region, they could create the conditions required for the emergence of life.

This conclusion is based on experimental laboratory simulations of the plasma torch produced by a hypervelocity impact. These experiments confirmed the possibility of the synthesis of complex organic compounds in an expanding impact-produced plasma torch. A comparative analysis of the new properties of an impact-produced plasma torch with the known characteristics of a plasma jet produced by a plasma-chemical reactor definitively demonstrated the great similarity between the physical processes that occur during adiabatic expansion-away of plasma formations in impact torch and in plasma-chemical reactor. In particular, in both cases the available conditions ensure highly efficient synthesis of new compounds in fast chemical processes where plasma acts as a catalyst. Therefore at the initial stages of the Earth's formation meteorite impact processes must have ensured the synthesis of complex organic compounds, whereas the presence of water under these circumstances could favor the formation of an environment required for the emergence of life with no additional conditions.

Thus in the "event—phenomenon—environment" chain considered above the event, i.e., the formation of a planetary system around a star, is the only process of low probability. Planets are nevertheless the most likely places for the emergence of life and, judging by the Solar System, nature has so far demonstrated such a possibility only once making use of the exceptional conditions on the primeval Earth to this end.

Given the presence of planets the other components of the chain – "phenomenon—environment" - can be viewed as inevitable.

Thus in the case of the Solar System the energy brought to the Earth by meteorite impacts during the first 20—30 million years was about twice the energy of solar radiation reaching our planet (but not its surface), and the characteristic sizes of planetesimals may have amounted to ~1000 km (Pechernikova and Vityazev, 2008). Hence the plasma environment created by an impact may have been of a truly global size.

The chemical composition of meteorites, like that of the Earth's surface layers, varied enormously from impact to impact because the latter were highly inhomogeneous. Thus while the Earth was bombarded by planetesimals whose composition was similar to that of carbonaceous chondrites, it was also subject to impacts of cometary nuclei, which consisted mostly of a mixture of ice and rocks. The Earth was also hit by stony meteorites, whose chemical composition varied over a wide range, and by iron meteorites.

The Earth's surface, in turn, was also far from uniform. Its nonuniformity was due to the distribution of large quantities of various rocks, water ice, seas, oceans, and, possibly carbon-rich areas. For this reason, the chemical composition of the products of synthesis could also vary greatly depending on that of the meteorite and of its impact site.

Not only the chemical composition, but also the velocities of the meteorites varied from impact to impact from 11 to 72 km/s (Bronshten, 1987). For more accurate average asteroid and cometary impact velocities see Nemchinov et al. (2008a) and Ivanov (2008a, 2008b). The above authors showed that the average impact velocities must have been of about 15 to 20 km/s for asteroids, 20 to 30 km/s for short-period comets, and up to 50 km/s for long-period comets. Such a wide range of impact velocities could result in different initial temperatures of the plasma torches produced, and hence in a considerable variety of the final composition of organic compounds synthesized during the plasma expansion-away. Variations of the characteristic sizes of projectile could also lead to similar consequences.

Hence because of the highly fluctuating nature of the initial conditions plasma torches produced by meteorite impacts must be extremely poorly reproducible in terms of many important parameters. This should have inevitably showed up in the final product and, in particular, in the emergence of an infinite variety of synthesized organic compounds, ensuring the formation of many different molecular structures, which, in turn, created the conditions for the origination of primary forms of living matter.

Thus the meteorite bombardment of the Earth, which continued for 500—600 million years, created the conditions for the synthesis of simple and complex organic compounds. These processes involved the realization of a countless number of trials to create macromolecular structures that differed radically in many parameters. All this could, owing to the exceptional properties of highly nonequilibrium and fluctuating torch plasma, favor the origination of primary forms of living matter with such characteristics that allow this event to be considered as the emergence of life on the Earth.

It is probable that because of the high catalytic activity of the plasma torch the primary forms of living matter formed as a result of a single impact of a large meteorite almost "simultaneously", in the process of plasma expansion-away. However, given the required temperature conditions and high concentration of organic compounds, life could have appeared later, after the expansion-away of the torch – in the "comfortable" (for the process considered) fine-dispersed, warm, and moist environment of fine-grained rocks of the impact crater.

Although this environment could be not as catalytically active as plasma, it would suffice 10 minutes or less for the simple structure like Altstein's protoviroidn to form – according to the estimates obtained by the author of the hypothetical being.

2.2. Characteristics of the Environment

A full-blown living cell is known to be an object of extremely complex structure and functional capabilities. As the very remote ancestors of the cell, primary forms of living matter must have been much simpler. However, for primary forms of life to emerge as a result of nonbiological processes, the environment where they could have originated must have possessed a number of unique properties. Note that these properties, like the processes that made possible the emergence of primary forms of life, are still not entirely known. Yet, we have certain information about some of the properties indispensable to the origin of life.

Let us consider characteristics of the plasma torch that could meet these requirements.

We briefly addressed this issue at the end of Chapter 1 and we will thoroughly discuss it in Chapters 3 and 5. In this paragraph we analyze those new properties of the plasma torch as a new environment that have remained unknown until now.

As we already pointed out above, some of the properties of the plasma torch have been well known as early as 1950-ies (Zel'dovich and Raizer, 2002). Back then, in 1960-ies, Q-switched laser emitters appeared, which were capable of producing ~10—20 ns long pulses with the radiation fluxes of 10^9 to 10^{12} W/cm^2. Such a flux proved to be sufficient to generate a plasma torch, and this, in turn, made it possible to study the electric and magnetic fields of the torch plasma (Stamper, 1991). These fields were found to be always oriented in one and the same direction. Their natural asymmetry, which is determined only by the polarity of electrons and ions, aroused no particular interest. It was, however, clear that the polarity of these fields can be reversed only by reversing the charge of electron from negative to positive.

At the time no one associated electromagnetic fields generated in the torch plasma with the processes favoring the emergence of live. However, at that period of time the researchers began to view torch plasma as the most promising environment for laser-driven fusion. Although these ambitious and long-term studies were not especially successful, they proved to be rather useful. Thus all the major properties of the torch and, in particular, electromagnetic fields, plasma instabilities, plasma radiation from the recombination region, etc., have been studied in the framework of this program, thereby substantially facilitating further studies of the torch in its new role.

The new approach to the electromagnetic fields of the plasma torch emerged only after the discovery of the properties of the synthesis of organic compounds in these hot plasma formations (Managadze, 2003). Later studies (Managadze, 2005a, 2007a, 2010a, 2010b) showed that these fields form in nonequilibrium processes and can therefore be viewed as local chiral physical fields supposedly capable of inducing symmetry breaking in the processes of enantiomer synthesis.

It is now well known that the detection of the symmetry breaking of organic compounds may be a natural marker indicating that the process considered is possibly related to other processes that led to the emergence of living matter. Therefore the answers to the questions of whether the mirror symmetry breaking of bioorganic world was determined by the natural asymmetry of the plasma torch, and how strong this asymmetry may have initially been, and whether a small initial symmetry breaking could determine the "sign" of the symmetry of bioorganic world – the answers that are of great scientific value – may clarify a number of difficulties and inconsistencies concerning the prehistory of life (Avetisov and Goldanskii, 1996a, 1999b; Keszthelyi, 1995; Mason, 1991; Goldanskii and Kuz'min, 1989).

A preliminary analysis of the expansion-away of the torch plasma showed that the configuration of the artificial plasma outburst generated by a laser is, like its natural analog – the impact-generated torch, is very similar to that of a modern plasma-chemical reactor. During its adiabatic expansion-away the torch produces an ideal environment for the synthesis of new chemical compounds and makes possible their further complexification.

Torch plasma has a number of important applications. One of them deals with analytical instrumentation technology. For almost half a century the torch plasma, irrespectively of its origin, has been the most popular and convenient pulsed ion source used in time-of-flight mass spectroscopy. The processes of ion formation have been thoroughly investigated in both laser and impact plasma. Many theoretical and experimental studies have addressed this problem (see, e.g., Avrorin et al., 1985; 1986). These studies were in great demand because of the sweeping development of laser and dust-impact time-of-flight instruments designed for the chemical analysis of samples on the Earth and in space. Given such an ambitious use and extensive studies of the plasma torch, it is difficult to explain why its other properties – those associated with its capability of synthesizing new compounds including the organic ones – have remained ingnored.

The discovery of the new properties of the plasma torch has an interesting history. Thus I discovered the properties of the plasma torch concerning its capability of synthesizing organic compounds as early as 1992, and published my results in the "Universal Multi-Purpose Transportable Mass-Spectrometric Complex" report of APTI (Washington D.C., USA). In these experiments the LASMA time-of-flight mass–reflectron recorded carbines (up to 300 a.m.u.), the fullerene peak at 720 a.m.u., and multiply ionized hydrocarbons shown in Figs. 18, 19, and 20. It was clear from the very beginning that the experiment discovered an interesting and hitherto unknown phenomenon related to abiogenous synthesis of organic compounds. However, further analysis of this problem had to be postponed because of the work on the development of a new-generation mass spectrometer for the measurements of heavy metals at the Los Alamos desert nuclear test site.

When I returned to the study of the torch in 1998, I found that no papers had been published on related subjects and that the priority had not been lost. Moreover, an analysis of earlier published literature showed that no one had considered plasma-chemical processes in connection with abiogenous synthesis of organic compounds in nature.

The research on the subject performed over the past ten years allowed us to find other, hitherto unknown properties of the torch plasma in addition to its ability to synthesize organic compounds. This has changed the perception of the problem viewed as the extent to which the combination of the currently known properties of the plasma torch is sufficient to explain the realization in nature of the extremely complex physical and plasma-chemical processes required for the emergence of the simplest forms of life. Today the importance of and interest in this problem have become obvious.

In Chapter 5 we show that the results of the experimental studies of plasma torch properties reported in this book, like the results of a number of other works (Korobkin et al., 1977; Stanper, 1991; Zhang et al., 1999) dedicated to laboratory reproduction of the processes similar to those that occur in the torch, indicate that the available "combination" of plasma environment properties may have been sufficient to create the conditions necessary for the emergence of primary forms of living matter.

However, such an important prerequisite needs bona fide experimental corroboration proving that the entire event and its individual, key fragments can be realized in nature. To this end, let us logically consider what properties of the environment are needed for the emergence of living matter and whether the properties of the plasma torch can meet these requirements.

Such a comparison should be done with the understanding that many scientifically founded scenarios and hypotheses have been proposed over the past 50 years where the emergence of life was explained by the presence of various natural mechanisms and environments. Neither of these hypotheses is known, however, to have been able to offer the mechanisms and ambient properties that would at least possess the potential capabilities for the creation of the required conditions for the synthesis of the simplest structures of living matter.

In this book we choose the torch plasma as the model environment for such an analysis. To determine its properties that are necessary for the formation of the simplest replicating macromolecule with a primitive genetic code, we use the experimentally found properties of this environment. We assume that these properties could have ensured the spontaneous symmetry breaking required for the formation of homochiralic molecular structures. The environment must meet the criteria of a strongly nonequilibrium system far from the thermodynamic branch of equilibrium with the realization of "dissipative structures".

Given the above factors, the desired environment must ensure that:

(1) All the needed constituent elements of organic compounds are available in the impact interaction region. If the synthesis region contains only C and H then simple and complex hydrocarbon structures form and the process involves the realization of self-assembly and ordering; if C, H, and O are available then sugars and fatty acids form; in the presence of C, H, N, and O the chemical processes involve the formation of amino acids, polypeptides, nitrogen bases, nucleosides, and polymer organic compounds such as acetylenic hydrocarbons and dendrimers, whereas in the presence of C, N, H, O, and P the reactions involve the formation of nucleotides and their oligomers, as well as fragments of RNA molecules.

(2) During plasma expansion-away, as "true" chiral factors emerge, local chiral physical fields form that have a stable asymmetry «sign». The emergence of highly nonequilibrium dissipative structures that result in a spontaneous symmetry breaking with the formation of homochiralic molecular structures. When subject to local chiral physical fields of the plasma torch, spontaneous processes must acquire and preserve the corresponding polarity "sign".

(3) Local chiral physical fields with a stable asymmetry "signature" are produced during plasma expansion-away by nonequilibrium and unidirectional electric and magnetic fields. The environment must also ensure the development of highly unstable dissipative structures whose evolution leads to spontaneous symmetry breaking and synthesis of homochiralic molecules. It must also confine spontaneously developed homochiralic structures with local chiral fields of the plasma torch and "fix" polarity of enantiomers.

(4) Organic compounds synthesized in the torch undergo self assembly and ordering via nonequilibrium processes. The environment must also ensure prebioic selection during the synthesis of organic compounds on a molecular level including inner sequences of monomers, and this imposes substantial restriction on possible stable combinations in polymer chains. The environment should provide extremely high

rates of plasma-chemical reactions due to the catalytic activity of hot plasma and irreversible nature of the synthesis processes.

(5) There are large amounts of high-density matter involved in the synthesis of organic compounds with unlimited source of energy and raw materials combined with the development of the conditions to ensure the protection of synthesized organic compounds against radiative destruction and make possible its preservation and accumulation.

All the above requirements, with few exceptions, have been experimentally confirmed as described in Chapters 4, 5, and 6 of this book. These requirements, which are formulated for the basic characteristics of a rapidly varying environment, must be all fully satisfied and reproduced in laboratory simulations of the natural phenomenon with no restrictions or additional effort. Such an environment would be optimal for fast and efficient synthesis of homochiralic polymer organic compounds.

As is well known, the properties of living matter were shaped in abiotic processes and were regulated by the laws of physics and chemistry. Therefore the research aimed at the study of how did the prototype of living matter develop from inorganic substance should be based on the laws and methods of physics and chemistry.

To ensure fast development of the first living forms from inorganic matter, other requirements should be satisfied in addition to the exceptional properties of the torch plasma mentioned above. These requirements concern the initial characteristics of the local substance that gets involved in the formation of the torch. This local substance consisted of materials that enter into the composition of the meteorite and the target. Innumerable meteorite impacts during the early stages of the formation of the Earth ensured the high repeatability of the formation of such a substance. This substance could have been characterized by a great variety of its initial chemical composition, which differed from one impact to another. High impact velocity ensured high density of matter and enormous concentration of energy. Such a varying substance was more suitable for the synthesis of a wide range of organic compounds than a stationary environment, and it could result from the great variety and inhomogeneity of the chemical composition of the meteorites and rocks at the surface of the planet. Only such an infinite variety of initial conditions could make possible, in some rare cases, the formation of an environment having the composition and properties required for the emergence of living matter. Note that the number of fast processes that produced various environments on the early Earth was, like the number of such processes on the modern Earth, rather limited.

It follows from a comparison of the hypothetical conditions required for the emergence of life that to attain this goal, the initial conditions of the environment and the properties of the mechanisms of its realization should vary over a wide range. Such a variety of input parameters is very difficult to achieve in nature, especially in stationary or equilibrium environments, e.g., when they are subject to intense solar radiation. The point is that the initial conditions of such external factors are determined by the density and composition of the atmosphere and the intensity of solar radiation. However, these are the parameters that do not change radically with time. Hence the environments that form within a narrow range of possible initial conditions are incapable of providing the great variety of conditions required for the emergence of life.

The situation is more or less the same with more dynamic and nonequilibrium environments, which form in nature as a result of lightning discharges, volcanic eruptions, and thermal heating of rocks.

Laboratory simulations of the corresponding environments showed that exposure to such factors resulted mostly in the synthesis of monomers, e.g., amino acids, whereas no organic compounds in the form of short polymers could be produced in this way. Organic compounds synthesized in these experiments exhibited no symmetry breaking that could be viewed as ambient property needed for the emergence of living matter.

It is important that after the expansion-away of the torch plasma produced by the meteorite impact other, stationary environments form, which could make possible the completion of the processes that occurred during the previous stage. The following of these processes could have been of greatest importance for securing further conditions required for the survival of the first organism and for the preservation of newly-emerged life:

- Generation of powerful ground-rock outbursts in the meteorite impact zone during crater formation and the conversion of these ground rocks into the fine-grained fraction of dense dust clouds. At the initial stage of the expansion-away of matter and after its deposition these dust clouds protected the organic compounds synthesized in the torch against the destructive effect of radiation, ensured the preservation and accumulation of the synthesis products;
- Thermal heating of the impact region and of the deep layers resulting in the production of water ice, and maintaining moderate positive temperature required for the evolution of the synthesis products;
- Impact-driven injection of complex organic compounds synthesized in the plasma torch deep down into the Earth crust layers, underground reservoirs, and interiors of planetary satellites.

Combined with the plasma-torch properties, the above impact-provided possibilities could radically extend the application domain of the proposed concept.

A joint analysis of the impact processes considered above shows that the environments that are generated in these processes and that have all the above properties could have made possible, when occurring in nature, the synthesis of molecular structures. These structures with all the necessary conditions typical of, e.g., protoviroids, could have emerged in accordance with the hypothetical scheme proposed by A.Altstein. It is important that except for the final stage the plasma torch could provide all the needed conditions and processes. And although both the process and the model are very clear, the assembly of the protoviroid remains the responsibility of the author. Such is life and the specifics of its emergence.

Thus the environment where life could have emerged on the Earth has long evaded the attention of researchers. And this was the case despite the fact that specialists in plasma physics were well aware of certain properties of the plasma torch. However, the unique properties of the torch concerning the synthesis of organic compounds remained unnoticed.

It is important that in the torch plasma the above processes, which may have been directly associated with the emergence of life, usually occurred with high reliability, with no additional preconditions, and with a substantial "safety margin". Hence there are grounds to expect that not only precursors of molecules of the biological level of complexity, but also primeval, simplest representatives of living matter could have formed under favorable conditions in a single meteorite impact.

2.3. Water – The Key Substance for the Emergence of Life

Water is believed to be the key component that makes possible the emergence and evolution of life on the Earth. The researchers studied the possibility of replacing it by other solvents in extraterrestrial life. However, water has become a much more preferred option compared to other solvents after it had been found to be inexplicably abundant in the Universe.

The high abundance of water beyond the Solar System has been established from the data obtained with ISO IR orbital telescope (Salama, 2004). These measurements showed with high degree of confidence that water is present everywhere in the Universe including the regions that had been hitherto believed to be absolutely water-free. Moreover, water was shown conclusively to be one of the most important ingredients in the processes of star formation.

The Solar System also contains much water. This became clear after a number of successful missions to Mars and giant planets – Jupiter and Saturn – had found new evidence supporting this claim. The results of these studies are indicative of the presence of thick ice fields under the surfaces of Martian polar caps (Mitrofanov at al., 2007), of water under the ices of Enceladus, possible existence of an ocean under the icy surface of Europa, as well as lakes, seas, and rubble ice deep down under the surface of Titan (Porco, 2005,2006). An analysis of these results combined with the fact that giant planets have silicate-ice satellites, such as Ganymede, Callisto, Mimas, Tethys, Rhea, and Iapetus, can describe the real ice situation in the Solar System only to a first approximation. The point is that future studies may discover water on other objects, which have yet not been explored from this viewpoint.

Particular interest in the abundance of water and ice in nature is due to the fact that large quantities of this compound found on a cosmic object simplify considerably the requirements for the initial conditions needed for the emergence of living matter, which is triggered in this case by a meteorite impact. The latter condition is critical because only the tremendous energy of major meteorite impacts is capable of heating the impact region, melting ice in polar regions, and ensuring the presence of water and heat throughout extensive regions for a long time amounting to several tens of million years on the Earth and other cosmic bodies. Thus it is possible that the injection of complex organic compounds into underground reservoirs of planetary satellites with low surface temperatures may create the necessary initial conditions for the emergence of extraterrestrial life.

Water is known to be one of the most abundant substances in the Universe, third only to diatomic molecules H_2 and CO (Bochkarev, 2009). Given the triatomic structure of this rather complex molecule, the probability of the synthesis of water in space should be lower than that of such diatomic molecules as CH or SiO. However, this is not the case. Water is abundant everywhere: it is found in large quantities both in dust and gas clouds and in the Solar System.

It can be expected that the synthesis of water in nature began after the first-generation supernova explosions and formation of oxygen. This could have happened about 300—500 million years after the Big Bang. It is a natural assumption, because oxygen is needed for the synthesis of water (Goldsmith and Owen 1980, Jones 2004, Ulmschneider, 2006).

Today the relative abundance of water in the Universe is estimated at ~ 0.01% of the hydrogen abundance. Oxygen ranks third in terms of abundance after H and He. Oxygen abundance may be underestimated (Bochkarev, 2009), and this element may actually be much more abundant in the so-called "cold showers" - heavy "rains" that fall onto many galaxies from the «desert» regions that surround them.

The earlier proposed mechanisms of water synthesis in space environment or mechanisms involving ion-molecular reactions in the gaseous state or in the process of the dissociation of atoms at the surface of dust particles (O'Neill and Williams, 1999; Dulieu et al., 2007; Goldsmith and Owen 1980) are not very efficient (Ioppolo et al., 2008). Other mechanisms included reactions in the gaseous state involving oxygen ions colliding with H_2 in molecular clouds of moderate optical depth (Bochkarev, 2009). Such mechanisms were used to explain the synthesis of a broad class of interstellar molecules discovered in early 1970-ies because no other mechanisms operating at ultralow temperatures were at the time known to be capable of explaining the abundance of polyatomic molecules including organic compounds in the cold tenuous interstellar gas.

However, the Universe proved to contain much more water and hence there must be another, more efficient mechanism in nature to produce the required quantity of this compound.

The discovery of a new, highly efficient mechanism for water synthesis in space could not only simplify substantially the requirements to the initial conditions of the emergence of life, but also be of a separate interest for understanding a number of important processes in astrophysics and cosmochemistry.

The possible existence of a highly efficient mechanism of the formation of water molecules is implied by the results of observations of cosmic masers with the brightest recorded H_2O, OH, and SiO lines (Bochkarev, 2009). These observations allowed I.S.Shklovskii to propose a model where the so-called maser nests are located around young massive protostars with very powerful stellar winds with the mass-loss rates amounting to ~ 10^{-3} Mo/yr. The matter blown away from the star may have produced gas-and-dust disk where maser radiation originated in different layers. Masers could be excited, e.g., by the stellar wind and shocks generated by the pulsations of the central star. Cosmic masers therefore served as indicators of water molecules in the vicinity of protostars, and the intensity of maser radiation could be used to determine the abundances of these molecules and observe the dynamics of their formation and destruction.

It is important that many maser sources are associated with late-type giants and supergiants that have strong stellar winds and that are surrounded by cold and extended circumstellar shells.

We point out, without going into the details of the complex mechanisms of formation, pumping, and radiation in cosmic masers, that the H_2O and OH molecules required for the generation of radiation were produced supposedly as a result of interaction between strong stellar wind and gas-and-dust disk. Initially, the researchers believed the most efficient water synthesis mechanism to be the interaction between the hydrogen contained in the stellar wind and the surfaces of interstellar-dust particles and large atmosphereless cosmic objects.

Stellar winds are known to be like the Solar wind and consist mostly of hydrogen ions moving with the velocities of 10^2 - 10^3 km/s. Ion density may vary over a wide range depending on the nature of the star and the distance from it. For example, the velocity of the Solar wind amounts to 3-4 · 10^2 km/s and the ion density in the vicinity of the Earth's orbit is 5 cm^{-3}.

To understand the physics of the proposed mechanism of the synthesis of water molecules under the conditions of interstellar clouds, let us make a simplified analysis of the interaction between the stellar wind and gas-and-dust disk. To this end, we can use the characteristic parameters of the Solar-wind plasma and, in particular, we assume that the energy of hydrogen ions is equal to ~ 1 keV.

Dust particles play important part in the analysis of the circumstellar shell and its interaction with the stellar wind (Goldsmith and Owen 1980). The initial sizes of these particles, which have the form of fine-crystalline or amorphous objects consisting of silicates, metal oxides, and graphite, do not exceed 10^{-6} cm. During ~ 10^8 years these particles develop a coat of water and ammonia and transform into 3 10^{-5} cm diameter ice balls.

Thus during the evolution of a star the hydrogen ions of its stellar wind like those of the Solar wind initially interact with "naked" particles of interstellar dust grain nuclei. The most conclusive results inferred from the absorption bands of dust grains at micron wavelengths imply that these bands correspond to silicate absorption and may belong to SiO_2, $MgSiO_3$, Mg_2SiO_4, and metal oxides, i.e., to oxygen-rich compounds.

During further evolution of the star, from the beginning of the formation of the compact gas-and-dust cloud, accumulation of planets from the swarm of intermediate bodies, and until the formation of planets, stellar-wind ions constantly collide with dust grains whose surfaces may be ~60% composed of oxygen atoms.

Such a process, namely the bombardment oxygen-rich dust particles by 500 eV to 5 keV hydrogen ions, occurs almost always and anywhere where a star shines in the Universe. What may be the results of such an interaction? These interactions may involve the synthesis of water in collisions between hydrogen ions and oxide molecules followed by the "release" of oxygen and formation of hydroxyl in subsequent processes of hydrogen addition and synthesis of water molecules. Let us now consider this process in more detail.

Bombardment of surfaces of celestial bodies devoid of atmosphere and dust particles is a very common phenomenon in nature. This is how solar-wind ions act on the surfaces of Mercury, Moon, Phobos, and Deimos.

This process is a natural analog of the laboratory technique known as secondary-ion mass spectroscopy (SIMS). In particular, this process makes it possible to knock out secondary ions by sputtering the surface of the specimen by a beam of primary ions, and then determine the chemical composition of the target by analyzing the mass spectrum of secondary ions (Cherepin, 1992). However, as we pointed out above, this process is also capable of synthesizing new chemical compounds.

Each incident primary ion knocks 10^{-2} or 10^{-3} secondary ions and a dozen neutral, so-called sputtered atoms out of the thin surface layer.

In laboratories bombardment is usually performed by Ag, Kr, Xe, and, sometimes, O atoms with energies of up to 10 keV. Except for some special cases, secondary ions and sputtered particles emerge in the form of atoms, although some secondary ions may also emerge as polyatomic molecular ions – the so-called clusters, and this property is of great importance for the case considered.

Numerous mass-spectrometric measurements made using this technique for over 95 years show that ion bombardment of oxides results in secondary atomic O ions and molecular O_2, OH, and H_2O ions appearing in the mass spectra. The production rates of molecular compounds increase substantially if hydrogen ions – the main component of the stellar wind – are used as primary agents.

The physical mechanism of this phenomenon can be described schematically as follows: a 1-keV hydrogen atom interacts with the surface layer of the molecules of the target. The energy of the hydrogen atom allows it to break the chemical link between the constituent atoms of the target, e.g., oxygen, and tear them away from the surface thereby creating, after ionization, a secondary ion. However, not all oxygen leaves the target in the form of atoms. Because of its high chemical activity, part of atomic oxygen manages to add one, and sometimes even two hydrogen atoms. In this way, water and hydrogen oxide can be synthesized and these molecules may leave the target as secondary ions or remain within it.

This is a simplified scheme of water synthesis driven by a primary hydrogen ion beam acting onto an oxide target.

It is important that this mechanism depends only slightly on the surface temperature and is equally efficient both at low temperatures typical for interstellar clouds and at the incandescent Mercurian surface. The efficiency of water synthesis in the cases where oxygen is in the solid state, e.g., in oxides, must be substantially higher than in the case where it is in the gaseous state. However, the water-synthesis efficiency could not be determined so far and laboratory calibrations are needed to estimate it.

Over the past 40 years the researchers have been trying to find evidence for such a highly efficient mechanism of water synthesis. One of the possible options is the synthesis of water as a result of ion bombardment of solid-state oxygen-containing substances at the surfaces of atmosphereless cosmic objects or interstellar-dust grains. Similar experimental studies were also performed in laboratories, some of them dedicated to simulating ion-collision processes in space. It is important that in a number of cases the bombardment of targets by primary ion beams resulted in the formation of new substances including organic compounds (Starukhina and Shkuratov, 1994, 1995). These results also indirectly indicate that water is a lilely product of such ion-impact processes.

Laboratory simulations of this process performed back in mid-1970-ies showed that only hydroxyls form when silicon (Mattern et al. 1976), aluminum (Gruen et al., 1976), and titan oxides (Siskind et al., 1977) are sputtered by H and D ions with energies ranging from 10 keV to 10 MeV.

The results of these very experiments may have largely affected further interpretation of many other measurements including those made onboard modern space probes. Thus the authors of a number of works avoided interpreting water absorption bands observed by onboard visual spectrometers in lunar surface studies as due to the ion-collision mechanism of water synthesis. The observed effects were explained by the presence of hydroxyl at the lunar surface (Starukhina and Shkuratov, 2000) or by the deposition of water on the optical elements of onboard instruments.

And only recently the possibility of water synthesis has been demonstrated in laboratory experiments involving the bombardment of solid-state oxygen molecules by H and D ions at temperatures from 12 to 28K (Ioppolo et al., 2008). However, this mechanism can hardly been widespread in open space.

In 1999 the onboard Visual and Infrared Mapping Spectrometer of the Cassini interplanetary probe detected surface absorption of lunar regolith at the wavelengths corresponding to hydroxyl and water. Clark (2009) explained these results by surface adsorption of the ancient water from lunar interiors. The above author did not rule out the possibility of the synthesis of water as a result of bombardment by Solar wind ions, however, no experimental evidence for the realizability of this process has been produced because it was unavailable.

A strange aspect of this situation consisted in the following. Lots of laboratory experiments have been performed over more than 90-years long history of the application of SIMS technique, and they have regularly produced mass peaks of molecular water ions in direct mass-spectrometric measurements. The researchers usually explained these peaks by the presence of water vapor in the residual gas in the vacuum camera. And this explanation persisted despite the fact that the observed peaks at 18 a.m.u. could be interpreted as rather solid evidence for the realization of the ion-collision scenario of water synthesis.

Later, such experiments were also confirmed by measurements made onboard space stations, where water absorption bands were observed using remote spectrometers during a number of space missions studying the lunar surface (Pieters et al., 2009; Sunshine et al., 2009). Another important evidence for the possibility in principle of water synthesis as a result of bombardment of oxides by hydrogen ions comes from the results of experimental studies that demonstrated the possibility of the synthesis of polycyclic aromatic hydrocarbons (PAHs) when a carbon target is bombarded by hydrogen, helium, and nitrogen ion flows (Starukhina and Shkuratov, 1994, 1995).

A comparison of these results may have pointed to the possible synthesis of water via the bombardment of oxygen-containing targets by hydrogen atoms. There appeared to be strong reasons for a more in-depth study of the mechanisms of water synthesis in the processes that accompany ion bombardment both in space and in the laboratory. However, such studies did not materialize for two reasons: (1) experiments of this kind are very specific and difficult to perform in a laboratory and (2) space stations lacked the instruments capable of directly detecting and measuring molecular water ions. The corresponding investigations have not been done until recently.

R.Sagdeev proposed to perform such direct measurements of molecular ions with onboard instruments of the lunar space probe of the LCROSS mission. Direct measurements of water ions during this mission could be made with the highly sensitive MANAGA time-of-flight mass analyzer, which we describe in detail in Chapter 7 of this book. This instrument, which was developed by the author of this book at the Space Research Institute of the Russian Academy of Sciences, was the perfect match for the task. Unfortunately, the proposal was rejected because of the lack of foresight on the part of the project directors.

The lack of mass spectrometers capable of directly detecting molecular water ions did not allow the researchers to associate water-line absorption observed during the recent Deep-impact (Sunshine et al., 2009) and Chandrayaan-1 (Pieters et al., 2009) space missions with the synthesis of this compound on the lunar surface.

That is why the experimental confirmation of the possibility of water synthesis as a result of the bombardment of metal oxides and other oxide compounds by energetic hydrogen atoms can be considered a result of especially high importance. This process was studied, at my request, under laboratory conditions by joint efforts of the Laboratory of active diagnostics of the Space Research Institute of the Russian Academy of Sciences and by the staff of the

Laboratory of mass spectroscopy of Kurdyumov Institute of Metallophysics of the National Academy of Sciences of Ukraine headed by V.T.Cherepin. The experiments were performed at the Institute of Metallophysics, which has years-long experience in the use of SIMS technique for solving a number of scientific tasks with primary beams consisting of hydrogen atoms and which has a facility and original mass-spectrometric instruments needed for such works.

To ensure high reliability of the results of laboratory experiments simulating ion-collision mechanism of water synthesis, we decided to reproduce the main parameters of interaction and natural conditions of the process as closely as possible. This means that the energy of ions should lie in the 1 to 4 keV interval. Vacuum must be at least as good as $5x10^{-7}$ mm Hg. Moreover, we also had to make our experiments very clean. Furthermore, it was equally important to achieve direct detection of molecular water ions with a substantially wider range of initial target substances including some lunar minerals and metal oxides. We decided to use deuterium ions to prevent eventual contamination by «laboratory» water.

As of now, results have been obtained that are indicative conclusively of the possible synthesis of heavy water as a result of the bombardment of oxygen-containing substances by deuterium ion beams. These results will be published completely in the nearest future.

Along with these experiments laser radiation was used to simulate micrometeorite impacts in the same laboratory of the Space Research Institute. Although they have only just started and despite the preliminary nature of their results, these experiments have already demonstrated the possibility of the synthesis of molecular water and hydroxyl ions as a result of a hypervelocity meteorite impact if the composition of both the projectile and target is close to that of glass. The only reasonable explanation of this experimentally observed effect is the possible synthesis of water. The amplitudes of the hydrogen and oxygen peaks in the mass spectrum lead us to conclude that these elements are highly abundant in the target. Glass was chosen as the target because such a sample is guaranteed to be free of bound water.

The observed effect requires a more in-depth study, which must involve a wider range of initial substances, and work is currently in progress in this direction.

The observed results, despite their preliminary nature, provide important information about the mechanism of the synthesis of the «first water molecule». Thus oxygen and metals including Fe were synthesized simultaneously in the interiors of the first-generation supernovae and therefore the explosions of these stars could have produced the first metal oxides (Goldsmith and Owen, 1992, Jones, 2004, Ulmschneider, 2006).

The synthesis of water in the process of an explosion was unlikely. It is therefore safe to suggest that the most plausible mechanism of water synthesis under such conditions could be the bombardment of metal oxides by hydrogen ions, and that it was this process that ensured the synthesis of water – the new chemical compound in the Universe.

The above results, which were obtained by simulating in a laboratory the phenomena that occur during meteorite and ion bombardment of solid-state substances, are of special interest. They allow us to argue that if such conditions are realized in space then the synthesis of organic compounds and water molecules should be possible. In the former case this may happen in the plasma torch of a hypervelocity meteorite impact (Managadze 2009, 2010), and in the latter case, as a result of ion bombardment (Starukhina and Shkuratov 1994, 1995). Whereas synthesis mechanisms in meteorite impacts involve plasma processes, the corresponding mechanisms in the case of ion bombardment should be associated with the synthesis of matter in the «hot spot» of the ion impact. As far as synthesis is concerned, the physical mechanisms of these processes differ, but the final result is the same. Therefore taking the above facts into account should help us correctly interpret the results obtained in various space experiments involving artificial or natural active influence on the environment.

Note that in open space water molecules are not only synthesized, but also destroyed, e.g., as a result of ion bombardment or photodissociation by ultraviolet or harder radiation. The consequences of the photodissociation of water molecules in the Solar System can be observed, e.g., in the form of hydrogen coronas of comets. Hence water available in the Universe is a result of competing processes of synthesis and destruction.

This highly efficient and experimentally confirmed water-synthesis mechanism is especially valuable because it explains the high abundance of this substance in the Universe including the neighborhoods of stars with planetary systems, like, e.g., in the Solar System. The high abundance of water can be viewed as the main prerequisite for the emergence of primary forms of living matter, a condition that is always fulfilled if there are planetary systems. Other prerequisites – provided that planets have formed - have already been listed above:

- The formation of meteorite bodies;
- The meteorite-carried delivery of the substances – including carbon – required for the synthesis of organic compounds;
- The acceleration of these bodies to higher-than-critical velocities to ensure the generation of a plasma torch;
- The formation of a crater and heating of a part of the impact zone to moderate temperatures.

If the conditions at the surfaces of cosmic bodies are unfavorable, e.g., if surface temperature is extremely low or high, the prerequisites for the emergence of living matter are:

- The possibility for complex organic compounds synthesized in the torch during a penetrative meteorite impact to be introduced into the interiors of cosmic bodies and into subsurface reservoirs;
- The availability of water or ice under the surface at different locations on the same planetary body and on several cosmic bodies.

The above conditions are likely to be provided by nature. Therefore organic compounds can be synthesized and accumulated, and favorable conditions for the formation of the simplest forms of extraterrestrial life can develop in the interiors of various cosmic bodies including those with extreme surface temperatures.

2.4. Meteorite Bombardment and Nightly Darkness on the Earth

The discovery of a living cell that existed 400—600 million years after the formation of the planet (Schidlowski, 1998) and the availability of liquid water on the Earth since the first 100—200 million years (Wilde et al., 2001) shifts substantially back in time the epoch when conditions appeared for the emergence of life on our planet. The new results of geological studies indicate that the conditions on the Earth at that time were different, much unlike those used in the early scenarios. However, these conditions may have been suitable for the emergence of primary forms of life if viewed in the framework of the new concept proposed in this book.

Studies of lunar craters showed that during its formation together with the Moon the Earth too was subject to strong meteorite bombardment, which, with varying intensity, continued for 500—600 million years (Safronov, 1972; Grieve, 1980; Bronshten, 1987; Vityazev et al., 1990; Hartmann et al., 1990). This result has been recently corroborated by ground-based studies. Researchers from Oxford and Queensland universities found traces of meteorite matter in samples retrieved from ancient rocks in Greenland and Canada. The results of computer simulations indicate that at the early stage of its formation and, in particular, during the first 200 million years, the Earth was subject to more than 20 thousand impacts of giant meteorites with sizes exceeding 20 km. These catastrophic impacts occurred, on the average, once in 10000 years (Pechernikova and Vityazev, 2008) and were unlikely to determine the "dust weather" on the Earth. However, regularly falling smaller meteorites could have determined the degree of dustiness and the duration of the period of high dust content.

The dustiness of the Earth atmosphere during the first 100—200 million years can be estimated based on the results of earlier published studies of the formation of fine-dispersed dust, which could have stayed in the atmosphere for long time periods amounting to several months. The corresponding papers report the data on the atmospheric dustiness produced by nuclear explosions and the results of numerical simulations of the generation of dust clouds during the formation of meteorite impact craters (Nemchinov et al., 2008a).

These results are consistent with the following estimates that I derived based on a natural experiment and, in particular, on the results of measurements of atmospheric dustiness produced by the fall of the Tunguska meteorite.

The fall of the Tuinguska meteorite is known to have been a unique natural phenomenon, which occurred during the night from June 30 to July 1, 1908. In mid-May of the same year, systematic observations of atmospheric transparency began at Mount Wilson observatory in California, 10000 km from the Tunguska impact site. The unprecedented global natural experiment involving the intrusion of a meteorite – purportedly a cometary nucleus – into the Earth atmosphere made it possible to perform the so far unique measurement of atmospheric dustiness after explosive disintegration of a cosmic body. Ch. Abott, a well-known astronomer, recorded a more than month-long decrease of atmospheric transparency by about 10% at 0.40, 0.45, and 0.75 μm 360 hours after the Tunguska meteorite explosion (Bronshten, 1987).

We now use the results of these observations to estimate the dustiness of the early atmosphere resulting from the meteorite bombardment about 4.3 billion years ago, i.e., during the first 100—200 million years after the formation of the Earth.

The Tunguska meteorite is known to have exploded about 5 km from the Earth's surface without producing any crater. According to various estimates, the mass and diameter of this body amounted to $\sim 10^6$ tons and ~100 m, respectively (Bronshten, 1987; Fesenkov, 1949).

Relative estimates combined with the results of observations of atmospheric dustiness can be derived without determining the mass of the meteorite matter that produced the fine-grained phase and "hovered" in the atmosphere. We can assume that the amount of matter in the form of submicron-particle dust is more or less the same for different initial impact conditions and is equal to ~10% of the mass of the meteorite provided that it has produced a crater. However, the Tunguska meteorite did not reach the Earth surface and therefore the resulting dust cloud contained no contribution from the impact crater, which could have been quite appreciable under different circumstances. This means that the estimate characterizes the case of minimum dustiness.

We use the results of Pechernikova and Vityazev (2008) to determine the number of Tunguska-size meteorites that could have bombarded the Earth during the first 100—200 million years since its formation. According to these data, at the final stage of the formation of the Earth the number of meteorite impacts was 10^{10}, 10^{7}, and 10^{3} for 1-km, 10-km, and 100-km large meteorites, respectively. According to this size distribution, during the first 200 million years of its history the early Earth should have been subject to 10^{13} impacts of 100-m (i.e., Tunguska-sized) meteorites. Hence during the time period considered the Earth was hit by ~5 x 10^{4} meteorites/year or at least 100 meteorites/day.

Such an intense bombardment could have produced global and long-lasting dust pollution and completely prevented solar radiation from reaching the Earth surface for a long time. Our estimate shows that during this period night darkness has reigned on the Earth for several million years with the properties characteristic of "nuclear winter".

An analysis of dust-cloud generation during impact-driven crater formation led us to conclude that in the case of a constant meteorite impact rate there may have been a peculiar mechanism that could be capable, to a rough approximation, of maintaining the dust pollution of the Earth atmosphere at a constant level irrespectively of the atmospheric density.

Qualitatively, the mechanism consists in the following: to retain water on the Earth, the primeval atmosphere must have had a density of at least ~0.01 its current value (Pechernikova and Vityazev, 2008). Could the density of the early Earth atmosphere have influenced substantially the level of its dust pollution? This is highly unlikely for the following reason.

In a low-density atmosphere dust settles faster and therefore dust pollution should also decrease faster. However, a low-density atmosphere allows more small meteorites to reach the Earth surface while moving with velocities sufficient to produce impact craters and generate dust clouds, which should not be the case for a dense atmosphere. With increasing atmospheric density the efficiency of dust formation must decrease because of the deceleration of small meteorites, but, at the same time, dust should stay longer in the air. Hence, to a rough approximation a variation of atmospheric density should have no effect on the degree of dust pollution because of automatic compensation.

An analysis of carbon isotope ratios as life markers in sedimentary rocks (Schidlowski, 1998) shows that widespread microbial life has existed on the Earth soon since the formation of the planet.

The time scale of the decrease of the meteorite bombardment intensity as determined from lunar craters combined with the above estimates of atmospheric dust pollution lead us to conclude that the "reign of darkness" could have lasted for at least 200 million years. And this means that the chemical stage of evolution involving the accumulation of organic compounds needed for the emergence of life must have proceeded without solar radiation and hence the first forms of living matter may have appeared on the early Earth during the "dark age".

The possible lack of solar radiation during the chemical stage of evolution is one of the most difficult-to-resolve inconsistencies in the traditional and generally adopted "warm scenario" of the origin of life. However, moderately positive temperatures on the Earth at that time, which ensured the presence of water on its surface, along with the conditions that allowed the formation and accumulation of complex organic compounds can be viewed as sufficient and capable of resulting in the emergence of living matter.

***

The above description of the processes of the formation of planets, initial conditions on the early Earth, and the new hypothesis about the origin of water in the Universe allows us to make the following important conclusion: if a Sunlike star had a system of planets then the most important initial conditions required for the realization of the proposed concept in nature existed automatically and without fail. These conditions include the presence of carbon and the constituent elements of organic compounds in the meteorites that bombarded the planets and their moons, the availability of water or ice, moderate temperature at the surface or in the interiors of these objects, and sufficiently high meteorite velocities to produce a torch.

When analyzing the dependence of the formation of impact plasma and, as a consequence, of the generation of a plasma torch, on impact velocity we should remember that earlier estimates made by some authors who stated that no plasma forms in impacts of 2 g/cm^3 projectiles moving at velocities ranging from 5 to 10 km/s or even from 10 to 15 km/s are wrong, need to be revised, and require substantial correction.

Thus the results based on experimental studies of this process carried out in various laboratories in a number of countries demonstrate conclusively that the generation of a plasma torch via surface ionization during an impact begins at velocities of 2 and 5—6 km/s for an iron or carbon projectile, respectively. In this case the pressure shock compression of interacting bodies may be as high as several units of 10^5 bar.

Volume ionization in an impact is achieved for a projectile density of $2g/cm^3$ and collision velocity of 15-20 km/s. Under these conditions, a large meteorite may produce a plasma torch of impressive dimensions and it may produce an ionospheric disturbance resulting in a global radio blackout for a considerable period of time.

The processes of plasma formation and the physical mechanisms of their realization during an impact are discussed in more detail in Section 1 of the Afterword.

The current lack of due interest in plasma outbursts among the experts in impact processes can be explained by the following factors. First, it takes, on the average, about 5% of the impact energy to produce a plasma torch and, second, the duration of the entire process does not exceed 0.25% of the time from the meteorite impact to the end of the formation of the crater. Hence the lack of interest can be explained by the small fraction of the impact energy spent for the torch and the short duration of' the process, as well as by the total lack of material evidence in the form of "lithic record". Thus many renowned authors of the book "Catastrophic effects of cosmic bodies" published in 2005 and translated into English in 2008, except Kovalev et al. (2008), do not consider at all the possibility of the generation of a plasma torch in hypervelocity meteorite impacts.

I must admit, however, that the above book is of great interest and contains useful information about the physics of impact processes including those that occurred on the early Earth; about the formation of impact craters; about the properties of heated matter ejected from the crater in the gas-vapor state, and about the main factors of asteroid danger and their elimination methods. I found the papers published in the monograph to be of great assistance in the development of my new concept and preparation of this book for printing.

Of great interest is also the fact that the papers published in the above book (Nemchinov et al., 2008a; Ivanov, 2008a, 2008b) show conclusively that the average velocities of the impacts of asteroids and cometary nuclei are equal to 20 and 50 km/s, respectively. If the diameter of the projectile exceeds 1 km then these velocities are sufficient to produce powerful plasma torches with all the ensuing consequences, which we consider above and in the rest of this book.

It would therefore be appropriate, based on the experience gained from numerical simulations of impact processes, to discuss with plasma physics experts the peculiarities of these interactions that show up in impacts with the velocities of 30—50 km/s. Such an analysis will allow us to start investigating the unknown consequences of impact effects in the generation of the plasma torch and ensure fast progress in this so far scantly explored field even in the case of low-velocity impacts.

The claims that the initial conditions for the realization of the new concept are extremely simple do not mean that these conditions should be satisfied for any configuration of the planetary system. To remove this restriction, we must analyze in more detail every particular available bona fide piece of evidence for planets orbiting individual stars. Only after the analysis of concrete data it will be possible to make conclusions about the occurrence in nature of conditions consistent with the above statements.

The point is that the extrasolar planets discovered in recent years move in highly elliptic orbits. The surfaces of such planets may heat up at the perihelion and freeze at the aphelion. The temperature difference may be quite substantial. Can life exist under the surface of a planet moving in such a trajectory? It turned out that this is still possible as evidenced by planets and moons of the Solar System.

Thus in Chapter 7 I try to show, based on experimental results, that in underground reservoirs of such "cold" Saturnian moons as Titan and Enceladus organic compounds can be found that participate in the processes similar to low-temperature chemical evolution. We cannot rule out completely the possibility of existence of the primitive forms of extraterrestrial life under such conditions. Hence if a planet has underground ice or underground water reservoirs then, according to our concept based on impact action, ultralow surface temperatures should not rule out the possibility of the emergence of extraterrestrial life.

The situation appears to be much more serious when such a planet passes its perihelion. However, in this case too we may assume that if the "zone of life" is located under a thicker than 100 m layer even a substantial heating of the surface should not result in appreciable change of the local temperature in deep layers.

In the Solar System this case is exemplified by Mercury. It is the closest planet to the Sun at a distance of ~60 million km from it. The maximum surface temperature at the perihelion amounts to ~400 C. However, radar observations indicate the presence of thick ice deposits in polar regions with the sizes of up to 150 km. It might be supposed that the ice that is protected there from direct solar radiation is additionally covered by a heat-insulating layer of fine-grained regolith.

The available observational results do not rule out the possible presence of ice and water in deeper layers of Mercury, and this fact can be used to expand substantially the possible "habitat" of extraterrestrial life for planets moving in highly elongated orbits.

Chapter 3

PHYSICS AND METHODOLOGY OF IMPACT STUDIES

3.1. CHARACTERISTIC FEATURES OF IMPACTS

To understand the main physical processes responsible for the synthesis of organic compounds in the plasma torch of a hypervelocity impact, we must analyze the most important stages of this phenomenon and the initial conditions that make possible the formation of the torch. The principal stages include the development of a hot spot, formation and expansion-away of a naturally asymmetric plasma torch, the synthesis and "hardening" of chemical compounds including organic substances.

The expansion-away of the torch is followed by crater formation processes. At this stage, ground matter is ejected, dense dust clouds form, and a large volume is heated in the impact region. Practically all the above processes that accompany a hypervelocity impact play an important and, some of them maybe even the crucial, part in creating the conditions needed for the emergence of life.

Impacts are among the most common events in nature. There is a great variety of impact processes and the characteristic feature of an impact is that it must involve at least two bodies. One of these bodies is considered to be at rest and is referred to as the target, whereas the other body, which runs into the first one, is referred to as the projectile. Hence the projectile has nonzero velocity relative to the target. Depending on the velocity and mass of the projectile, energy is released, which may result in the deformation, destruction, or heating of the colliding bodies. The consequences of the impact depend on the physical characteristics of the projectile and the target and on the relative velocity of their interaction.

Impact processes realized on the Earth have a common feature – they all have low relative velocities. The main barrier that prevents the achievement of high velocities is, first and foremost, the Earth's atmosphere. Therefore impacts of bodies that occur on the Earth may only result in deformation, destruction, and thermal heating.

Generally, on the Earth the maximum relative velocity of an impact can be achieved if the projectile and the target move head on toward each other.

Thus the maximum impact velocity for such a head-on collision in the Earth's atmosphere involving high-velocity supersonic drones (or other aircraft) may amount to ~2 km/s. Such processes can be conventionally classified as high-velocity impacts.

The phenomena that accompany impact processes have threshold-type behavior. Thus according to Zel'dovich and Raizer (2002), interaction remains elastic and does not result in substantial temperature increase needed for plasma formation up to an impact velocity of about 5 km/s in the case of a $2g/cm^3$ projectile. However, even a small extra increase of impact velocity may result in a substantial and irreversible compression of the lattice and significant increase of the ambient temperature. Such a temperature increase is sufficient to ensure surface ionization of interacting bodies and formation of a plasma torch.

An increase of the impact velocity to 15-20 km/s would result in volume ionization of both the projectile and the target and in a factor of 10—30 increase of the plasma density.

The minimum impact velocity that results in the formation of a plasma torch can be used to separate two types of impact processes. Thus if no torch is produced, the collision can be called a high-velocity impact, and the process resulting in the formation of a plasma torch can be referred to as a hypervelocity impact. A similar delimitation of impact interactions can also be based on such a parameter as the Mach number.

The Mach number is equal to the ratio of the velocity of aircraft in the Earth atmosphere to the velocity of sound and is denoted by letter M. Thus if the velocity of an aircraft is five times greater than that of the propagation of sound in the atmosphere then the Mach number for such a process is equal to 5. In this case, a shock develops in the atmosphere and propagates at a velocity equal to that of the aircraft.

For impact interactions between solid bodies the Mach number is, like in the case of atmospheric processes, equal to the ratio of the velocity of the projectile to the velocity of the propagation of sound in the matter of the projectile and the target. Note, however, that this process now occurs not in the atmosphere, but in condensed-state matter. One should further bear in mind that the velocity of sound in a solid body is about 10 times higher than the velocity of sound in the atmosphere.

Thus in the case of a *high-velocity* impact, where $M < 1$, the velocity of the propagation of perturbations in the constituent materials of the projectile and the target is, like the velocity of the outflow of heat from the interaction region, higher than the projectile velocity. This means that under such conditions significant heating of the projectile and target materials is prevented by the intense heat outflow from the interaction region. In this case, the matter may be heated up to several thousand degrees and transform into the gas-vapor state. Heated matter is then ejected into the ambient space as vapor consisting of neutral atoms and molecules. Therefore the impact energy would then be insufficient to dissociate and ionize the projectile and target materials, i.e., to produce ions and electrons. Hence because of the lack of charged particles no electromagnetic fields or charged particle flows will be generated in the vaporized matter.

Unlike the above scenario, in the case of a *hypervelocity* collision between condensed-state bodies with $M>2$ or $M>3$ the velocity of the projectile is, by definition, several times higher than that of the propagation of sound waves in the colliding bodies. For example, in the case of the Earth the maximum velocity of a meteorite impact may be as high as 70 km/s and the Mach number for such interactions may be 10 or higher.

Under such conditions, the state of the medium changes radically during the impact. The point is that in this case heat outflow from the interaction region cannot keep up with the fast energy input, whereas the shocks produced in the projectile and target materials are "carried" with the flow in the process of the destruction of the colliding bodies and therefore remain inside the interaction region. The velocity of the shock propagating through the target is then always higher than the velocity of the "projectile—target" interface and hence the distance between the shock and the interface increases progressively with time. It is important that in the case of low impact velocities this distance increases faster than in the case of high impact velocities.

This results in intense heating of the matter in the interface region and leads to the formation of a fully atomized and ionized plasma environment. The plasma temperature and pressure in this region may be as high as 10^5—10^6 K and 10^6 atm, respectively. It is the explosive expansion-away of a current-carrying medium with such characteristics that produces the plasma torch.

In the case of a *hypervelocity* impact a hot spot forms in the contact region between the projectile and the target and propagates depthward until the shock moving through the projectile body destroys it completely. Only then, the plasma produced during the impact is able to freely expand into the ambient medium. Until that moment the projectile body, acting as a piston, pushes the hot spot deep down into the target. In the process, a toroidal zone forms between the projectile and the target, plasma is partly ejected outward through this zone to produce a torch, which also has a toroidal configuration.

If the projectile crosses a finite-thickness target undestroyed, then a part of the plasma produced in the hot spot is "pushed out" by the projectile piston into the domain located under the target. We show in Chapter 7 that such an impact configuration may be of special interest from the viewpoint of eventual introduction of organic compounds synthesized in the torch into the subsurface cavities fully or partially filled with the atmosphere or water.

Therefore if the impact velocity exceeds the critical threshold then a domain filled with the matter in the fourth aggregate state, i.e., with plasma, forms in the projectile—target interaction region. It is a fully atomized and ionized, highly overheated, high-density substance consisting of the atoms that until that were part of the projectile or the target.

The above can be summarized as follows: at the initial stage of hypervelocity impact the so-called hot spot forms where the projectile material and part of the target material with the mass comparable to that of the projectile transform into the plasma state with the temperature and pressure reaching one million degrees and one million atmospheres, respectively. These conditions ensure explosive ejection into the ambient medium of a flow of high-temperature, highly irregular, high-velocity plasma, which is initially 100% atomized and ionized. The researchers call such a plasma formation the plasma torch (Delone, 1989), and refer to the event that produces such a plasma torch as a hypervelocity impact. Note that formations like the plasma torch with their exceptional characteristics can form in nature only as a result of hypervelocity impacts. However, under laboratory conditions a plasma torch similar to a impact-produced torch can be generated using Q-switched laser radiation with a pulse duration of 0.3 to 10 ns and with the radiation fluxes of 10^9 to 10^{12} W/cm^2. The characteristics of the laser pulse can be varied so as to make the physical characteristics of the plasma torch as close as possible to those that arise as a result of a hypervelocity impact.

The formation of a crater is believed to be an important property of a hypervelocity impact. Crater forms after the expansion-away of the torch plasma, and it is the process that always accompanies every impact. It consists in the destruction of the target while the projectile penetrates into it, and in the explosive ejection of molten fragments of the projectile and the target along with the impact-ground matter of the interacting bodies into the ambient space, and ends up with the formation of dense dust clouds in the impact region.

The size of the crater produced by a hypervelocity impact depends on the impact velocity and the physical properties of the interacting bodies. To more conveniently and accurately compare the characteristic sizes of the projectile and the crater, the concept of intermediate crater has been introduced. Intermediate crater is the maximum-sized cavern produced in the target by the projectile for a very short time during the formation of the crater. The volume of the intermediate crater may usually be greater than that of the projectile by two orders of magnitude. Thus an impact of a 1-km meteorite produces an intermediate crater with a diameter and depth of 5 and 2—3 km, respectively, which, after rock displacement ceases acquires a stable shape with the diameter of rock destruction of about 10—12 km. By the end of the crater formation, the thermal heating zone around the impact center is 10 times wider and three times deeper than the diameter of the projectile. Because of the low thermal conductivity of the rocks the heat released during the impact is retained for a long time. Thus, according to simple estimates, the rock cooling time for a 10-km diameter meteorite impact may be as long as ~10 million years.

3.2. HYPERVELOCITY IMPACT PROCESSES IN NATURE

Thus an impact of a $2 g/cm^3$ projectile moving at a velocity exceeding the critical threshold of V_{CR} = 15—20 km/s (Zel'dovich and Raizer, 2002) results in volume ionization and produces a plasma outburst or a plasma torch in the interaction region (Friichtenicht and Slattery, 1963) (see Figure 4). For a torch to form, the time scale t_{IMP} of the interaction of the incident body with the target must be shorter than the time scale t_H of heat outflow from the impact region. Under these conditions, within a very short time energy is released that exceeds the atom binding energy in the matter and the ionization energy. As a result, the incident body and a part of the target transform into high-temperature plasma, which is ejected into the ambient space by the gas-kinetic pressure.

Depending on the physical properties of the interacting bodies the temperature of the plasma produced in impacts with velocities $V_{IMP} > V_{CR}$ reaches 30—50 eV and the plasma is, at the initial state, fully atomized and ionized. However, the plasma begins to cool down rapidly during its adiabatic expansion-away. The decrease of the plasma temperature triggers neutralization (recombination) and condensation, and the ionization degree decreases by a factor of $\sim 10^2$—10^3. Note that the ion composition of such plasma is approximately representative of that of the interacting bodies. At a certain stage of the torch expansion-away, some – mostly atomic – low-energy ions recombine into more complex chemical structures and produce new compounds in the form of neutral molecules or polyatomic ions (Managadze, 2003). If the projectile or the target contains constituent elements of organic compounds then such compounds can be synthesized in the plasma torch during its expansion-away. Below we pay a special attention to the synthesis of molecular ions including organic ions during the expansion-away of the impact plasma.

Hypervelocity impact is a typical cosmic phenomenon (Figure 5). Thus the surfaces of atmosphereless objects of the Solar System are subject to permanent hypervelocity bombardment by micrometeorites with masses ranging from 10^{-15} to 10^{-12} g. Micrometeorites with diameters less that 1 μm are accelerated to velocities V_{IMP} ~50 km/s by solar radiation pressure. Such bombardment plays important part in shaping the surface regolith layer of atmosphereless objects (Sagdeev et al., 1986a).

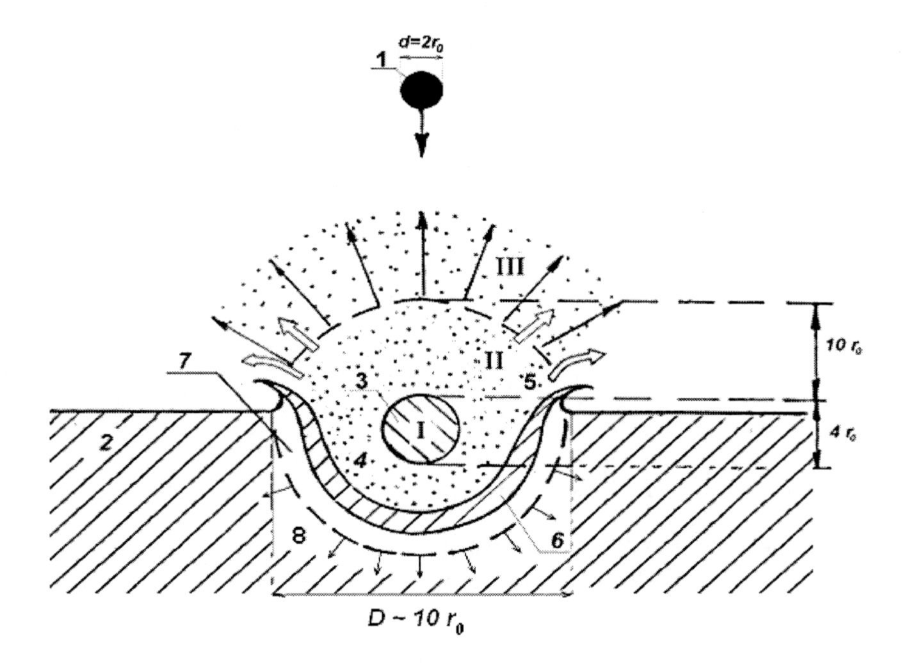

Figure 4. After a hypervelocity collision of projectile 1 with target 2 hot spot 3 of the plasma torch forms that occupies domain I. The adiabatic expansion-away of the hot plasma produces domain II of the plasma torch, where plasma ions accelerate and thermal energy of the plasma transforms into the energy of directed gas-dynamic motion. In domain III the velocity of plasma particles reaches its asymptotic value and the thermal energy of the plasma becomes substantially smaller than the energy of its directed motion. These processes run ahead of the formation of crater 4 and involve the ejection of the ground material of the target in the direction shown by arrows 5; formation of melt 6, high-pressure zone 7, and shock 8 that propagates deep down into the target. In this figure we also show the relative characteristic sizes of the physical zones and features that formed as a result of a collision with a projectile of diameter d. The diameter of the intermediate crater produced by the impact is about twice its depth.

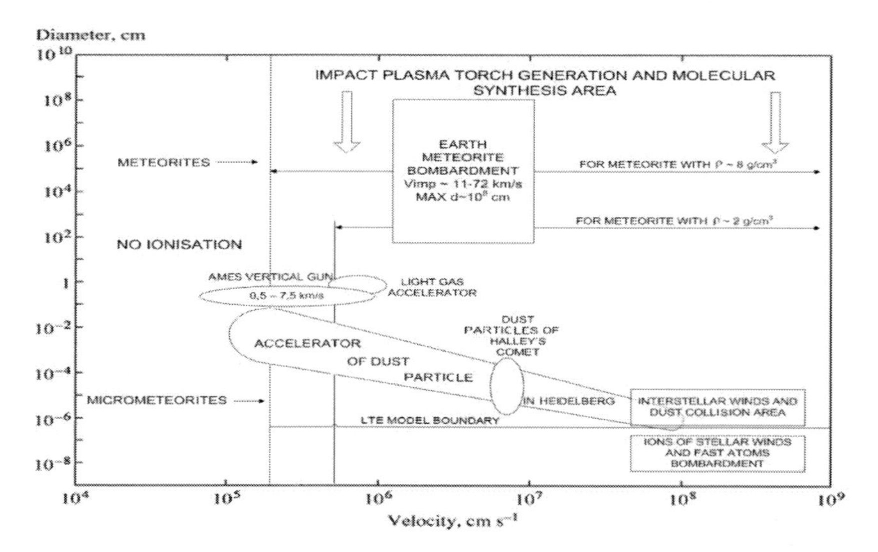

Figure 5. Overall scheme of the generation of impact plasma in various processes in nature and under laboratory conditions depending on the size and velocity of the body.

Traces of catastrophic hypervelocity impacts of large celestial bodies that occurred mostly during the formation of planets remain well preserved in the form of craters on many planets and moons of the Solar System. During this period of time, 4.6—4.0 billion years ago, the Earth too was subject to high-energy bombardment. The hypothetical influx of impact energy was enormous and, according to various estimates, amounted to 10^{22} to 10^{24} eV/m^2 year (Grieve, 1980; Kobayashi and Saito, 2000). At the initial stage of the Earth's formation, i.e., during the first 20—30 million years, the energy of meteorite impacts was about twice the incident energy of solar radiation (Pechernikova and Vityazev, 2008). During this period the velocity of meteorite body impacts onto the Earth's surface depended on the type of the meteorite, its origin or its formation region in space, and varied from 11 to 70 km/s.

The following are two useful factors in the physics of impacts:

- The probability distribution of impacts of meteorites with sizes ranging from a fraction of a micron to several tens of kilometers can be described by a single power law;
- The processes of the formation and expansion-away of the plasma torch in the meteorite size range mentioned above are described by uniform physical laws. Crater formation processes are characterized by analogous similarity.

These factors make it possible to precisely scale meteorite fluxes in the case if meteorite sizes have to be changed, and also allow the results of laboratory simulations of hypervelocity impacts of micron-sized meteorites to be used to reconstruct the processes accompanying the impacts of large bodies.

Many different processes occurring in the interstellar space are also accompanied by the acceleration of dust particles up to above-critical velocities. These processes are mostly associated with explosions that happen during the formation of new stars. However, even slower, but long-lasting processes also contribute to the acceleration of dust particles to rather high velocities.

Thus in gas-and-dust clouds dust particles can be accelerated by stellar radiation pressure or shock fronts. These processes may accelerate dust particles to 10^2—10^3 km/s and allow them to further participate in hypervelocity impacts (Spitzer, 1978). A more detailed analysis shows that hypervelocity impact is a common phenomenon at almost any stages of the evolution of the Universe from the formation of new stars and planetary systems to the death of these bodies (Voshchinnikov, 1986; Bochkarev, 1992; Tielens, 1994).

3.3. Specifics of the Expansion-Away of the Plasma Torch

Plasma torch can be produced not only by hypervelocity impacts, but also by Q-switched laser radiation (Figure 6). Ultrafast input of energy into the interaction region heats up the medium instantaneously. This results in the formation of the so-called hot spot both in the process of the impact and in the case where the medium is subject to pulsed laser radiation. The necessary condition for the formation of a hot spot consists in the following: the target heating time scale, which is determined by the velocity v_L of the projectile, must be substantially shorter than the heat outflow time scale, which is determined by the velocity v_h of heat propagation from the interaction region, or $v_L << v_h \sim c_s$, where c_s is the speed of sound in the target material (Bykovskii and Nevolin, 1985; Kostin et al., 1997).

The formation of the hot spots produced by an impact and laser radiation differ by the input of energy into the target. This difference determines that impact produces a crater whose volume is about two orders of magnitude greater than that of the projectile or the crater produced by laser radiation (Kissel and Krueger, 1987a; Guring, 1973). The latter means that laser radiation must reproduce the impact kinetics rather poorly.

That is why laser simulation is not widely used for the studies of crater formation. However, plasma processes, especially plasma expansion-away and formation of the torch, in the case of impact and laser-radiation produced phenomena are highly similar and almost identical. The plasma processes that occur inside the torch and that are responsible for its formation are also similar. Hence, provided appropriately chosen initial conditions and characteristics of laser radiation, the resulting torch can be made indistinguishable from an impact-produced torch.

Therefore laser-produced torch is believed to be a good model for the plasma torch generated by a hypervelocity impact.

The differences between the processes of the formation of the hot spot in the two cases should have little effect on the final characteristics of the torch and on the dynamics of its expansion-away.

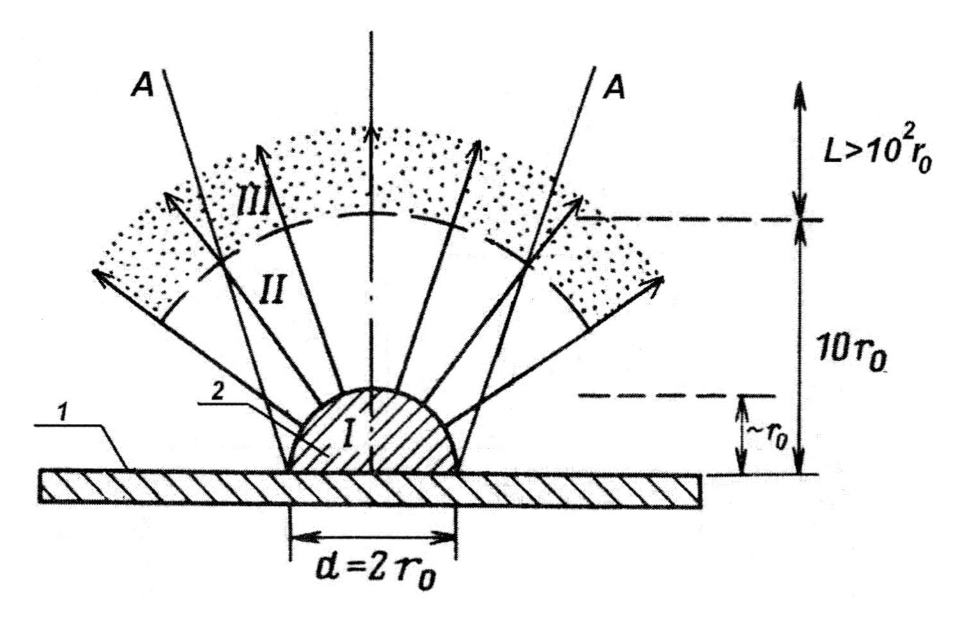

Figure 6. Interaction of laser A operating in Q-switched pulsed mode focused onto target 1 onto a spot of diameter d $=2r_0$ with a power flux density of $W_L>10^9$ W/cm^2 produces hot core 2 that fills zone I. According to Bykovskii and Nevolin (1985), in the case of adiabatic expansion-away of dense hot plasma, like in the case of impact action, a plasma torch forms with zones II and III. The processes of impact- and laser-produced torch formation are similar. In zone II plasma acceleration is powered mostly by its thermal energy, and at its boundary with zone III radiating, recombining plasma forms, where new compounds may presumably be synthesized. In zone III the velocity of plasma particles reaches its asymptotic value. Laser-radiation driven crater formation is several hundred times less intense than the process of impact-driven crater formation.

As is well known (Knabe and Kruger, 1982; Ratrcliff et al., 1997), the atoms and molecules of the target and projectile are heated directly at the initial stage of hypervelocity impact, and this heating is due to the transfer of impact energy to the projectile and the target. The process results in atomization and ionization of the entire projectile material and of a target volume equal to about 5 to 10 projectile volumes.

Plasma formation consumes only a few percent of the total impact energy. The remaining part, about 95%, is spent on the destruction and grinding of the target material, formation of the crater, and generation of the shock that causes the fragmentation of deep-seated rocks. The ejection of the torch produced by a meteorite impact runs ahead of the processes of rock ejection, i.e., crater formation, and part of the fine-dispersed material ejected from the crater forms dense dust clouds (Figures 6 and 7).

Thus the peculiarity of the formation of the hot spot in hypervelocity impact processes is that in this case fast heating of the matter involves direct interaction between atoms and molecules of the colliding bodies (Petrovtsev et al., 1998; Avronin et al., 1985, 1996).

When the target is subject to laser radiation (Bykovskii and Nevolin, 1985; Tonon, 1965; Basov et al., 1970), plasma forms as a result of the interaction of laser radiation both with the solid body and with the products of evaporation. In this case radiation energy must be sufficient to provide for local melting, evaporation, and ionization of the target material and its subsequent transformation into the state of fully ionized gas, i.e., plasma. Thus the target material transforms into plasma in the hot-spot region. It is important that in both cases the heating of matter proved to have no effect on its further development and does not change the physical characteristics of the plasma torch irrespectively of its nature.

The dynamics of the formation and expansion-away of the plasma torch in impact processes is of special interest. Experimental studies of the impact-produced plasma torch under laboratory conditions on the Earth appear to be possible only for micron-sized particles. The problem with the use of natural phenomena for the study of meteorite impact processes on the Earth or on other cosmic bodies is first and foremost due to the impossibility to predict when and where such an event should occur. Therefore the proposed new experiment involving a hypervelocity collision of artificial objects in the near magnetosphere with the characteristic projectile size of up to 20 cm and the most important parameters of the impact consequences recorded during the test may play the crucial part in the eventual discovery of hitherto unknown properties of impact processes.

The properties of a laser-produced torch localized in space are much easier to study. Laser plasma is now widely used in practice in a number of important fields including controlled fusion the generation hypervelocity microparticles – the so-called “flyers”. Plasma torch is, irrespectively of its nature, used as a source of ions in dust-impact and time-of-flight laser mass spectroscopy, and as a source of the synthesis of chemical compounds like fullerenes, linear-chain carbon, and nanotubes. Within the framework of the concept proposed in this book plasma torch began to be used for the first time as an environment for abiogenous synthesis of organic compounds and for the study of physico-chemical laws of this phenomenon in hypervelocity impact processes.

Because of the high pressure in the hot spot, plasma expands freely into space above the target irrespectively of whether this expansion-away occurs into vacuum, gaseous or liquid medium. This provides a realization of the gas-dynamic mechanism of plasma torch. However, the electrodynamic mechanism plays the dominating part in the acceleration of plasma ions (Kon'kova et al., 2009).

This acceleration process is realized because of the breaking of quasi-neutrality during the plasma expansion-away, or, more precisely, because of the “escape” of hot electrons from the plasma torch and the generation of strong electric fields.

During its free expansion-away, plasma begins to cool down, its temperature decreases, and recombination processes start at a certain stage of the torch expansion-away. This results in a decrease of the plasma ion density by two to three orders of magnitude. Ions interact with plasma electrons and form neutral atoms. Recombination processes may continue during further expansion-away, and the already formed low-temperature, low-density plasma may move away from the target with atomic ions interacting only slightly with neutral atoms. This motion in the vacuum chamber ceases when the flux reaches the chamber walls or the detector surface, or, if the plasma expands in a neutral gas or in the atmosphere, the motion ceases as a result of collisions in a neutral gas or in the atmosphere – due to the deceleration of the flux because of collisions with ambient atoms and molecules.

At a certain stage of the plasma torch expansion-away, presumably in the recombination region, individual atomic ions and neutral atoms may combine into molecular ions or neutral molecules (after the recombination of molecular ions), thereby making possible the synthesis of new chemical compounds.

To understand the plasma processes that occur in the torch, let us describe the generally agreed structure of the torch and its characteristic zones (Delone, 1989).

Let us assume that the characteristic size of the hot spot – in particular, its diameter D – approximately coincides with the diameter of the projectile or that of the focal spot in the case of the hypervelocity impact and laser radiation, respectively. The height of the hot spot then does not exceed half its diameter, i.e., the boundaries of the hot spot should have the form of a half-sphere of radius D/2. The ion density at the boundary of the hot spot can be estimated at $\sim 10^{21}$—10^{22} cm^{-3}. Beyond the hot-spot boundary the luminous zone or recombination processes begins, where the particle density is $\sim 10^{18} cm^{-3}$. Beyond the recombination zone the zero-potential surface is located, which separates the domains of positive and negative volume charges, and, finally, beyond this surface the conventional electron-cloud front is located with the particle density equal to $\sim 10^{11} cm^{-3}$.

Beyond this zone the acceleration of the plasma torch particles ceases and the so-called inverse "hardening" takes place. However, atomic and molecular ions continue to move away in the vacuum if they do not collide with the residual ambient gas.

We so far do not know the location of the zone where new compounds are synthesized during the torch expansion-away. It may be located between the hot spot and the domain where the principal recombination processes occur. The relatively high density of plasma ions ensures the combination of atomic ions into molecular ions. It must be a highly efficient process for relatively slow ions with the energy of ordered motions less than or equal to 30 eV.

3.4. Study of Hypervelocity Impacts under Laboratory Conditions

The studies of the physical processes of impact interactions under laboratory conditions are performed at special accelerators of microparticles capable of accelerating particles with the masses of 10^{-11} to 10^{-17} g to velocities of 30—100 km/s (Knabe and Kruger, 1982; Ratcliff et al., 1996; Ratcliff et al., 1997; Roybal et al., 1995; Drobyshevsky et al., 1995). Laboratory experiments with accelerators showed, among other results, that the mass composition of the projectile, e.g., microparticles, can be determined from the spectrum of plasma ions produced as a result of the impact. To study the mass and isotopic composition of microparticles in field experiments in space dedicated mass-spectrometers have been developed, where the source of ions is the plasma produced as a result of a hypervelocity impact of a micrometeorite or a dust particle onto a high-purity metal target (Kissel et al., 1986). Thus, for example, dust-impact instruments PUMA and PIA of VEGA and GIOTTO space missions made it possible for the first time to determine the elemental and isotopic composition of microparticles in the gas-and-dust cloud of comet Halley (Sagdeev et al., 1987; Mendis, 1988; Reinhard, 1988). The microparticle impact velocities in these experiments amounted to ~80 km/s. Figure 5 shows the summary picture of the collision processes mentioned above as a plot of the projectile size versus its velocity.

Crawford and Schultz (1988, 1991, 1993, 1999), Göller and Grün (1989), and Srama et al. (2009) have made an important contribution to the study of the physics of impact processes under laboratory conditions. The above studies showed convincingly that the formation of a plasma torch and the synthesis of new compounds begin simultaneously during plasma expansion-away starting from the impact velocity of 2 and 6 km/s for iron and carbon projectiles, respectively. P. Schultz managed to record the processes of the generation and expansion-away of the plasma torch in NASA's Ames Vertical Gun Range during the preparation of the Deep Impact mission. We show these processes in Figures 7 and 8 and thoroughly discuss the results of the mission in Section 1 of the Afterword.

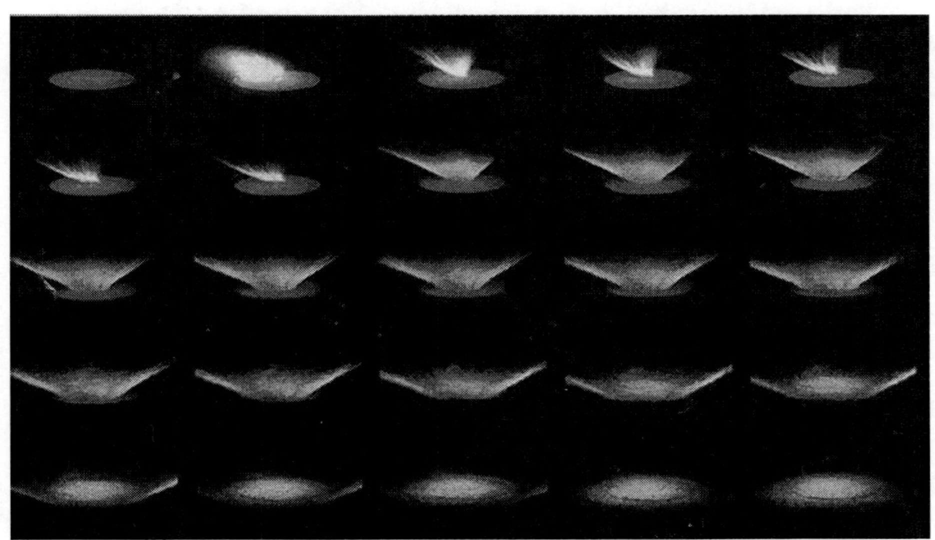

These images were obtained by P.Schulz (Brown University) and are published with his kind permission.

Figure 7. Frame-by-frame photography made under laboratory conditions featuring the impact of a pyrex sphere onto a silicon powder target (the diameter of the projectile is shown on the first frame in the right-hand part of the figure and that of the future crater, at the left-hand side of the figure.) Frames 2—6 show the formation and flight of the plasma torch, and frames 7—23, the process of impact-driven crater formation. The last frame shows the crater at the end of its formation.

The laboratory and space experiments on the study of impact interactions or the determination of the mass composition of microparticles can be classified as direct impact experiments with the hypervelocity impact of a microparticle studied in real time. Such experiments can also be viewed as unique or exotic on the basis of their cost and the possibility of realization of the initial conditions. However, modern science also has other, simpler and more easily available methods of the study of impact processes. These include, first and foremost, laboratory modeling and numerical simulations.

Numerical simulations play an important part in the study of hypervelocity impact processes. This becomes clear from the results obtained in the series of works on the simulation of the destruction of spacecraft protection (Sagdeev et al., 1983) or ionization processes for dust-impact instruments (Hornung et al., 1996; 2000) used in the field experiment during the study of comet Halley. Another experimental direction is associated with laboratory modeling of the hypervelocity impact plasma torch using laser radiation. This technique (Kostin et al., 1997) is based on the capability of Q-switched laser radiation to generate pulses with the duration of t_L~1—10 ns and provide a power density of up to 10^{13} W/cm^2 in a 10^2—10^3 μm diameter spot. Model experiments of this kind do not reproduce the kinematics of the impact (Kostin et al., 1997), however, they reproduce rather reliably such plasma processes as the formation of the torch, adiabatic expansion-away of plasma, and recombination provided the correct choice of the parameters of laser exposure (Managadze, 2003, 2005b; Managadze et al., 2003b, 2003c).

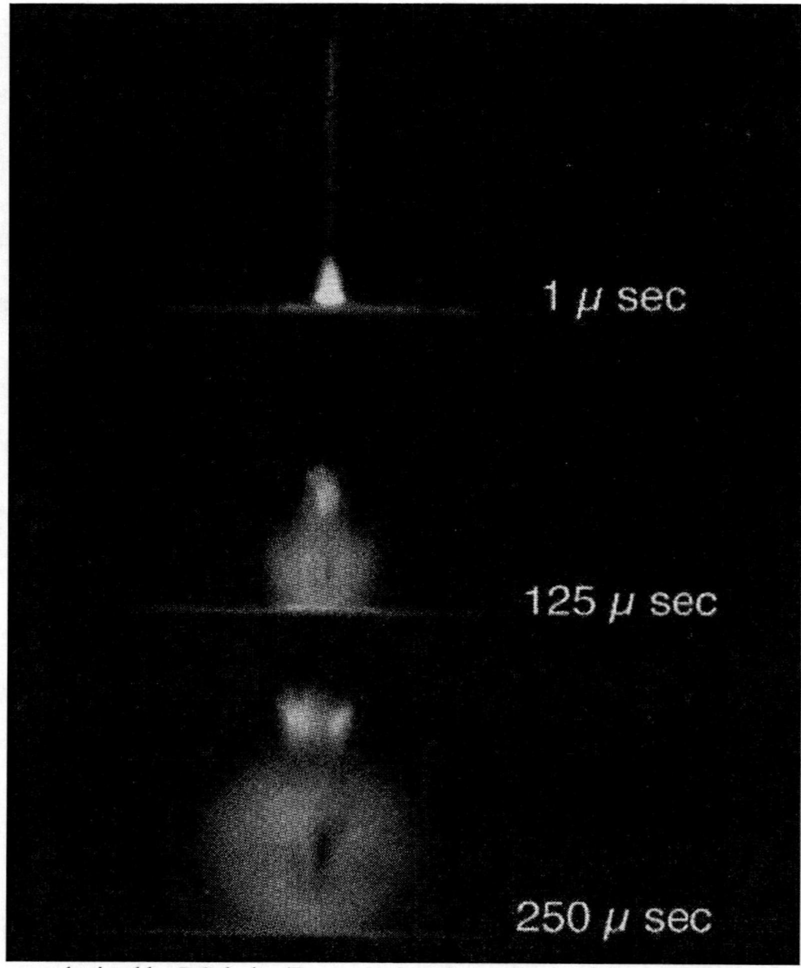

The images were obtained by P.Schultz (Brown University) and are published with his kind permission.

Figure 8. Three photographs taken at different times demonstrating the development of interaction between the projectile and a dry-ice target. This interaction results in the formation of a plasma torch seen on the first image and the ejection of fine-dispersed material of the target during impact-driven crater formation seen on the last two images.

The choice of laser exposure parameters simulating the impact process is based on the use of experimental results obtained in field dust-impact experiments in space or in direct impact experiments performed on dust particle accelerators. To this end, one must know such initial parameters of the impact as the velocity and size of the projectile and the density of both the projectile and target material. Thus in this book we choose laser exposure parameters based on the results obtained by VEGA mission (Kissel et al., 1986; Mendis, 1988; Reinhard, 1988).

According to these results, in the case of impacts of cometary dust particles with the diameters of 0.01—1 μm and masses of 10^{-18}—10^{-12} g moving at a velocity of ~80 km/s the ionization degree and energy of plasma ions was α~1 and E_i~50—70 eV, respectively. The properties of plasma ions were corroborated in studies involving numerical simulations of the impact process. These simulations modeled mass-spectrometric measurements of cometary dust made with onboard dust-impact mass analyzers PUMA-1 and PUMA-2 (Petrivtsev et al., 1988; Avronin et al., 1985; 1996). These authors also showed that with the above initial conditions of the field experiment plasma is produced by particles with the diameters as small as 0.01 μm.

To correctly determine the parameters of laser exposure to simulate an impact, let us consider the peculiarities of plasma formation in the impact process. The volume of the intermediate impact crater is known to exceed, depending on the projectile velocity, the volume of microparticles by a factor of ~100 (Kissel and Krueger, 1987; Guring, 1973), and plasma forms in a volume that is 5 to 10 times greater than that of the projectile microparticle. It follows from this that the energy contribution to the formation of plasma in the case of an impact may be equal to several percent of the meteorite energy, whereas the bulk of it is spent for formation of the crater, deformation and destruction of the target. This explains the rather low ionization degree and low energy of plasma ions generated as a result of the impact despite the fact that the power density computed by the diameter of the particle is high and amounts to ~10^{13} W/cm^2 (Hornung et al., 2000).

The dependence of the ionization degree and ion energy on the power density of laser exposure has been thoroughly studied (Bykovskii and Nevolin, 1985; Tonon, 1965; Basov et al., 1970). Thus the increase of power density from 10^9 to 10^{11}W/cm^2 increases the ionization degree for Al from 1 to 5 and particle energies from ~100 eV to 2 keV. Hence these parameters can be used to approximately estimate the energy contribution to the process of plasma formation and make it possible to choose the laser exposure parameters that would correspond to an impact with the given parameters. Such a procedure makes it possible to achieve the similarity of the characteristics of impact- and laser-produced torch plasma and choose the correct simulation parameters.

The criteria for the choice of the parameters of laser exposure for reproducing an impact torch can be briefly formulated as follows. Laser exposure must ensure the formation of a plasma torch where the ionization degree and the energies of plasma ions should match the corresponding parameters for an impact-produced torch. Under these conditions laser exposure must also ensure that the plasma formation volume matches approximately the plasma formation volume for the impact-produced torch.

The first condition imposes restrictions on t_{ION}, in particular, $t_{ION} < t_L$, where t_{ION} is the plasma-generation time scale. A laser with t_L in the 0.1 to 10 ns interval satisfies this condition with a good safety margin.

To satisfy the conditions of the similarity of the parameters of impact and laser plasmas, we must ensure that energy contributions are identical in the two cases. Thus if the power density of laser exposure is W_L ~ 10^9 W/cm^2, then plasma ions are mostly singly ionized and their average energy is ~50—70 eV. This result agrees rather closely with the impact plasma properties for the impact velocity of ~80 km/s.

The number of newly formed ions for laser exposure for the given power density depends on the diameter and depth of laser crater: these parameters can be used to determine the volume of plasma formation. These data, in turn, allow one to compute the diameter of a microparticle for which the simulation was performed. The procedure can be inverted, i.e., the diameter of laser exposure can be chosen based on the diameter of the particle. Thus an impact of a 10-μm particle onto a target produces plasma with an equivalent volume of 10^4 μm^3. The same plasma volume can be obtained by laser exposure with a diameter of 5-μm and a crater depth of 3μm for the case of a carbon target.

Thus by varying the power density and diameter of the laser-produced crater we can reproduce the principal parameters for the given diameter and velocity of a microparticle.

The above procedure for the choice of the parameters of laser exposure simulating the impact is nothing but the determination of the similarity parameters (Managadze and Podgornyi, 1968, 1969) with elements of limited simulation principles without changing the spatial scale length of the experiment. These parameter values should be used to the full in order to extend the results obtained in microparticle impacts to the results of impacts of large meteorites.

The identical nature of ion-formation processes in laser and impact plasmas is evidenced by the fact that such plasma is used as the source of ions in laser (Managadze and Shutyaev, 1993) and dust-impact (Kissel et al., 1986) time-of-flight mass analyzers. In the former instrument the generation of ions in laser plasma is used to determine the mass composition of the target and in the latter, generation of ions in impact plasma is used to determine the mass composition of both the incident particle and of the target. Another common feature of these instruments, which are similar both in the principle of operation and often in design, is that ions are provided by high-temperature plasma produced as a result of ultrafast concentration of energy.

Laser plasma has often been used as the source of ions in dust-impact instruments during their testing and laboratory calibration. In particular, I used a laser ion source to test the laboratory prototype of PUMA flight instrument. These tests allowed me to discover and analyze the effect that decreases the mass resolution of the instrument in the case of high plasma density, which corresponded to the impact of a relatively big particle onto the target (Sagdeev et al., 1987). The results of laser simulation served as the basis for further improvement of the flight instrument – the mode of narrow energy window was introduced, which allowed unique mass spectra of large dust particles to be obtained.

The possibility of the synthesis of molecules or molecular ions during the expansion-away and cooling of impact plasma is of great interest for the study of the processes of the synthesis of organic compounds in meteorite impacts at the early stage of the evolution of the Earth and, given the identical nature of these processes, also in hypervelocity impacts of dust particles in interstellar gas-and-dust clouds.

3.5. "Warm Scenario" of the Origin of Life on the Earth

The generally agreed early hypotheses about the stage of chemical evolution on the Earth formulated the necessary conditions that make possible the emergence of life in terms of the so-called "warm scenario" (Goldanskii and Kuz'min, 1989). According to one of the versions of the "warm scenario", the synthesis of organic compounds – from monomers to polymers –

was made possible by the action of solar radiation on the primeval Earth atmosphere (Dickerson, 1978). According to this concept, formation of the living cell was a several-stage process. The first, necessary stage was the formation of the atmosphere, or the stage of the creation of the material for organic synthesis. At the second stage monomers were supposed to form under the action of solar radiation on the Earth atmosphere. The third stage was dedicated to the synthesis of biopolymer precursors from the already available monomers. The fourth stage was thought to consist of the emergence of precursors of cellular structures, microspherules, which must have been capable of separating protobiopolymers from the external environment (supposedly water), locating and concentrating them. Thus the hypotheses considered assumed that living matter was created from inorganic substance, and that during its further evolution living matter could have developed the capability for replication, for transferring this capability to successive generations, resulting in the formation of a prototype of the simplest primeval organism.

According to expert estimates, the total time needed for all these stages must have been quite substantial and amounted to about 1 billion years.

Various scenarios have been proposed for the emergence of life. Thus one of the most popular scenarios of the time was the so-called "small thermal reservoir" concept (Folsome, 1979). According to this concept, there was no need for large water expanses like the ocean for a protocell to form, and a small reservoir would suffice where in the surface water earlier synthesized organic compounds could have later become increasingly complicated and new complex polymer structures could form under the action of solar radiation. Folsome believed that the newly formed polymers must have been sinking into deeper layers of the reservoir to avoid destruction by solar radiation and emerge again for the process of further complexification to continue. This concept was rather popular because of the simplicity of its realization and clarity of its main idea.

For several decades after the publication of the first results of the experiments of S.Miller and H.Urey (Miller and Urey, 1959), the scenario proposed by these researchers for abiogenous synthesis or organic compounds required for the emergence of life also became very popular. However, with time, material evidence began to appear that indicated that in a number of cases the presumed initial conditions on the early Earth somewhat or even radically differed from reality. Thus the vulnerable points of many generally agreed scenarios proved to be:

- early emergence of the first living cell on the Earth;
- unknown composition of the primeval atmosphere;
- low density of abiogenically synthesized monomers of organic compounds;
- lack of natural mechanisms ensuring the synthesis of polymer organic compounds or, at least, polymerization of monomers;
- the fact that solar radiation did not reach the Earth surface;
- lack of a mechanism of isomer symmetry breaking.

The above and other difficulties and inconsistencies have so far prevented the development of a satisfactory theory of the emergence of life based on real processes that occur under natural conditions.

In view of this, it is of interest to assess the potential of the new concept proposed in this book. In particular, this concerns identifying the weak points, difficulties, and inconsistencies of this concept and assessing the degree of its "unsinkability".

We already discussed in the preceding chapters the possible advantages of the natural processes that occur during a hypervelocity meteorite impact and that are associated with the properties of the torch plasma. In this chapter we consider the important physical properties of the emerging plasma environment, because it is very important to understand whether the proposed plasma-chemical concept may play the crucial part in the realization of the "warm scenario" in nature.

One of the important properties of impact processes is that they originated or occurred simultaneously with the formation of the Earth. Hence if impact processes were capable of ensuring the emergence of life then this could happen during the earliest period of the formation of our planet. This factor removes the important inconsistency due to the early origination of the first living cell.

A meteorite impact under terrestrial conditions provides the realization of the following important processes: generation of the plasma torch where organic compounds are synthesized; formation of a crater with the generation of dense dust clouds that protect the products of synthesis and ensure their accumulation, and thermal heating of the impact zone where moderate ambient temperature is maintained for a long time after the impact.

As an environment that ensures the synthesis of organic compounds, plasma torch is enormously superior to any other natural process. The point is that the projectile material and part of the target material become completely atomized and ionized at the early stage in the hot spot, thereby preparing the active environment for the synthesis of new compounds in plasma-chemical processes.

Then during the adiabatic expansion-away the following processes occur that are important for the synthesis:

- plasma-chemical reactions of the synthesis of organic compounds become irreversible, and intermediate products are rapidly removed from the reaction site;
- products of synthesis are moved to cooler layers where they enter new reactions, which result in further complexification of these compounds;
- catalytic activity of the plasma and prebiotic selection of organic compounds on a molecular level ensure high rate of self-assembly and ordering;
- breaking of the mirror symmetry occurs in the synthesis of isomers, resulting in the formation of homochiralic structures of organic compounds;
- fractionation and grouping of light elements during the expansion-away of the torch, which ensures high efficiency of the synthesis of organic compounds.

Some of the above processes are highly reliable and occur without fail during the expansion-away of the plasma torch. The remaining ones have preliminary experimental corroboration and require a more thorough analysis. However, such analyses are not always needed. Thus preliminary results of direct measurements of organic compounds synthesized in the torch are indicative of a weak breaking of the mirror symmetry of enantiomers. Given their preliminary nature, these results would not be so impressive were it not for the experimentally discovered factors in the plasma torch that are capable of breaking the enantiomer symmetry. When it was proposed, the hypothesis of the symmetry breaking in the torch was by no means a purely theoretical speculation because of the already available electromagnetic field measurements in the torch plasma, which met the requirements for local chiral physical fields.

One may have the impression that the above properties of the plasma torch make possible the synthesis of simple organic compounds exclusively, while proving to be insufficient to ensure the high efficiency of the synthesis and the required high complexity of these compounds. This is not the case.

Hence to understand and demonstrate the potential of hypervelocity impact and plasma torch processes one must take into account the following points. During the first 200 million years in the process of the Earth formation a total of 10^{10}, 10^{7}, and 10^{3} ~1-, ~10-, and ~100 km large meteorite bodies, respectively, fell onto the surface of our planet. Laboratory simulations of hypervelocity impacts whose results we report in Section 4.4 show that a proportional relation exists between the characteristic size of the projectile, mass of the organic compound produced, the complexity of the structure of this compound. This result indicates that the synthesis process considered here has, even at the initial stage of its analysis, no alternatives and shows great promise for increased characteristic sizes of the torch region.

Other properties of the plasma torch of a hypervelocity impact, besides those mentioned above, include hitherto little known, but very productive and bona fide established physical parameters of this phenomenon that may play the crucial part in the preparatory processes for the emergence of life. These properties include:

- Power density exceeds that of incident solar radiation by a factor of 10^{13}—10^{14};
- Initial mass density in the torch at the time of its formation exceeds the atmospheric density by a factor of 10^{6};
- The rates of plasma-chemical reactions in the torch are 10^{7}—10^{9} times higher than the rates of chemical reactions that occur in the liquid state under laboratory conditions without catalysts;
- The initial temperature in the hot spot may amount to 10^{5}—10^{6} K and decreases rapidly down to the ambient temperature during the free expansion-away of the torch;
- Practically infinite number of environment formation trials may be realized with an enormous range of the parameters of the initial conditions of synthesis;
- The organic compounds synthesized in the torch can be injected into deep layers of the Earth crust and into underground reservoirs.

If analyzed jointly with the results of laboratory simulations of impacts, the above properties of impact effects and of the plasma torch suggest that the plasma environment produced by the impact of a greater than 1-cm meteorite is capable of synthesizing complex organic polymer compounds with no extra conditions. Organic compounds synthesized under such conditions may have a polymer structure and, according to extrapolated relations, they have molecular masses ranging from 200000 to 500000 a.m.u. It is possible that some of the molecular structures with such masses produced in the torch plasma will be incapable of replication and translation, i.e., they will be unable to produce a protoviroid-like object.

The generation and expansion-away of the plasma torch run ahead of the process of crater formation, which is of a separate interest. The consequences of this process may also play an important part in the formation of living matter and in the protection of organic compounds produced in the torch against destruction.

To have a general understanding of the impact dynamics, let us consider the sequence of events of the formation of the Popigai model crater as shown in Figure 9 (Ivanov, 2008a, 2008b). This crater was produced by an impact of a stony meteorite about 36 million years ago near the Northern border of the Anabar shield. The figure shows six "snapshots" of the formation of the crater, which was produced by a 8-km diameter projectile moving at a velocity of 15 km/s. About 20 s after the impact the depth of the crater reached its maximum value of 18 km and this was followed by the immediate collapse of the crater. The diameter of the intermediate crater was about five times greater than that of the projectile. It is evident from the figure that organic compounds synthesized in the plasma torch and fallen onto the lateral surface of the intermediate crater may have been buried under a several kilometers thick layer of ground rock. The sequence of crater formation shown in the figure is only an illustration; however, it clearly demonstrates that the final diameter or the crater was about 10 times greater than that of the projectile.

As is well known, about 95% of the meteorite impact energy is spent for the destruction and heating of rocks located at the surface and in deep layers of the Earth crust. The maximum temperature at the initial stage is achieved in the intermediate crater region, where rocks melt. This is followed by the collapse of the transition cavity accompanied by the formation of the "transitional hill" of considerable height, and the rise of deep-seated rock toward the surface. During the "spread" of the hill mixing of rocks occurs and the crater acquires its final, stable form. The temperature in the newly formed crater decreases from the center toward the periphery and, depending on the size of the meteorite, remains constant over a considerable time interval because of the low heat conductivity of the impact-ground rock.

Meteorite impact experts usually estimate the crater zone cooling time scale by the simple formula, which I learned from A.Vityazev.

According to this formula, time t of rock cooling in the crater can be computed as $t \sim D^2/\sigma$, where D is the diameter of the meteorite in cm and σ is the average heat conductivity of rocks, which is equal to 10^{-2} kall/cm s K. For a ~10 km diameter meteorite $t \sim 10^{14}$ s, or 10^7 years. These estimates are extremely simple and therefore very reliable and conform to the processes that occur in nature.

For the case considered the most important processes among those that occur during the crater formation are:

- Generation of dense dust clouds in the impact zone;
- Thermal heating of the impact zone;
- Injection of organic compounds synthesized in the torch into deep-seated layers of the Earth crust.

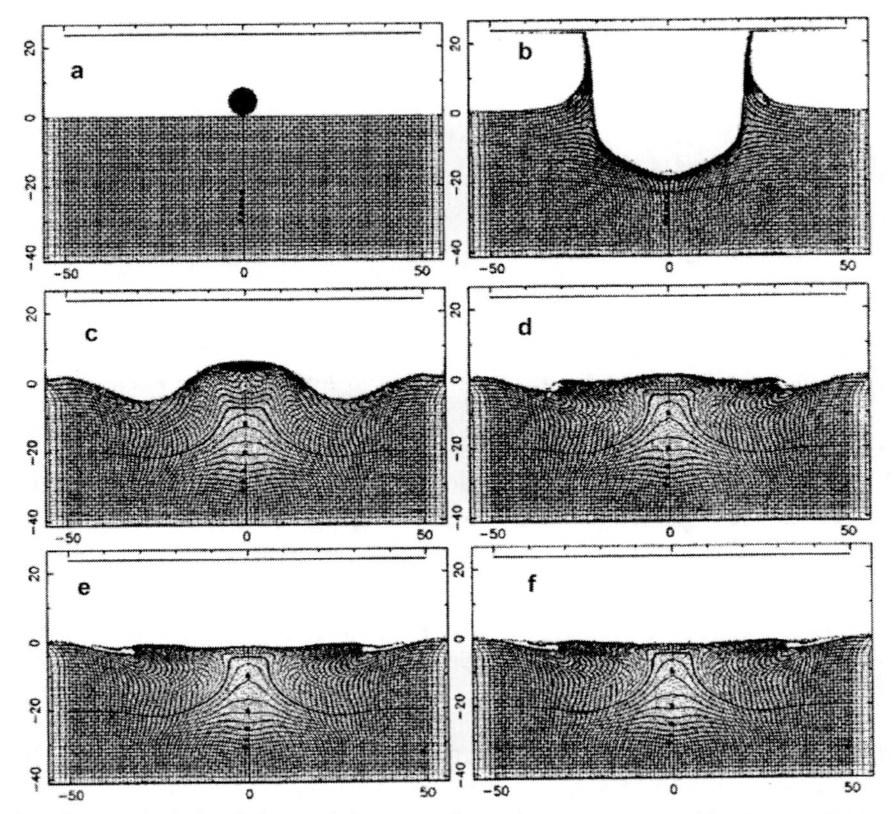

The results of numerical simulations of the crater formation were reported by Ivanov (2008a, 2008b) and we print them in this book with the kind permission of the author.

Figure 9. Sequence of events during the formation of the Popigai model crater: the diameter of the meteorite and that of the intermediate crater were equal to 8 and 40 km, respectively. Panel (a) shows the initial positions of the spherical projectile and of the layered target; panel (b) shows the snapshot computed for the time 23 seconds after the impact when the transition region reaches its maximum depth of 19 km; panel (c) shows the snapshot taken 115 seconds after the impact when the collapse of the transition cavity (rise of rocks at the center resulting from the subsidence of the walls) results in the formation of the 5-km high "transition hill" and deep-seated rock under the crater rise above their initial depth. Panel (d) shows the snapshot taken 200 seconds after the impact, when the "transition hill" spreads in the gravity field, whereas deep-seated rocks stopped moving because internal friction had returned to its normal value. The velocity of near-surface spread reaches 200 km/s. Panel (e) shows the snapshot taken 300 seconds (five minutes) after the impact, when motions have almost ceased. Panel (f) shows the snapshot taken 400 seconds after the impact, when the crater acquires its final, stable shape.

We perform a more detailed analysis of the possible consequences of the above processes in Chapter 7. According to these data, in the impact zone favorable conditions are created for both the emergence and evolution of living matter. These conditions include moderate temperature, high density of organic compounds synthesized in the torch, and sufficiently high ambient humidity provided the availability of ice or water reservoirs in the impact zone. The dense dust clouds generated during the impact are capable of protecting organic compounds synthesized in the torch from eventual radiative destruction.

The exceptional properties of the plasma torch concerning the synthesis of organic compounds combined with the conditions created after the formation of the crater may have provided all the prerequisites for the realization of the "warm scenario" and the emergence of living matter.

3.6. "Cold Scenario" of the Synthesis of Organic Compounds in Interstellar Clouds

Interstellar gas consists mostly of hydrogen (~70%) and helium (~28%) atoms and ions. Other elements and molecules account for less than 1% of all atoms and ions and are mostly contained in dust particles. Organic constituent elements, such as C, N, O, S, and P, are three to four orders of magnitude less abundant compared to hydrogen. For each dust particle in these formations there are $\sim 10^7$—10^8 hydrogen and helium atoms.

In recent years, about one hundred molecules and molecular ions including organic compounds have been discovered and identified using radio astronomical methods. Clouds of this type are dominated by organic compounds among which molecules as HCN, CH_2NH, CH_3NH are known to be the prime material for the formation of amino acids (Spitzer, 1978; Bochkarev, 1992).

According to modern concepts, two processes are responsible for the formation of molecules: chemical reactions in gaseous media and reactions at the surfaces of dust grains in clouds (Bochkarev, 1992).

Dust in interstellar clouds has the form of small crystals or, possibly, amorphous formations consisting of silicates, graphite, and metal oxides, and have the sizes of $\sim 10^{-2}$ μm (small particles) or up to 1 μm (larger particles). These larger particles formed as a result of contaminated water freezing over the surfaces of smaller particles. Dust particles form mostly in slowly outflowing atmospheres of red giants, although other hypotheses have also been proposed to describe the processes of the growth, acceleration, and destruction of dust grains. Thus, for example, dust particles may accelerate up to $\sim 10^3$ km/s when they cross the shock front produced by a cloud—cloud collision or under the action of stellar radiation pressure (Spitzer, 1978). Collisions of dust particles are one of the main and generally agreed mechanisms of their destruction.

Low ambient temperature does not favor efficient synthesis of organic compounds in interstellar clouds. Thus because of the low density and temperature we must now admit that the synthesis of molecules in gaseous media cannot be as efficient as reactions at the surfaces of dust grains, which have become quite popular among the researchers. However, scenarios involving reactions at the surfaces of dust grains also face certain problems.

In particular, temperature conditions there proved to be so unfavorable that various mechanisms of low-temperature synthesis had to be developed to explain the observed abundances of organic compounds and especially those of more complex molecules. A new synthesis mechanism has been proposed that ensured the association of atoms into molecules and involved the low-temperature quantum limit of the rate of chemical reactions discovered in 1970-ies as well as a realization of the mechanism of molecular tunneling that is capable of synthesizing organic compounds at ultralow temperatures (Goldanskii, 1979; 1993). The hypothesis that at ultralow temperatures in interstellar clouds the rates of chemical reactions are not equal to zero has been confirmed not only theoretically, but also experimentally. These results made it possible to develop, along with the "warm" terrestrial origin-of-life scenario, also the concept of "cold" cosmic scenario where at least some of the stages of chemical and prebiotic evolution can occur in space.

Later the applicability limits and the yield efficiencies for molecules and polymer chains of amino acids have been analyzed for the case of molecular tunneling or the so-called "cold prehistory of life". It was shown, for example, that the time scale of the formation of an M~130 a.m.u. molecule is of about 10^{10} years, and that reactions of the polymerization of nucleotides to a molecular mass of ~350 a.m.u. should be forbidden because the time scale of this process amounts to 10^{55} years. Hence the probability of the formation of complex organic molecules in cosmic conditions is practically equal to zero (Gerasimov and Mukhin, 1978).

The synthesis of the polycyclic aromatic hydrocarbons anthracene and pyrene with the masses of 178 and 202 a.m.u., respectively, discovered in recent years in the protoplanetary nebula known as the "Red rectangle" cannot be explained in terms of the generally agreed mechanisms, which assume that these compounds must form in the gaseous state or in the process of adhesion of atoms onto the surfaces of dust grains. The results mentioned above and also those reported in numerous other publications indicate that the formation of relatively complex polyatomic organic molecules in the processes of the synthesis of organic compounds at low temperatures in interstellar clouds, where matter is in the low-density gaseous state, is too unlikely to be considered even slightly probable. The observed abundance of organic compounds in the interstellar medium indicates that in nature there must be another mechanism that should be capable of providing efficient synthesis of organic compounds under extremely hard conditions of this environment.

Let us now consider the formation of H_2 molecules during the adhesion of atoms onto dust grains in the case of typical properties of interstellar clouds.

Interstellar gas is known to consist mostly of H and He with an admixture of O, N, and C whose combined abundance is close to 10^{-3} (Bochkarev, 1992). Dust grains, which are an important component of the interstellar medium, make up for only 1 to 2% of the total mass of gas. The masses of the main components of gas – hydrogen and helium – and the mass of dust grains relate as 75:24:1, whereas the number densities of H and He relate as 10:1. If the diameter of dust grains is ~10^{-5} cm then the ratio of the number densities of gas-forming atoms, n_A, to that of dust particles, n_D, should be n_A/n_D ~10^{11}. The average relative velocity of dust particles may amount to ~3—5 km/s, however, ~10% of dust particles may have velocities ~5 10^6 cm/s, and this population forms as a result of the acceleration of dust grains in the interstellar medium by stellar radiation pressure or as a result of charged dust rains crossing the shock front.

In the case of interstellar gas we can assume that if a hydrogen atom reaches the surface of a dust grain, it remains there. The probability of adhesion is then equal to unity. We further assume that after a short time interval an atom moving over the surface should combine with another hydrogen atom to form an H_2 molecule, which then leaves the surface under the action of external factors. Given the number densities of dust grains ($n_D \sim 10^{-10}$ cm^{-3}) and that of hydrogen atoms, one can easily obtain and order-of-magnitude estimate for the time required for an H atom to enter into a reaction at the surface of a dust grain. This time is equal to $t_A = 1/(V_A\ n_D\ \beta_D) = 2\ 10^{14}$ s, where D is the cross section of dust grains. The same formula can be used to determine the time of a single collision of dust grains, which is equal to $t_D \sim 2$ 10^{13} s.

Thus during ten collisions of dust particles there will be one act where an atom enters into chemical reaction resulting in the formation of a single H_2 molecule. Given that $n_A/n_D \sim 10^{11}$

and that only 0.1 adhesion act occurs during a single collision between dust particles, the number of H_2 molecules must amount to N_H~10^{10}. To make the comparison of the estimates more convenient, let us determine the number of diatomic carbon clusters produced in 2 10^{14} s. Given the cosmic abundance of carbon, which is equal to ~3.3 10^{-4}, a total of N_C ~ 3.3 10^6 C_2 molecules must form during the synthesis of 10^{10} H_2 molecules. Longer time intervals are needed to synthesize more complex molecules or molecular ions consisting of three, four or more atoms.

To compare the data obtained in the experiments to the estimated molecular yields in hypervelocity dust impacts, we have to determine the effective yields of the molecules containing constituent atoms of organic compounds for a single act of a dust particle impact. These quantities can be determined rather accurately from the results of dust-impact experiments performed on the dust particle accelerator in Heidelberg (Germany). According to these results (Stubig, 2002), a 2.4 10^{-13} g latex particle (213 PANi-PS) with a diameter of ~0.75 10^{-4} cm and up to 80% carbon content accelerated to 20 km/s produces ~6 10^6 atomic and molecular ions when it hits a pure rhodium target. Given that for an impact velocity of 20 km/s the ionization degree for carbon plasma is equal to 10^{-4}, i.e., that for every ion there are 10^4 neutral atoms or molecules, their total number should be ~6 10^{10}. Increasing the velocity from 20 to 50 km/s results in a two-to-threefold increase (to 2 10^{11}) of the number of neutral particles produced. However, this quantity should be adjusted to account, in particular, for the following fact: the mass of dust particles in the above estimates of the synthesis of molecules in the process of adhesion was ~300 times smaller than the mass of the latex particle; only 10% of dust particles moved at a velocity of 20 km/s and the efficiency of the production of molecular compounds for a carbon-rich projectile or target was equal to 50%. In this case, the total number of molecules produced should be smaller, N_{IMP} ~3 10^7 mol. Hence in dust particle collisions the efficiency of the synthesis of molecules should be about one order of magnitude higher compared to the efficiency of the formation of molecules as a result of adhesion of atoms onto the surface of dust particles.

The result obtained applies mostly to diatomic molecules; however, an analysis of the effective yields of polyatomic molecules in the plasma torch generated by hypervelocity impacts of dust particles implies that the number of such molecules should be appreciably greater than that of the molecules produced in adhesion processes. This is due to the specifics of the synthesis of new compounds in plasma formations because plasma torch is characterized by approximately equal yields of simple and complex molecular ions.

Thus, given sufficiently high carbon content in the nucleus or in the ice cover of dust particles, the synthesis of rather complex organic compounds during a hypervelocity collision of dust particles may become a dominating process.

This result can be explained by the high efficiency of interaction between the carbon atom and other constituent atoms of organic compounds during the expansion-away of the plasma torch. In particular, dust-impact laboratory experiments showed that, other conditions being equal, the ionization degree of carbon plasma should be 10^{-4}, which is two orders of magnitude lower than the ionization degree of plasma consisting of iron or aluminum atoms. This result is indicative of the high reactivity of carbon ions, which in the process of the torch expansion-away participate more efficiently in the synthesis of new compounds including organic molecules.

***

The new proposed mechanism of the synthesis of organic compounds in a plasma torch of arbitrary nature is capable, owing to the unique physical properties of hypevelocity impact processes, of providing the conditions for the realization of the chemical evolution stage on the Earth and on other cosmic objects including the possibility of the formation of molecular structures meeting the requirements to the simplest forms of living matter. This process, which is a natural phenomenon, becomes explainable and understandable; however, numerous and time-consuming experimental, theoretical, and dedicated long-term studies are needed to confirm it.

In these investigations we use all the available data about the ambient and process parameters. These data allow fast progress to be achieved in the study of the problem. Thus the plasma torch processes that reproduce the processes that occur in the well-known industrial facility –plasma-chemical reactor – may explain quite a lot. In particular, the presence of high-velocity, high-temperature, and high-density plasma in the torch implies fast development of chemical synthesis reactions catalyzed by the plasma. The dynamics of adiabatic plasma expansion-away ensures, as plasma-chemical reactions reach equilibrium state, fast removal of the products of synthesis from the active zone, their preservation and further participation in the processes of complexification in the cooler layers of the torch. Chemically active products of synthesis are represented by molecular ions in excited state, ion radicals, and neutral radicals. After "hardening" in the process of free expansion-away, such compounds leave the torch region and, remaining chemically active, may enter into further reactions with ambient atoms and molecules and synthesize new, more complex compounds. Thus the specific nature of the free expansion-away of the plasma torch is a unique feature that makes possible the synthesis of complex compounds including organic molecules. The fact is certain that such or similar process must also develop, albeit with certain complexifications, in the impact torch.

The possibility of the synthesis of inorganic chemical compounds and carbon structures was demonstrated in experimental studies where plasma torch was generated by laser radiation. In these experiments simple chemical compounds, such as molecular ions of rare-earth monoxides and hydroxonium (H_3O^+) (Ramendik et al., 1979), and a number of carbon structures, e.g., fullerenes and their derivatives (Zhang et al., 1999), have been synthesized.

The realization of the synthesis of organic compounds in the plasma torch became a new landmark in the creation of chemical compounds whose formation processes in nature had been previously unknown. In the plasma torch such synthesis processes were favored by anomalously high catalytic activity of the plasma and the reactivity of atomic carbon, which was available there in ionized or excited state. In the presence of other constituent elements of organic compounds these properties ensured high effective yields of the products of synthesis.

The investigation of the synthesis of organic compounds in the plasma torch has just started, and it is therefore difficult to imagine how the initial conditions and physico-chemical properties of the process may reflect on the complexity and specificity of synthesized organic compounds. Note, however, that according to the results presented below, plasma-formation regions smaller than 0.1 mm obey a proportional relation between the characteristic size of the interaction region and complexity of synthesized organic compounds. Obtaining such results became possible owing, among other things, to the approach where the discovered possibility of the synthesis of inorganic compounds in the plasma torch was immediately applied to the synthesis of organic compounds. These results have laid down the basis for the new idea about the possible relation between the synthesis of organic compounds in the impact-produced plasma torch and the processes of the emergence of life on the Earth.

The undeniable capability of the plasma torch to synthesize new compounds in free expansion-away mode follows from the following excerpt from the Physical Encyclopedic Dictionary published in Moscow in 1983 (in Russian). The "Plasma" section composed by B.Trubnikov based on the works of L.Artsimovich, D.Frank-Kamenetskii, V.Ginsburg, and B.Trubnikov (Artsimovich, 1969; Frank-Kamenetskii, 1968; Ginsburg, 1970; Trubnikov, 1978) concludes with the following description of the plasmatron – industrial analog of the plasma torch: "Plasmatrons, which produce jets of dense low-temperature plasma, are widely used in various branches of technology. In particular, they are used to cut and weld metals and apply coatings. In plasma chemistry low-temperature plasma is used to produce some chemical compounds, e.g., halogenides of inert gases, which are impossible to synthesize in another way. Moreover, the high temperature of plasma implies high rates of chemical reactions – both direct reactions of synthesis and inverse reactions of breakdown. Inverse reactions of breakdown can be suppressed, resulting in a substantial increase of the yield of the desired product, if the synthesis is performed during "flyby" of the plasma jet by expanding and thereby cooling it down in the downstream jet portion (such operation is referred to as "hardening")".

Chapter 4

EXPERIMENTS ON THE SYNTHESIS OF ORGANIC COMPOUNDS IN THE PLASMA TORCH

4.1. SPECIFICITIES OF THE APPROACH TO THE PROBLEM

We already pointed out that the bottleneck of the classical scenario of the chemical stage of evolution is the lack of bona fide natural mechanisms capable of ensuring highly efficient synthesis from inorganic substance of complex polymer organic compounds possessing the required functional and structural properties. Without addressing this task it is impossible to find a solution for the problem – that of the emergence of life on the Earth. Therefore the problem of abiogenous synthesis of polymer organic compounds has for some time become the focus of attention of the research community. This problem continues to be relevant, possibly, because no natural phenomena and mechanisms have been found so far that would, when reproduced exactly in laboratory conditions, ensure the formation of at least the precursors of the simplest fragments of biological molecules.

The lack of proper natural synthesis mechanisms is not the only difficulty that we face when addressing the problem. Once the much needed natural synthesis mechanism of polymer organic compounds is found and reproduced in a laboratory, an enormous challenge will have to be faced. We will have to extract from the organic medium produced by the said mechanism the key compounds that are directly related to the emergence of living matter and to identify them.

The severity of the task is also due to the extreme "contamination" of the present-day Earth by organic compounds of biological origin, and this process has been going on for at least 3.5 billion years. Therefore even if abiogenous synthesis is successfully reproduced in laboratory experiments, the artificially produced organic compounds will be rather difficult to analytically distinguish from organic compounds of biological origin. The researchers specializing in the field consider the identification of the newly produced organic compounds to be the most challenging part of the experiments on abiogenous synthesis, and argue that the results of this identification must be very reliable and that they will determine the reliability of the final conclusions.

The research community has been long aware of the above issues, which, however, still remain without proper solution, especially in the case of small quantities of the synthesis products.

It is difficult to imagine that the synthesis of organic compounds on the early Earth could be a result of purposeful processes. These processes were most likely unpredictable and stochastic, although they may have possessed some properties of prebiotic selection. Therefore the production of the "right" organic compounds may have been accompanied by the synthesis of very large quantities of "trash" or "background" organic material.

The separation of the "right" organic compounds from "trash" is a daunting analytical task whose complexity increases with molecular mass. The situation is further compounded by the fact that modern analytical chemistry has neither techniques nor instruments to address it. We have neither chromatographic, mass-spectrometric, spectroscopic, nor other equipment that would allow us to unambiguously and securely identify the "key" macromolecules needed for the emergence of life in a situation where they are mixed with other organic compounds, especially if the target compounds have small concentrations. This is usually the case in experiments involving the synthesis of organic compounds.

Such studies are almost as time consuming as a tangled criminal investigation with the difference that in the case considered we have to study the cause of the origin of life and not the cause of the death of a living organism.

The studies of abiogeneous synthesis of organic compounds at the early stage usually include an analysis of and a search for the natural phenomena suitable for the development of such a process, and also the choice of the appropriate mechanisms of its realization. In the case of a successful conclusion of this stage, a laboratory simulation of the process of synthesis is performed followed by the analysis of the final product. Many researchers have been doing such simulations, and many continue to do them now. The years long work experience in this direction crystallized into standard approach, which determines the priority requirements to the phenomena and mechanisms in order for them to ensure the synthesis of organic compounds in nature.

According to these requirements, every newly found mechanism of the synthesis of organic compounds should be reproducible in laboratory conditions with no additional efforts that could potentially affect the physics of the process, and with minimum intervention on the part of the experimenter. The laboratory action that ensures the synthesis of organic compounds must be identical and as close as possible to its natural analog. If the number of initial conditions to be realizable in nature for the newly found mechanism to operate is very small, this should be viewed as a positive factor indicative of the high probability of the operation of this mechanism.

We now illustrate these requirements for the case of a meteorite impact. The proposed mechanism of the synthesis of organic compounds allows, with no additional complexifications or tricks, an exact analog of the natural "impact" process to be reconstructed in laboratory conditions. Hence a meteorite impact is an example of a natural phenomenon with the minimum number of requirements for its realization. Therefore in the case of an impact both the required velocity and initial substance are provided by nature.

In the next chapter we report the results of original experiments that demonstrate that both simple and complex chemical compounds including organic molecules are synthesized in a plasma torch irrespectively of its nature. Various analytical instruments, measurement techniques, and approaches are used to record the data, interpret the results, and identify the synthesized organic compounds.

To perform a preliminary detection of products and control the synthesis process, we used LASMA time-of-flight mass-reflectron combining time-of-flight analysis with laser ablation of the medium studied. This instrument, which I developed back in 1986 (Managadze, 1992b, Managadze and Managadze, 1997), proved to be the optimum device for the study of abiogeneous synthesis of simple organic compounds. Laser exposure made it possible to simultaneously simulate the impact torch in the mode of free plasma expansion-away, which is typical of natural interactions, and measure the mass composition of the molecular ions of organic compounds synthesized in the torch with no restrictions on the range of masses measured. These measurements allowed us not only to "weigh" the synthesized organic compounds, but also determine their structural characteristics from the distribution of mass peaks. This procedure made it possible to interpret the synthesis products to a first approximation.

To reliably identify the products of synthesis, we also used modern high-precision chromatographs and new-generation mass-spectrometric instruments, combinations thereof, as well as MALDI-TOF and MALDI-TOF-TOF like instruments.

4.2. Laser Mass-Spectrometer – An Instrument for Synthesis and Analysis

LASMA instrument was developed and manufactured by the Space Research Institute of the Russian Academy of Science and has the form of a compact laser time-of-flight mass-reflectron with a fully axisymmetric configuration (Managadze, 1992b; Managadze and Shutyaev, 1993; Managadze and Managadze 1997, 1999).

The instrument combines high analytical, technical, and operational performance required for the determination of the elemental and isotopic composition of samples, and provides a mass resolution of 300—600, a sensitivity of about 10^{-5} – 10^{-6} g/g per single exposure, and high reproducibility of the spectra if used on a homogenous sample. It is capable of determining the masses of molecular ions produced in the case of a laser exposure. The periodicity of the mass peaks in the spectrum may be a source of additional information about the structural characteristics of the molecular ions produced.

Figure 10 shows the design of LASMA instrument. This device is axisymmetric both in terms of laser radiation and relative motion of the plasma ions of the torch. Laser radiation reaches the surface of the sample after crossing reflector grids and the central hole in the detector. This configuration ensures high reproducibility of the spectra and makes it possible to perform layer-by-layer analysis of the sample practically with no restrictions in terms of its depth. In this configuration an optical microscope coupled with a TV camera allows both the target surface and the site of laser exposure to be observed. A variable neutral filter was used to change the power density of laser radiation. Such a layout allowed the mode of laser operation to be chosen arbitrarily in order to vary the ionization degree and the energy of newly produced ions over a wide range.

The quantitative measurement technique employed is based on the physical principle that allowed the correct results to be obtained in the registration of all generated ions, from monovalent to polyvalent, in the energy interval from 0 to E_{max}.

Laser ion source is known to ensure not only fractionless evaporation of the sample, but also approximately equiprobable yields of different elements. In the case considered, with the laser focal spot diameter and exposure density equal to 50—60 μm and 10^9 W/cm^2, respectively, ions of most of the elements are concentrated in the energy interval from 0 to 150 eV. Hence the results obtained in this energy interval in the case of the registration of atomic ions should be correct and the ratio of the measured elemental concentrations to those of the corresponding isotopes must match their concentrations in the sample, thereby making it possible to perform an analysis without a standard. However, measurements performed within a narrow energy window proved to be more promising for accomplishing the task, and, in particular, for investigating the process of the synthesis of new compounds in the plasma torch.

To analyze the properties of the instrument when operating in a narrow energy window, various highly homogenous samples were used with the elemental concentrations ranging from 50 to 0.05%: e.g., NBS-612 standard glass with the atoms of 61 elements “trapped” as admixtures in its matrix. However, the mass spectra of this sample showed the yields of adjacent elements to be the same in the narrow-window mode. After the integration of the mass peaks the resulting quantitative elemental abundances in the matrix proved to be close to their standard values listed in the standard-sample passport or reported in the literature.

An analysis of the analytical properties of LASMA instrument showed that the device provided reliable data within the narrow energy window from 10 to 30 eV for the analysis of the elemental and isotopic composition of samples made of light elements including organic compounds. In this energy interval the principal constituent elements of organic compounds, such as H, C, N, and O, had the same yields. Such measurements performed within a low-energy interval exclusively were sufficient to correctly determine the elemental and isotopic composition of the samples including those that consisted of organic compounds, and also made it possible to infer the masses of molecular ions and their fragments in the cases where they formed in a wide range of masses. LASMA instrument was used in various research organizations and institutes to perform a number of correct mass-spectrometric studies with the aim to determine the elemental compositions of simple and complex organic compounds and reveal marker and tracer elements and materials contaminants.

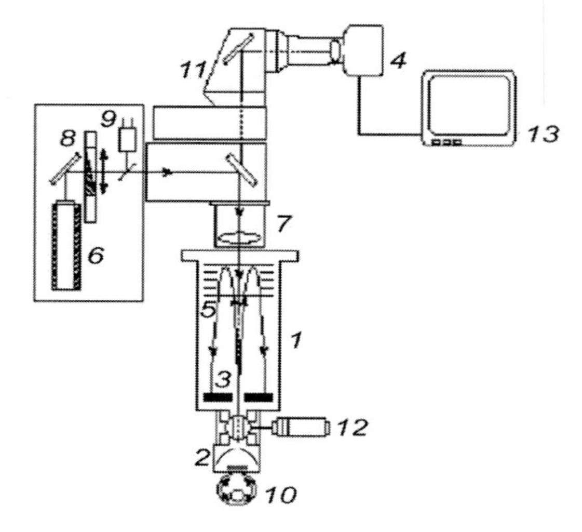

Figure 10. A scheme of the LASMA- TOF laser mass-reflectron: 1—vacuum chamber; 2—lock-chamber with sample and holder for collecting the products of synthesis; 3—detector; 4—TV camera; 5—reflector; 6—laser; 7—focusing lenses; 8—neutral filter; 9—radiation power meter; 10—system for press lock-chamber; 11—microscope; 12—motorized vacuum valve; 13—monitor.

An in-depth analysis of the potentialities of the instrument when operated in a narrow, low-energy window was performed with targets made of pure carbon or materials with high carbon content. The analysis of such samples revealed mass-spectrum peaks that could not be identified either with elemental or isotopic peaks. The characteristic interpeak spacing amounted to 12 a.m.u., indicating the presence of low-energy carbon clusters in the plasma torch. An increase of the power density of laser radiation up to 10^{11}-10^{12} W/cm^2 resulted in the emergence, along with multiply ionized carbon atoms, of the peaks that could be interpreted only as due to multiply ionized hydrocarbon atoms. These results were obtained back in 1990, when nobody suspected the possible existence of such compounds. The certainty that the mass peaks found corresponded to real compounds played the decisive role because the only possible explanation of such compounds was the combination of multiply ionized carbon with hydrogen, i.e., the synthesis of new compounds.

Later, a more detailed analysis of the possibility of the synthesis of new compounds in the laser plasma torch showed that a similar process may also occur in the impact-produced torch because of the identical behavior of the plasma processes in these formations. However, at that time these «strange» results stimulated more in-depth studies of this previously unknown phenomenon and brought about the discovery of a new mechanism of the synthesis of organic compounds in the impact-produced plasma torch during the formation stage of organic molecules under severe conditions of primeval Earth.

4.3. Initial Stage of Research

At the initial stage we searched for carbon-containing solids, which, when subject to laser radiation, could ensure the synthesis of carbon clusters and hydrocarbon compounds. LASMA time-of-flight laser reflectron was the base instrument for simulating the impact process. As we pointed out above, LASMA included a laser used as an agent that generated plasma torch similar to the impact-produced plasma torch, and an analyzer for the determination of the masses of atomic and molecular ions including organic compounds synthesized as a result of laser exposure.

In the first series of experiments a Q-switched laser provided a power density of ~ 10^9-10^{11} W/cm^2 in a 30-150 μm diameter spot with a pulse duration of 7-10 ns. The plasma torch expanded into vacuum without additional acceleration of ions – i.e., in the so-called «free expansion-away» mode. The plasma torch generated by such laser radiation had the same parameters as the plasma torch generated by impacts of 5-10 μm diameter microparticles moving at speeds from 30 to 80 km/s.

We also took into account both the special role that carbon plays in the synthesis of organic compounds (Goldsmith and Owen, 1980; Miller and Orgel, 1974) and the capability of this element to combine into polyatomic structures. That is why we used carbon-rich targets. In this way, we also increased the probability of the synthesis of organic molecules.

In this series of experiments we used a target made of high-purity carbon meant for spectroscopic analytical instruments with impurity content no greater than 0.1 %. The high purity of the sample was confirmed by the measurements made with LASMA instrument, which is capable of simultaneously recording all elements of the periodic table including hydrogen, which was found to be highly abundant in the mass spectra obtained. Auxiliary experiments made it possible to determine the possible locations of the hydrogen source. They showed that hydrogen was located inside the sample volume including the lattice if any. However, hydrogen was also found to be present in the monomolecular surface layer of the target as a surface contaminant that forms as a result of adsorption of vacuum-oil vapor. The presence of hydrogen in the target had no negative effects on the measurements. It played a positive role because it increased the probability of the synthesis of hydrocarbon compounds along with that of carbon structures.

The mass spectra obtained with a carbon target and shown in Figure 11 revealed singly ionized molecules of methane, acetylene, ethylene, propylene, propadiene, and C_NH_M type hydrocarbon polymer structures containing one to 40 carbon atoms and one to four hydrogen atoms. These spectra proved to be highly reproducible. Such hydrocarbon structures and polyatomic molecules were also later obtained with targets made of other carbon-containing substances: technical graphite, heavy oil fraction and bitumen.

To determine whether silicon is capable of producing structures similar to carbon structures in abiogeneous synthesis processes, we performed experiments with high-purity silicon targets under similar initial conditions. The mass spectra obtained showed that silicon is also capable of producing polymer structures, but the maximum number of Si atoms in the chain did not exceed 11.

Of special interest were the above mass spectra obtained with heavy oil fraction targets and exposures with a power density of ~ $5 \cdot 10^9$-10^{10} W/cm^2 (Managadze, 1992a, 1992b) (Figure 12), where along with the mass peaks corresponding to ions with ionization degrees from 1 to 4 other peaks were found that were identified with multiply ionized hydrocarbons. The latter, in particular, included twice ionized CH2, CH_3, CH_4; triply ionized CH, CH_3, CH_4, C_2H, C_2H_3, and fourfold ionized CH_4, C_2H, C_2H_5, C_3H_8. There are grounds to believe that at the time we were the first to observe such ions in laser plasma. Six years later, we found publications whose authors associated the generation of such ions in plasma with the processes of recharge and stripping, which could also occur in laser plasma.

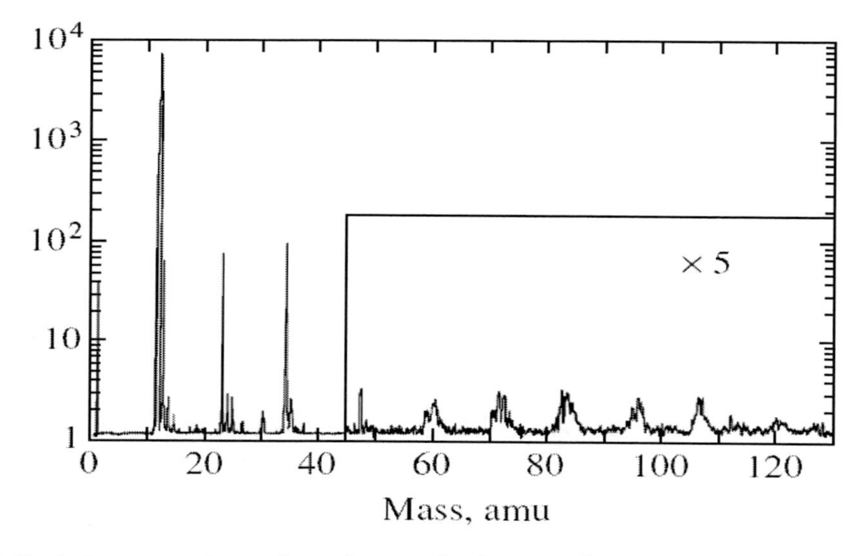

Figure 11. Typical mass spectrum of cumulenes and polyenes or linear-chain carbons synthesized in a laser-generated plasma torch acting on a pure carbon target. This compound represents the simplest case of ordering of matter.

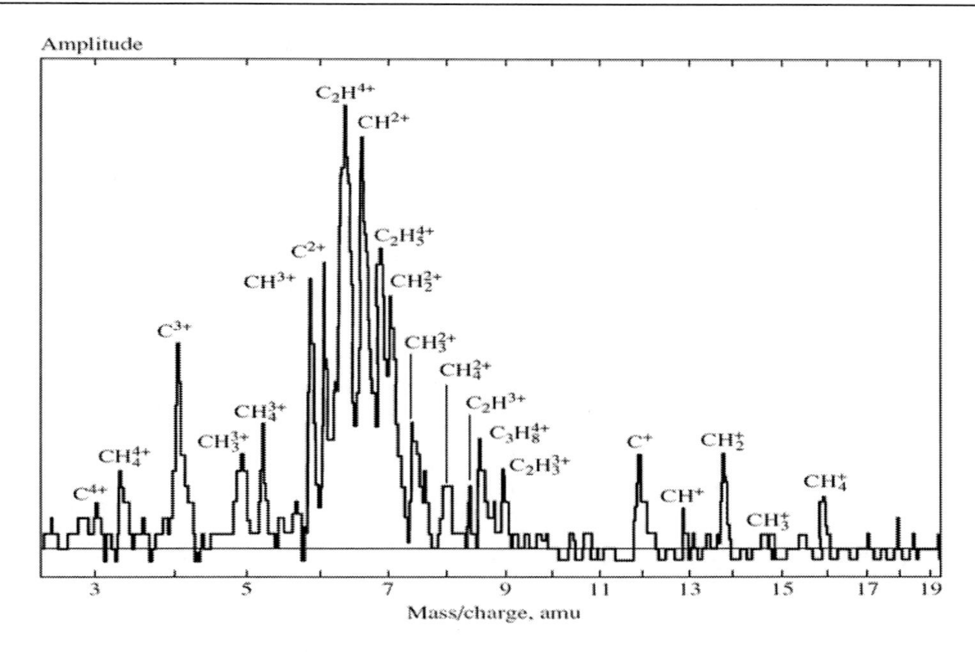

Figure 12. Mass spectrum of multiply ionized carbon and “strange” compounds - multiply ionized hydrocarbons.

According to theoretical computations and experimental results, the lifetimes of these ions were sufficiently long for them to be detected with a time-of-flight mass spectrometer (Andrews et al., 1992).

To synthesize more complex organic molecules containing the main constituent elements of organic compounds, such as H, C, N, and O, we performed a series of experiments where pure inorganic substances containing these elements were mixed with carbon powder. We used high-purity substances: NH_4NO_3, $Ni(NO_3)_2$, $NaNO_2$. The primary aim of these experiments was to demonstrate the possibility of the synthesis of (1) gases that form organic compounds; (2) intermediate reactive compounds; (3) precursors of organic molecules – e.g., those of amino acids, and amino acids themselves, and, possibly, (4) dimers and trimers of these compounds.

The identification of the peaks in the mass spectra proved to be the most difficult task because of the superposition of the mass peaks of chemical elements and their isotopes onto those of synthesized organic compounds. Thus the mass peaks of CN^+ (m = 26) blended with those of $C_2H_2^+$ (m = 26), and that of HCN^+ (m = 27), or hydrocyanic acid, which is a very important compound for the synthesis of organic compounds, coincided with the peak of Al (m = 27). The lack of ultrapure nitrogen-containing salts and oxides and the low productivity of the use of stable isotopes C_{13} and N_{15}, which could shift the mass peaks of organic compounds, complicated substantially the selection of «pure» peaks. Therefore to identify molecular peaks, we considered only those domains of the mass scale, where no elemental peaks could be found or where there where strong deviations from isotopic ratios. Figure 13 shows the results of the measurements of a series of these spectra and Table 1 lists the observed mass peaks. In these spectra we recorded, along with atomic ions H, C, N, O, Na, Ni, and the carbon structures of the type $C_NH_M^+$ mentioned above, also the mass peaks corresponding to molecular ions of gases and intermediate reactive compounds CH_4^+, NH^+,

OH^+ or NH_4^+, H_2O^+ or NH_4^+, H_3O^+, CO^+, N_2^+. The use of these very compounds allowed amino acids to be obtained for the first time in laboratory experiments simulating the synthesis of organic compounds (Urey, 1985; Miller, 1982). The combination of these substances with hydrocarbons may have been the main factor that could explain the synthesis of organic substances whose mass peaks were also observed experimentally. Thus the following compounds have been recorded: CH_2NH^+ (m = 29), H_2CO^+ (m = 30), $C_2H_2O^+$ (m = 42), $HNCO^+$ (m = 43), $NHCHO^+$ (m = 44), and $CH_2O_2^+$ (m = 46). It is probable that some of these mass peaks may correspond to other compounds with the same molecular mass, such as N_2O^+, NO_2^+, and CO^+.

To confirm the plasma origin of the polyatomic ions, we performed an experiment where a silicon-carbon or tungsten-carbon powder mix was subject to laser radiation. In such a mix the synthesis of polyatomic molecules containing Si and C or W and C can be ruled out at the stage of powder mixing. As a result, we recorded the mass peaks of polyatomic ions shown in Figure 14, which corresponded to: Si_m^+ with m up to m = 11 and $Si_mC_n^+$ and $Si_mC_nH_p^+$, where m, n, and p vary from 1 to 7, 1 to 3, and 1 to 3, respectively. In a series of similar experiments WC and WC_2 molecules were synthesized. The presence of the mass peaks of molecular ions in the spectra can be explained only by the processes of the synthesis of new compounds in the plasma torch. These results also indicate that the synthesis of organic compounds in the plasma torch should be viewed as a highly probable process.

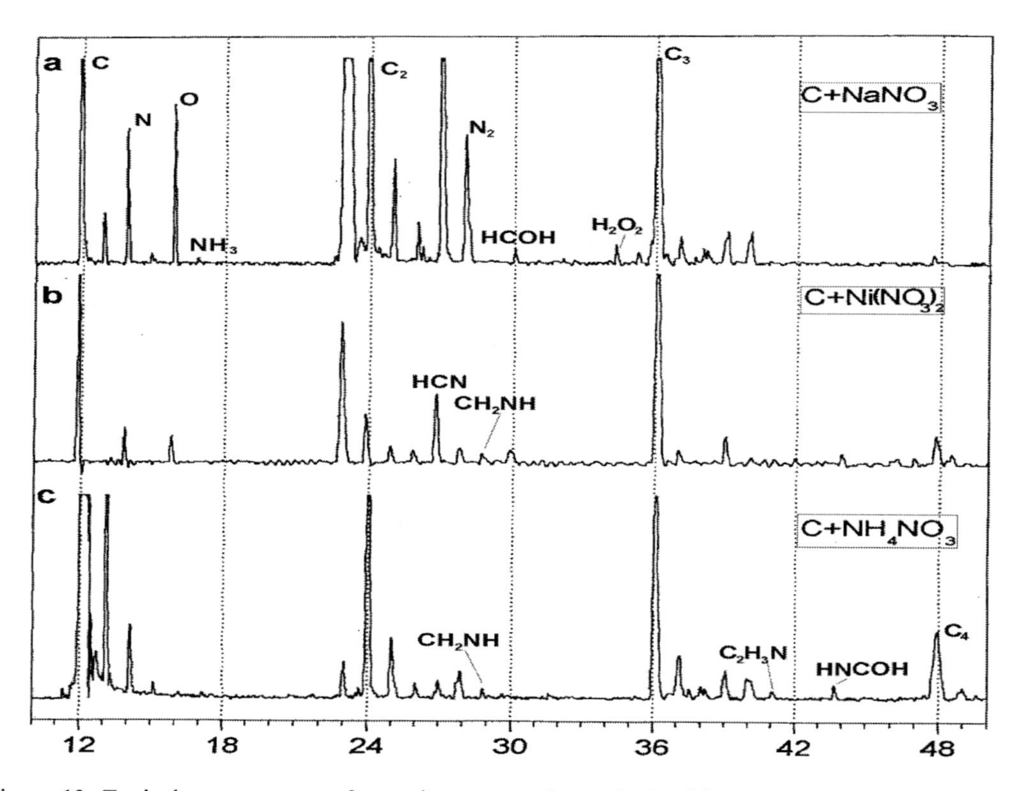

Figure 13. Typical mass spectra of organic compounds synthesized in the plasma torch produced by exposing a mechanical mix of carbon and salts to laser radiation. The composition of the salts is indicated for each spectrum.

Table 1. Mass peaks of experimentally detected polyatomic ions, and their interpretation

Mass, a.m.u.	Atomic ions	Molecule is based on			Name
		C	N	O	
12	C	-			
13	C	CH			
14	N	CH_2			
15	-	CH_3	NH		
16	O	CH_4	NH_2		Methane
17	-	-	NH_3	OH	Ammonium
18	-	-	NH_4	H_2O	
19	F	-		H_3O	
24	Mg	C_2			
25	Mg	C_2H			
26	Mg	C_2H_2	CN?		Acetylene
27	Al	C_2H_3	HCN?		
28	Si	-	N_2	CO	
29	Si	-	CH_2NH		Carbamide
30	Si	CH_2O		NO	Formaldehyde
34	S	-	-	H_2O_2	
36		C_3			
37	Cl	C_3H			
38	-	C_3H_2			
39	K	C_3H_4			
40	Ca	C_3H_4	C_2H_3N?		Propadiene
41	K	-	C_2H_3N		
42	-			C_2H_2O	Ketene
43	-		HNCO		
44	Ca	C_2H_4O	HNHCO	CO_2	Acetic aldehyde
46	Ti	C_2H_6O	NO_2	CH_2O_2	Ethanol/formic acid
48	Ti	C_4			
49	Ti	C_4H			
50	V	C_4H_2			
51	V	C_4H_3			
60		C_5			Also found were mass peaks corresponding to C_NH_M, where $N \leq 10$ and $M \leq 4$
72		C_6			
84		C_7			
96		C_8			
108		C_9			
120		C_{10}			

Although it is possible to determine the contribution of each of these compounds to the amplitude of the corresponding single peak, addressing this problem was beyond the scope of the preliminary stage of investigation.

We analyzed the mass spectra obtained to determine the relative yield of synthesized polyatomic ions. We found it to range from 0.1 to 1% of the total number of atomic ions of light elements (from H to Ca).

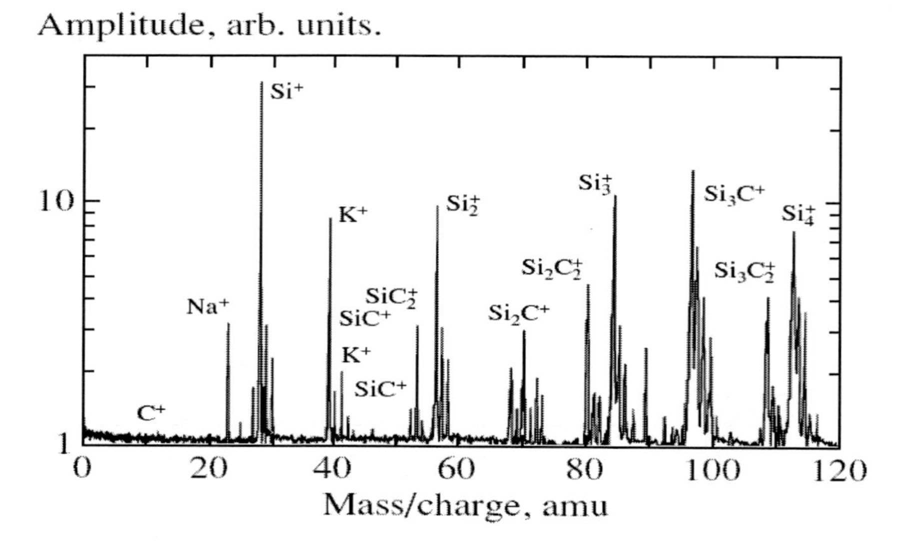

Figure 14. Mass spectrum of a mechanical mix of C and Si powders.

We determined the effective yield of polyatomic ions per ~ 1 eV energy assuming that the energy required for the dissociation and ionization of a single atom is equal to 25—30 eV. It follows from this that the generation of a single molecular ion in the case of a ~0.1% relative yield and 1 eV energy input must be accompanied by the synthesis of $\sim\eta = 10^{-3}$ $(30)^{-1} = 3\ 10^{-5}$ molecules.

The relative yield of 0.1-1 % should equally apply to the neutral component. After recombining plasma cools down, its ionization rate decreases down to 1 to 0.1 %, and hence the contribution of neutral atoms must be significant. The studies performed by Bykovskii et al. (1974) showed that the neutral component of laser plasma has the same nature as the ion component. Hence the neutral component of recombining plasma is also very likely to contain molecules synthesized during the expansion-away and with the same relative abundances. This means that the total number of synthesized molecular ions and neutral molecules must be comparable to the total number of atomic ions.

Laser experiments on the simulation of a hypervelocity impact have therefore demonstrated the possibility of the synthesis of organic molecules from inorganic substances based on carbon and silicon atoms. The power density of laser radiation was 10^9 W/cm^2 in most of the experiments (or somewhat higher in the case of the generation of multiply ionized hydrocarbons) with the spot diameter of 30—50 μm and, according to the similarity criteria, the properties of the resulting plasma must have been identical to those of impact plasma for impacts of 5—10 μm diameter microparticles moving with velocities of 30—80 km/s.

4.4. Experiments with Greater Diameter of Laser Spot

The results of the first experimental simulations of hypervelocity impacts showed that the complexity of synthesized compounds increases proportionally to the characteristic size of the plasma-formation region. However, the limited capabilities of LASMA's laser emitter allowed the diameter of its focal spot to be increased by a maximum factor of three, from 50 to 150 μm. Despite the paucity of the data obtained, the scientific value of these results was evident, and we therefore decided to repeat these experiments using technical facilities that allowed us to produce a plasma-formation region with an order of magnitude greater size.

For these experiments a laser was used operating in the Q-switched and single-mode generation regime, with a pulse energy of up to 600 mJ and a pulse duration of ~10 ns at a wavelength of ~ 1.06 μm. The laser was made as a three-stage system consisting of a master clock and two amplifier stages. It was the prototype of the onboard laser emitter earlier developed for LIMA-D laser remote sensor (Sagdeev et al., 1986b) of the «Phobos» mission.

Mass-spectrometric measurements were made using LASMA time-of-flight mass-reflectron (Managadze, 1992b, Managadze and Managadze 1997, Managadze et al., 1997). Laser emission was focused onto a carbon target into a 1.5 to 2 mm diameter spot. Radiation power density varied from 10^8 to 10^{12} W/cm^2.

Some of the earlier obtained results were reproduced during the initial stage of the experiments of this series. Thus with an increased laser exposure power density multiply ionized hydrocarbons were observed like in earlier experiments, whereas singly ionized carbon polymer chains were found in the case of lower power densities (10^9 W/cm^2). The results of repeated experiments were obtained with a laser exposure diameter of 150 μm.

An increase of the exposure diameter to 1.5—2 mm resulted in the emergence of intense peaks at 720 and 840 a.m.u., i.e., of typical fullerene features. The reliability of the results obtained was confirmed by the carbon "comb" with a step of 12 or 24 a.m.u. that showed up in the spectra with the mass peaks smoothly decreasing by 200 a.m.u., reappearing at higher masses, and reaching maximum amplitude at 720 and 840 a.m.u., which corresponds to 60 and 70 C (carbon masses), respectively. This pattern shows up conspicuously in the mass spectrum shown in Figure 15.

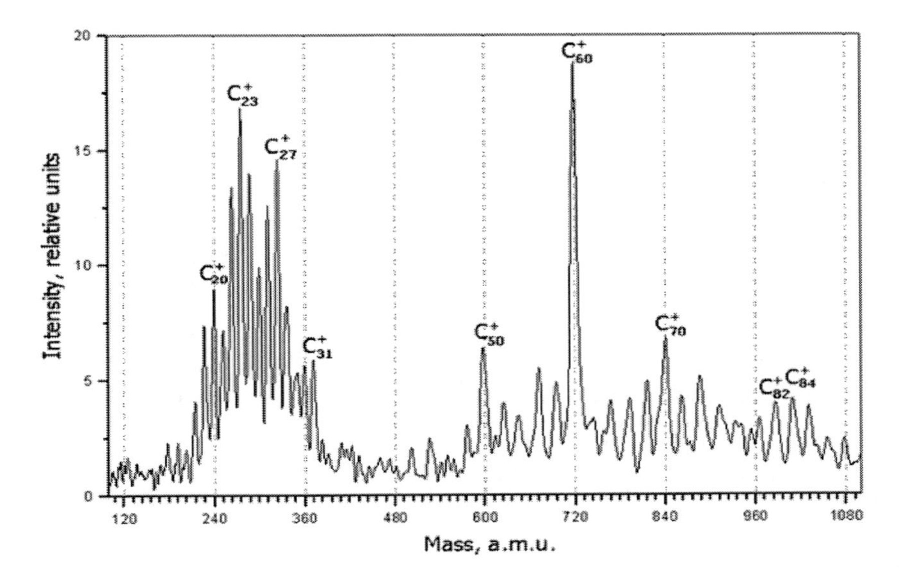

Figure 15. Mass spectrum of fullerenes and their fragments that was obtained in laboratory experiments with d_L ~ 2 mm and W_L ~ 10^9 W/cm^2. The peaks corresponding to the masses from 19 to 31 C with a

step of 1C must be due to individual fragments of fullerenes, and those corresponding to 50, 60, 70, and 82—84C with a step of 2C are most probably due to fullerenes proper.

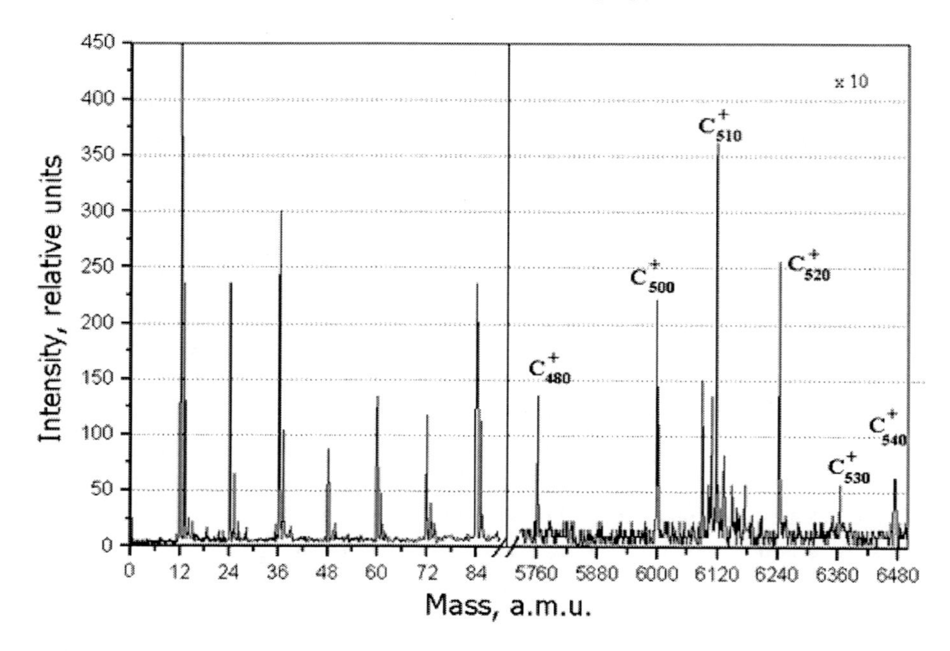

Figure 16. Mass spectrum of giant fullerenes. It is a typical mass spectrum of carbon and hydrocarbon formations obtained with $d_L \sim 2$ mm and $W_L \sim 5 \cdot 10^8$ W/cm^2. The mass peaks at 480 to 540C correspond to typical masses of fullerene onions consisting of nested quasi-spherical structures containing the following quantities of carbon: 60, 240, 540, 960.... The mass peaks at 12—96 a.m.u. form the carbon comb.

In similar, albeit softer, laser exposure modes with a power density of $W_L \sim 5 \cdot 10^8$ W/cm^2, heavy carbon structures have been detected with the masses as high as 6500 a.m.u. Figure 16 shows a typical mass spectrum of these formations.

The high-amplitude peaks in these spectra are ~10 carbon atoms, or 120 a.m.u., apart and, according to the distribution of mass peaks in the spectrum, the compounds found must have been new, hitherto unobserved, structures synthesized in the plasma outburst.

This conclusion was consistent with the initial composition of the target as determined from Raman spectroscopy. These data showed that the target, which, according to its certificate, consisted of pure graphite exclusively, contained no similar structures. In view of these results, the experimental detection of large carbon formations cannot be explained by the "rise" of clusters from the target under the action of laser radiation. A reasonable interpretation was suggested that the increase of the characteristic size of plasma formations created the initial conditions needed for the enlargement of the structure. The synthesis of these formations may have involved already available single hexagonal graphite structures. These structures could have regrouped during the expansion-away of the torch, and some of them could have restructured from hexagonal into pentagonal structures required for the synthesis of fullerenes.

Let us now consider in more detail the carbon structures produced under the conditions where $v_{IMP} < v_{CR}$ and shown in Figure 16. In this case, the energy of laser radiation could have been insufficient for the complete atomization and 100% ionization of the sample.

However, this energy may have been large enough to evaporate the sample, dissociate it into individual molecules, and ionize these molecules, i.e., to produce the so-called molecular plasma and a plasma torch consisting of formations. The expansion-away of such a torch may be accompanied by the combination of molecular ions into larger formations. Such a synthesis process can be conventionally referred to as «self assembly» as opposed to «synthesis», which occurs if $v_{IMP} > v_{CR}$.

Such self-assembly may have resulted in the formation of new, more complex compounds, which could include condensed-state fragments and germs. The product of self-assembly may have differed from the initial substance by its structure, chemical composition, molecular weight, and other characteristics. Experimental studies of soft evaporation and expansion-away in vacuum of molecular ions obtained from pure graphite subject to laser exposure with the power density of $W_L < 10^9$ W/cm^2 and a focal-spot diameter of ~ 1.5—2 mm made it possible to synthesize carbon structures that could be with high degree of confidence classified as hyperfullerenes with the masses amounting to ~ 6500 a.m.u. and with a characteristic period of mass peaks equal to 10C.

A joint analysis combining the results obtained with those reported in earlier published studies demonstrates that laboratory simulations of hypervelocity impacts using laser exposure of pure graphite targets may produce various molecular carbon and hydrocarbon structures.

An analysis of the mass spectra obtained in the above experiments shows that a change in the initial exposure conditions, such as the focal-spot diameter and the radiation power density, translates into a change of the spectra and, in particular, of the localization and distribution of the mass peaks, which, in turn, reflects a change in the structure of synthesized compounds.

Thus, if, for the purpose of a joint analysis, we combine the mass spectra (Figs. 15 and 16) into a single spectrum, the resulting pattern acquires the following form: in the resulting «carbon comb», which starts at M=12 a.m.u., the first rise of carbon peaks begins in the mass interval from 36 to 48 a.m.u. and is then followed by a smooth decrease of the peak amplitudes up to the mass of ~ 120 a.m.u. (C_{10}) (Figure 16). The next, second, rise of peak amplitudes is, according to Figure 15, observed in the mass interval from 228 (C_{19}) to ~ 372 a.m.u. (C_{31}). The period of the peaks in this spectrum is, like in the previous spectrum, equal to 12 a.m.u. Then the third rise of peak intensities begins at the masses of 600 a.m.u. (C_{50}), reaches its maximum at the fullerene mass of 720 a.m.u. (C_{60}), and decreases at about 1080 a.m.u. (C_{90}). Note that the C_{50}, C_{60}, and C_{70} peaks show up conspicuously and the period of mass peaks increases to 24 a.m.u. The next (fourth) rise of peak intensities (Figure 16) is observed at the masses of ~C_{480}, reaches its maximum at ~C_{510}, and decreases down to the noise level beyond the mass of C_{540} or 6480 a.m.u. The interpeak distances increase to C_{10} or 120 a.m.u.

An analysis of the results obtained allows us to associate the observed rises of peak maxima with the following carbon structures:

- The domain of the "carbon comb" and the first rise of the peak maxima in the mass interval from 36 to 48 a.m.u. must correspond to chain-linear type carbon formations, and the feature observed at the mass of ~ 72 a.m.u. may also be due to graphite structures.

- The domain of the second rise of the peak maxima agrees well with the masses of carbon structures referred to as prefullerene formations, which according to Kent et al. (2000) may have masses from C_{24} to C_{32}. These structures correspond to rings of various types – the lateral walls of fullerene. It is important that in mass-spectrometric measurements these structures appear simultaneously with the structures of fullerenes.
- The domain of the third rise of the mass-peak maxima agrees well with and can be explained by the synthesis of fullerenes consisting of 50, 60, 70, and 82—84 carbon atoms.
- Of special interest is the fourth domain of the rise of mass-peak maxima in the mass interval C_{480}—C_{540}. These peaks are very likely due to the synthesis of hyperfullerene-type formations. This hypothesis is supported by both the masses of the formations and the period of the main mass peaks.

Thus a factor of 40 increase in the characteristic size of the plasma-formation region – i.e., the laser focal spot diameter – from 50 μm to 2 mm, while maintaining the power density W_L at ~ 10^9 W/cm^2, results in the proportional increase of the masses of carbon structures produced from 180 to 6500 a.m.u. There are reasons to believe that an increase of the characteristic size of the plasma-formation region should result in a similar complexification of the structure and in the increase of the mass not only for carbon and hydrocarbon compounds, but also for organic compounds in general provided that the target contains N and O.

The effect is very likely to remain because the observed complexification is a characteristic feature of plasma formations including those that form in plasma-chemical processes. In the case considered, an increase of the transversal size of the torch results in the proportional increase of its height and, consequently, in the increase of the volume of the domain where organic compounds may be synthesized. Hence organic compounds synthesized at the initial stage of the torch expansion-away have extra time for interaction in the «synthesis domain», which may result in further complexification of the final product. It is therefore safe to suggest that the observed complexification effect is very likely to ensure not only the complexification pure carbon and hydrocarbon structures, but also that of organic compounds containing other elements, e.g., O and N. There are grounds to believe that the synthesis of monomers at the initial stage of the torch expansion-away and the experimentally found tendency for their complexification may also result in the synthesis of polymer structures. Note that one would expect organic compounds synthesized in the torch to become more complex with increasing size of the plasma-formation region because the results obtained are consistent with the main concepts of modern plasma physics where the increase of the characteristic size of plasma formations results in the complexification of synthesized compounds.

However, an experimental discovery of the signatures of self-assembly and ordering or self-organization of matter in the plasma torch during its adiabatic expansion-away should also be viewed as a result of great importance. This result is especially valuable because the processes of self-organization of matter constitute a sine qua non factor for the realization of the new concept. The mass spectra obtained demonstrate convincingly and unambiguously that self-organization processes occur in the plasma torch, that there is a great variety of final products, and that these processes should be viewed as the properties or salient features of the emerging plasma environment.

4.5. EXPERIMENTS WITH SHORT LASER PULSE

Unlike the earlier experiments, in the series considered here the time of laser exposure was reduced from 10 down to 0.3 ns, i.e., by a factor of about 30.

We did it initially in order to ensure equal interaction times for the two processes: micrometeorite impact and laser exposure. Actually, a 10 μm or greater diameter micrometeorite moving at a velocity of $\sim 5 \cdot 10^6$ cm/s must «sink» into the target during time $\sim$ 0.2 ns. To correctly simulate the plasma processes triggered by the impact of such a meteorite, we must ensure the same interaction time. However, when the study was at its initial stage only laser emitters with $t_L \sim 7$—10 ns were available. No emitters with a short pulse that would be suitable for laboratory simulation could be found at the time. And only recently did we find a small laser emitter with a short pulse duration of 0.3 ns, to ensure a power density of $\sim 10^9$—10^{11} W/cm^2. This diode-pumped laser based on a JAG:Nd crystal was developed in Russia by "ELS-94" research and production center. The emitter provided an energy output of up to 0.3 mJ at $\lambda \sim 1.06$ μm with a pulse frequency of up to 100 Hz.

The unique, miniature active element of this emitter was 6 mm long and 3mm in diameter, and consisted of a crystal with a passive phototropic shutter applied to one of its end surfaces. The small size of the emitter allowed such a short pulse duration to be achieved. Energy was transferred from the diode array to the emitter via a flexible optical fiber, thereby facilitating substantially the necessary manipulations during the assembly of the emitter.

Laser radiation was focused onto a 25 to 50 μm diameter spot. We chose single-exposure mode for the first experiments. The high stability of emission energy allowed us to achieve high reproducibility of the spectra.

The most interesting results were those obtained while exposing a mechanical powder mix of ultrapure carbon and chemically pure ammonium nitrate NH_4NO_3 to laser radiation. As is evident from a typical mass spectrum shown in Figure 17, new peaks appeared between the mass peaks of the "carbon comb" in the mass interval from 108 to 144 a.m.u. These peaks cannot be attributed to elements and their isotopes, and both the amplitudes and numbers of the peaks are significantly higher than those of the mass peaks of polyatomic compounds synthesized in earlier experiments involving a 10-ns long laser radiation pulse focused onto a 50-μm diameter spot.

The spectra obtained were found to show mass peaks up to 180 a.m.u., which corresponded to chemical compounds containing mostly the constituent elements of organic compounds and were therefore interpreted as due to molecular ions of organic compounds. The main problem in the identification of the mass peaks found was that each molecular mass can be associated with several isomers of organic compounds.

We developed the following simple selection criteria for preliminary identification of chemical compounds synthesized in the torch: precursor molecules of organic compounds required for the emergence of life on the Earth were such compounds as, e.g., amino acids. Life could not originate without these substances, and therefore they must have been most easily synthesized and most abundant under natural conditions. We therefore expected that in laboratory experiments simulating the synthesis of organic compounds in plasma processes, such as, e.g., the plasma torch, a simple and reliable mechanism must operate to synthesize the "key" amino acids. This possibility was indicated by the results of numerous earlier laboratory experiments aimed at the study of abiogenous synthesis of organic compounds in the past 50 years.

Therefore the mass peaks of molecular ions observed in the spectra were primarily interpreted as the "key" organic compounds of prime importance for the origin of life. This approach was also based on the working idea suggested in this book that impact processes must have constituted the mechanism that made possible the synthesis and accumulation of organic compounds necessary for the emergence of life.

We identified the mass peaks with particular organic compounds synthesized in the torch as follows: a mass peak at 119, if found, was primarily identified with the threonine amino acid, which we preferred to all the other possible 15 or 20 compounds having this molecular mass. Note that we did not assume that the corresponding peak was due to threonine exclusively. However, we considered the presence of threonine among other isomers of this mass to be likely or even very likely.

We used ChemFinder program to identify the possible organic compounds corresponding to a particular given mass peak. This program found the names of all existing chemical compounds having the given mass, and provided their physical-chemical properties. We used the above criterion to choose the particular organic compound from the list provided by the program.

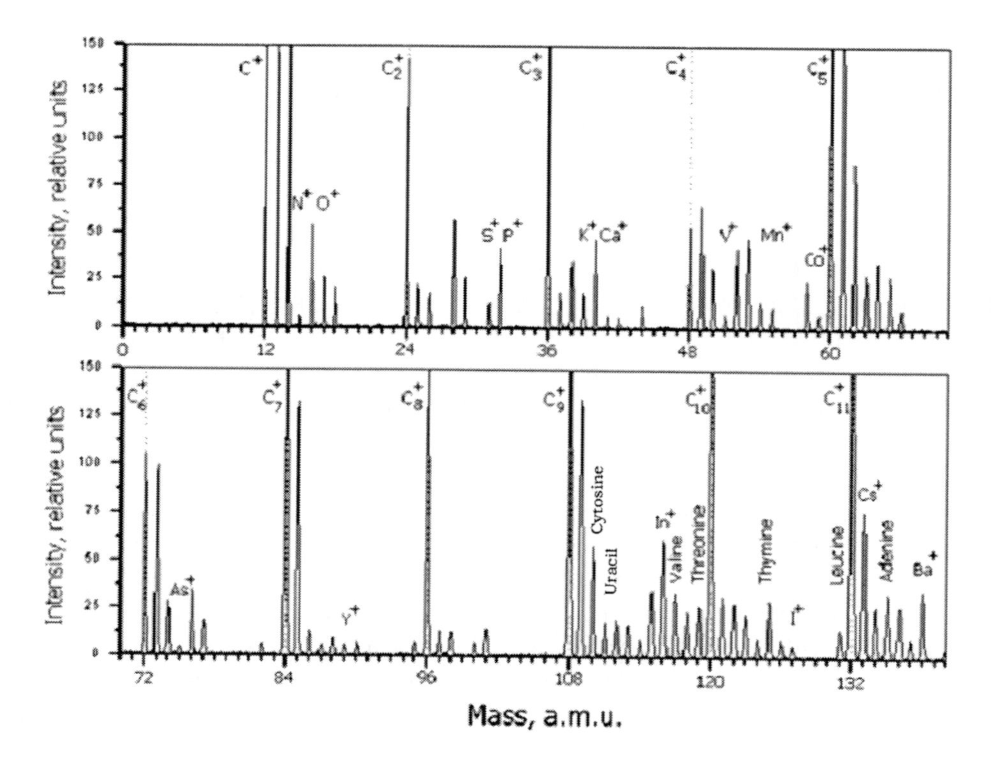

Figure 17. Mass spectrum of a mechanical mix of carbon and NH_4NO_3 obtained with a laser exposure with $W_L \sim 10^9$ W/cm^2 and $t_L \sim 0.4$ ns. We ignored the peaks of the elements and isotopes of the matrix, trace contaminants, and C_M and C_MH_N (N=1—3) type compounds. The remaining peaks were

interpreted as due to synthesized compounds consisting of C, H, O, and N. The presence of these elements in the target and the application of strict peak selection rules lead us to conclude that the emergence of new peaks was due to the complexification of synthesized organic compounds.

We did not analyze the mass peaks coinciding with multiples of carbon atomic mass (12, 24, 36 a.m.u., etc.) - i.e., those that belong to the so-called carbon comb. Neither did we analyze the next adjacent peaks within four a.m.u., and did not list them in the tables.

After processing the spectra in accordance with the above criteria we interpreted the resulting mass peaks as indicative of important prebiotic organic compounds – valine, proline, threonine, thymine, cytosine, uracil, cysteine, asparagine, aspartic acid, adenine, glutamine, lysine, etc., – synthesized as a result of laser exposure. However, we did it only at the initial stage. We then used chromato-mass-spectrometric techniques to reliably identify the resulting mass peaks.

These results, which were of great interest at the time when they were obtained, needed independent confirmation combined with exact identification of organic compounds. However, this was impossible at the time because the scanty amounts of organic compounds synthesized in these experiments prevented any kind of chomato-mass-spectrometric measurements. Such measurements became possible after a radical improvement of measuring instruments and after the development of the technique that allowed fast accumulation of organic compounds synthesized in the plasma torch.

The above results concerning the complexification of synthesized organic compounds with decreasing laser exposure time were, unlike the results reported in the previous section, difficult to forecast. In particular, no one expected to observe more organic compounds with a more complex structure to be synthesized as a result of a shorter laser exposure (0.3 ns instead of 10 ns) with the same focal-spot diameter and power density.

To explain the results obtained, consider the processes triggered by laser radiation with t_L ~ 10 ns. The processes needed for the formation of a plasma torch are known to last ~0.1 ns. It follows from this that in the case of a ~10 ns long laser exposure the plasma torch remains, during its expansion-away, subject to extra laser radiation for a relatively long time. This may also affect substantially the process of ion formation as a result of additional heating of the torch plasma, which, in turn, may result in late recombination and thereby slow down the synthesis of new compounds. Additional heating of the torch plasma may also result in the thermal dissociation of earlier synthesized organic compounds.

There is another possible explanation for the complexification of synthesized organic compounds in the case of short exposures. Such an important reduction of laser exposure results in a simultaneous change of two important properties of the process: (1) the energy spent for the production of ions decreases and (2) the exposure becomes too short for plasma to reach local thermodynamic equilibrium (LTE). These two factors result in incomplete ionization and atomization of the plasma, which, in turn, result in its early recombination. Such conditions may favor the synthesis of more complex organic compounds.

4.6. Synthesis of Organic Polymers from Amino Acids

The results described in the previous sections suggest that there are two independent ways to synthesize more complex organic compounds in the torch: (1) by increasing the characteristic size of the exposure domain or (2) by reducing the exposure time. Under the circumstances, we considered it most interesting to subject a target containing C, H, N, and O to a short exposure with a large focal spot. Such a combination of initial parameters was believed to produce maximum effect. However, the experiment has not been performed because of the lack of a laser emitter with a 0.3-ns pulse duration and a pulse energy sufficient to achieve the nominal power density in a ~ 1.5-2 mm diameter focal spot. That is why we used the laser emitter described above, which provided an energy of 600 mJ with a pulse duration of ~10 ns and a working diameter of the exposure region of ~ 1.5-2 mm, to further investigate the synthesis of organic compounds in the plasma torch with a target consisting of a mechanical mix of graphite and ammonium nitrate (NH_4NO_3) subject to laser radiation.

By subjecting such a multicomponent target to laser radiation with high power density (or high emission pulse energy) we obtained the mass peaks of the constituent elements of the target, i.e., H, C, N, and O. These results indicated that the excess energy in the emission pulse and its 10-ns long duration could have resulted in the disruption of organic compounds earlier synthesized in the plasma torch.

A decrease of the laser exposure energy by about 30—40% resulted in new peaks characteristic of chain-linear carbon with a maximum mass of 30—40C emerging along the H, C, N, and O peaks.

Further reduction of the power density of the exposure brought about a radical change of the mass spectrum: in particular, the elemental mass peaks decreased by a factor of more than 100. The peaks of chain-linear carbon became higher. Note that in the mass interval from 70 to 180 a.m.u. numerous peaks with the masses of amino acids, purines, and pyrimidines emerged between chain-linear carbon cluster peaks. They were followed by mass peaks of carbon structures with an interpeak separation of ~2C and with a well-defined amplitude modulation of carbon peaks with a step of 8C. This modulation can be seen out to the masses of 600—700 a.m.u. In the interval from 700 to 4000 a.m.u., individual, chaotically located peaks and formations consisting of closely spaced high-amplitude mass peaks located in the vicinity of 800, 1200, 2200, 3200, and 3900 a.m.u. are observed.

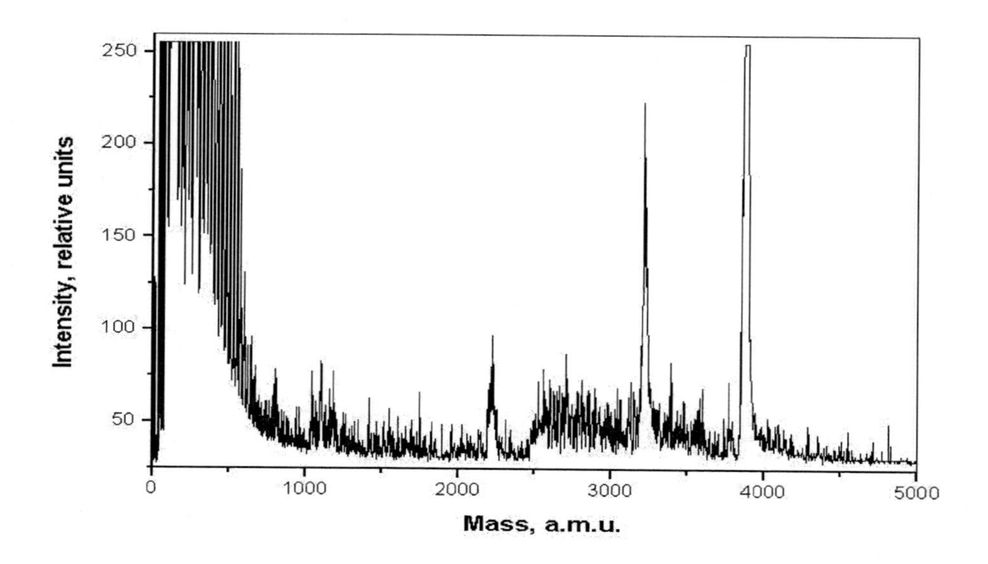

Figure 18. The spectrum of a carbon and $^{15}NH_4{}^{15}NO_3$ mix obtained via laser exposure with a spot diameter of 1.5-2mm. The mass peaks at 2250, 3250, 3900 a.m.u. are interpreted as dendritic-type, hyperbranched polymer organic compounds, and the mass peaks in the 1000-2000 and 2500-3100 a.m.u. domains, as peptides. (The full spectrum).

Figures 18—21 show typical spectral fragments, which we use below when analyzing the results. Note that these results were obtained with $W_L = 5 \cdot 10^8$ W/cm^2, which approximately corresponds to an impact velocity of 5-7 km/s for a carbon projectile, i.e., to surface ionization of the target.

To speed-up the analysis of the new compounds synthesized in the torch, let us begin with a preliminary selection of mass peaks earlier identified in the spectrum using various analytical techniques.

They include the mass peaks corresponding to chemical elements, i.e., C, H, N, and O, peaks of chain-linear carbon structures, and the peaks of amino-acid monomers, which are immediately apparent in Figure 19 by their location between the carbon cluster peaks in the mass interval from 70 to 150 a.m.u. and by a characteristic increase of their amplitudes.

It is now beyond question that the above new formations can be synthesized abiogenesously under the action of various analogs of natural factors. Hence no detailed analysis is needed for the initial portion of the mass spectrum.

Of special interest are organic compounds hitherto unobserved in the plasma torch, which we discuss below in more detail. Consider the fragments of the mass spectra shown in Figs. 19—21, which exhibit regular carbon peak structures recorded in the mass interval from 150 to 700 a.m.u. with a characteristic step of 2C and with a modulation amplitude of 8C. Such a distribution of peaks corresponds to an acetylene carbon chain where carbon atoms are linked by alternating single and triple bonds. This structure determines the interpeak distance of 2C. Such structures are also known to be maximally stable in the case of 8C chains, and this fact explains the observed modulation of peaks. In the spectra shown here interposed peaks of other elements can be seen between the main carbon peaks. These interposed peaks are shifted with respect to the main peaks by a step that is not a multiple of C. This means that the spectrum contains not only the signatures of "pure" carbon polymer structures, but also peaks of other high-molecular organic compounds whose structure is close to that of polymer organic compounds, but, in addition to carbon, also contain O and N atoms. Hence complex – so-called spider-web – polymer organic compounds could have been synthesized in the plasma torch on the basis of acetylene carbons.

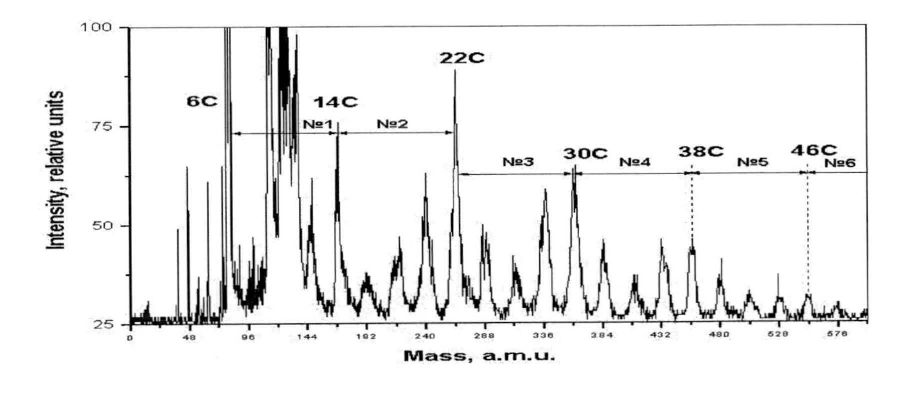

Figure 19. The 1—600 a.m.u. fragment of the spectrum shown in Figure 18 obtained by subjecting a carbon and $^{15}NH_4{}^{15}NO_3$, mix to laser radiation with a spot diameter of 1.5-2mm. The mass peaks (in domains nos. 1-6) were interpreted as branched polymer OC and cyclo-olefines (C8) based on linear acetylene carbon (-C2-C2-) or «sp» carbon allotropes. The periodic structure of the pattern typical of spider-web acetylene carbons is immediately apparent.

Acetylene carbon chains – an allotropic form of carbon based on sp-hybridization – were discovered and studied by Kasatochkin et al. (1967). Long chains were studied in detail by Lagow et al. (1995), who showed the new compounds to be sufficiently stable provided that they have "end caps". These compounds were synthesized in a special chemical reactor where graphite was evaporated in a gaseous mix consisting of He (90%) and C_2N_2 (10%) at a pressure of ~0.2 atm.

These conditions ensured the stability of synthesized compounds because the latter had "caps" put on their ends. The above authors also report the results of early, pioneering works of A.Kaldor (Rohlfing, 1984), who synthesized some linear carbon chains of various lengths (116C, 325C) in the gaseous state using Ar as the gaseous medium. It is important that the mass spectra that the above authors obtained for carbon chains were in a number of cases totally identical to the spectra shown in Figure 15, which were obtained using a carbon target during the expansion-away of the plasma torch into vacuum.

It was shown that high-resolution spectrum of acetylene carbon has a resonance at the masses of ~3900 a.m.u. This mass corresponds to ~325C and coincides with the characteristic size of the acetylene carbon chains earlier found by A.Kaldor and D.Cox (Kaldor et al., 1988).

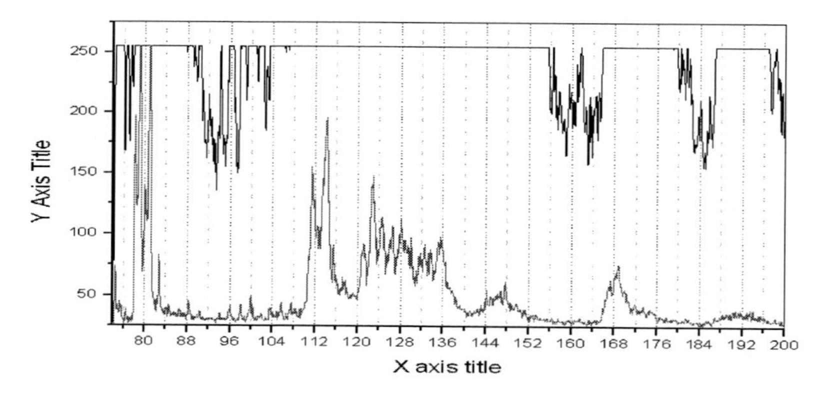

Figure 20. Fragment of the spectrum shown in Figure 18. The zone of amino acids. This zone is marked as domain No. 1 in Figure 19.

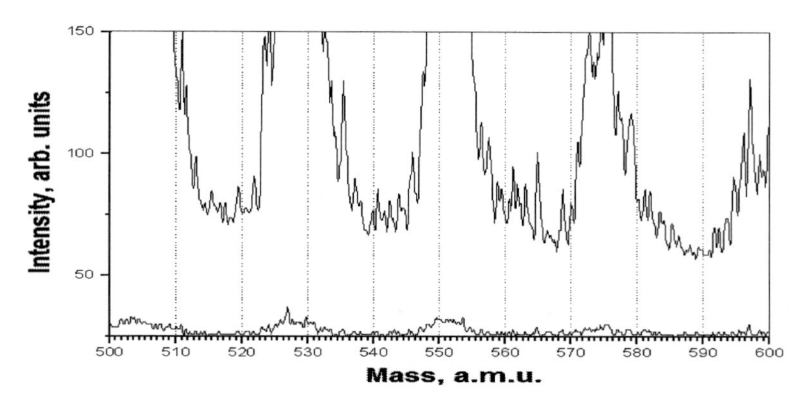

Figure 21. Fragment of spectrum shown in Figure 18. These mass peaks are interpreted as hyperbranched polymer organic compounds based on linear acetylenic carbon.

Such a coincidence may be due to the fact that 325C is the "magic" mass for linear structures considered. An analysis of the results obtained led Lagow et al. (1995) to conclude that linear carbon chains may be parts of the helical-chain structure shown in Figure 22.

The capability of laser plasma for "instantaneous" synthesis of extended acetylene carbons is indicative of the great potentialities of the plasma torch in the synthesis of complex polymer compounds. Note that the distribution of masses and peak amplitudes of acetylene chains produced in the plasma torch is very easy to identify and difficult to mistake for other similar structures.

The conditions of their synthesis are equally important: thus the synthesis of carbon structures by exposing pure carbon to laser radiation resulted in the production of carbines, fullerenes and their fragments, and hyperfullerenes. The ends of acetylene carbon chains became protected once O and N appeared in the plasma torch as a result of the introduction of NH_4NO_3 into the target. This circumstance may have made possible not only the synthesis of short, but also that of long chains.

The results described above give ground for a feasible interpretation of one of the high-amplitude peaks, namely the one located at 3900 a.m.u. This well-isolated ~24 a.m.u. wide band of peaks corresponds to the masses of 323—325C, or to the "magic" mass of acetylene carbons and with the resonance domain. The previous peak band located at 3230 a.m.u. can also be conventionally viewed as a "magic" band because it approximately corresponds to 245C.

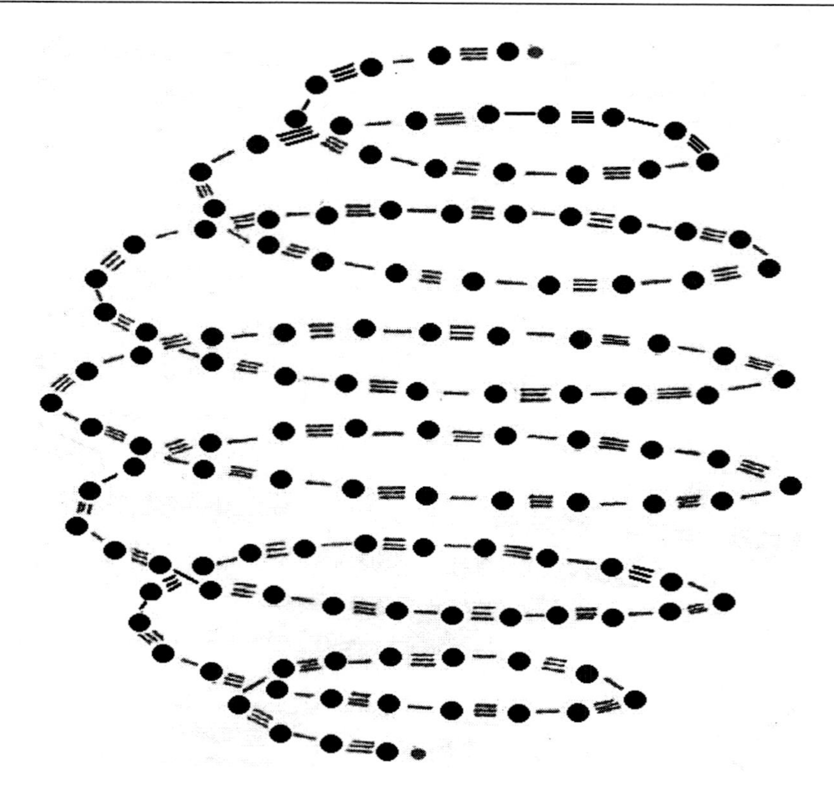

Figure 22. Hypothesis about the possible development of a spiral structure in the process of the synthesis of acetylenic carbons was first proposed by (Lagow et al., 1995). We show such a structure in this figure. The black circles denote carbon atoms.

These peaks, which are shifted by 80C with respect to the mass of 325C, make up 10 portions consisting of 8C large stable fragments. Such lucky coincidences do not occur for the band of peaks at 2220 a.m.u. because the location of bands corresponds to 180C and the 60C difference is not a multiple of 8. The band of peaks at 1100 a.m.u. corresponding to an average of 92 carbon atoms is almost equally "inconveniently" located. A comparison of various bands of high-amplitude peaks possibly indicates that the bands located at 3900 and 3200 a.m.u. have common nature associated with acetylene carbons, whereas the other peak bands located at 2200 and 1100 a.m.u. are of a different nature, or some bands of different nature may blend together.

The mass peaks observed in the initial part of the spectrum and corresponding to the masses of amino acids suggest that above-background irregular peaks recorded in the spectrum must correspond to peptides synthesized in the torch after their polymerization in the process of plasma expansion-away. This hypothesis needs experimental confirmation. To obtain such a confirmation, sufficient amount of organic compounds synthesized in the torch should be accumulated, and complex mass-spectrometric analysis should be performed with a time-of-flight mass spectrometer consisting of two serially connected instruments. Such a device is capable of obtaining the mass spectrum of amino acids that contribute to a single peak of the chosen peptide after its fragmentation. Currently, also of interest are other structures, which form amino acids and which may show up in the spectrum as mass peaks, - the so-called dendrimers.

As a class of compounds having the form of regular spider-web polymer structures shown in Figure 23, dendrimers synthesized based on amino acids form molecules with a certain mass ratio depending on their structural type. Thus depending on the degree of branching dendrimers synthesized from a single amino acid have masses that may relate as 500:1000:2000:4000:8000 in the case of $D_1D_2D_3D_4D_5$ generation (Vlasov et al., 2005). These ratios do not always show up in the measurements, but the overall pattern remains the same.

For homodendrimers synthesized, e.g., in a solution as a result of the classical process (which takes more than 10 hours), the molecular peaks of the compounds of different generations appear as monopeaks in the mass scale of the spectrometer.

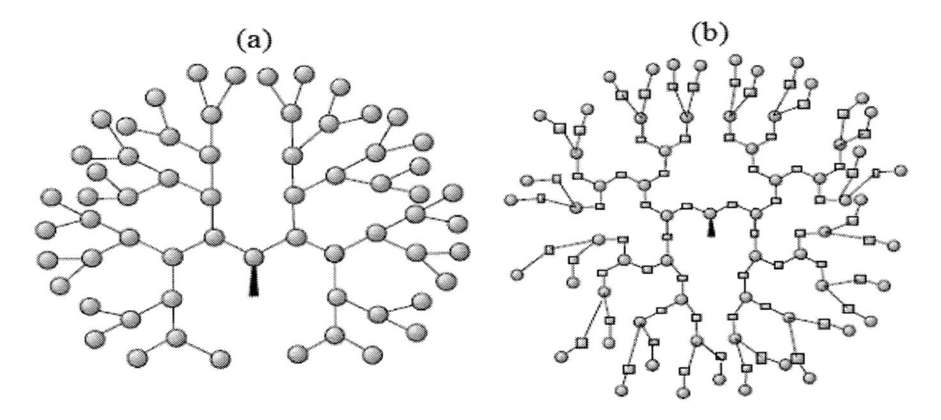

Figure 23. Schematic sketch of the fifth-generation homolysine (a) and heterolysine (b) dendrimers. The circles show lysine at "branching" points and rectangles, the residues of alanine, glutaminic acids, histidine and diglycine between the residues of lysine at the "branching" points (Vlasov, et. al., 2004, 2005).

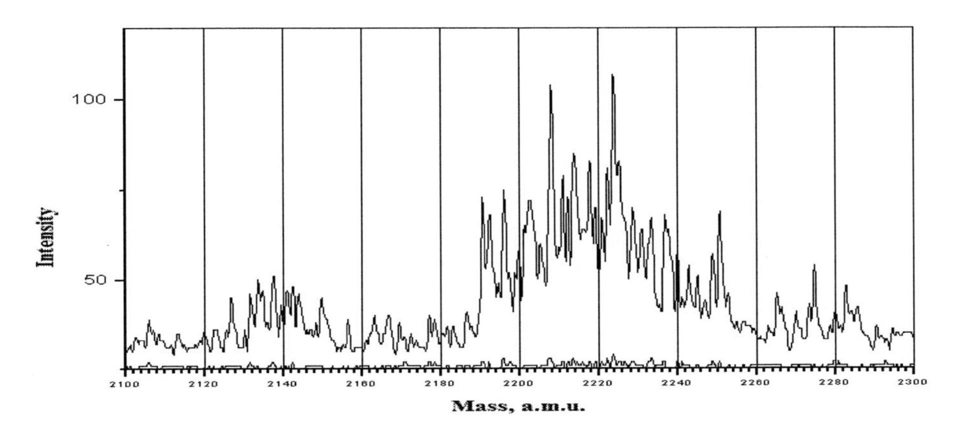

Figure 24. Fine structure of the magic peak of organic compounds is interpreted as a heterogeneous amino-acid dendrimer.

However, no one knows how inhomogeneous heterodendrimers synthesized in the plasma torch, presumably via the mechanism of “instantaneous” self assembly, may appear in the mass spectrum, and what should be the mass ratios of the dendrimers of different generations because no such compounds have been so far synthesized in a torch.

We will nevertheless try to imagine how heterodendrimers of different generations synthesized in a plasma torch would "appear". In the mass spectrum they should most likely show up as bands of closely spaced individual peaks (Figure 24). A possible scatter of the masses of amino acids that make up the structure may also result in the deviations from the exact ratios of the masses of dendrimers of different generations.

Let us now return to the mass spectrum shown in Figure 19 and analyze the large bands of peaks and their distribution. It is immediately apparent from the figure that each band consists of many closely spaced mass peaks, and if we interpret the band at 3900 a.m.u. as the location of the fourth generation, or D_4, then the band at 2200 a.m.u. can be conventionally interpreted as belonging to D_3. Furthermore, the band at ~1100 a.m.u. should be classified as belonging to D_2. Given the abrupt decrease of the peak amplitudes toward smaller masses, D_1 may be lost against background noise of, purportedly, polypeptydes. Hence the 1100:2200:3900 a.m.u. ratios may be due to heterodendrimers D_2, D_3, and D_4. One may also discern the bands of peaks corresponding to the 800:1600:3200 a.m.u. ratios, however, they are not very conspicuous.

Thus in experimental laser simulations of hypervelocity impact processes with a 1.5—2 mm large plasma-formation region and a target consisting of C, H, N, and O, the following compounds may be synthesized in the plasma torch: chain-linear carbon, amino acids and their precursors, acetylene carbon chains, and also, supposedly, peptides and heterogeneous dendrimers.

One of the important results of this series of experiments is that they demonstrated the possibility of the synthesis of polymer organic compounds. This conclusion follows from the facts listed below:

- The target used was a 50%:50% mechanical mix of ultrapure graphite and high-purity ammonium nitrate (NH_4NO_3) and hence the compounds synthesized in the plasma torch contained only C, H, O, and N atoms;
- Mass-spectrometric analysis of the synthesized compounds based on the periodicity of carbon peaks with a period of 2C and the modulation amplitude of these peaks with a step of 8C indicates that acetylene carbon chains - chain carbon inorganic polymers – were synthesized in the plasma torch;
- The mass-spectrum peaks found between the main carbon peaks, but shifted by an offset that was not a multiple of 12, could have formed in the following cases: (1) if N or O atoms became incorporated into the carbon chain; (2) if H, N, or O atoms were appended to the ends of the chain or if in the case of branching with these elements or their derivatives side attached to the chain. Thus the compounds synthesized in the plasma torch contain C, H, N, and O and should be classified, based on their composition, as organic compounds. Structurally, these materials may have the form of high-molecular organic compounds similar to irregular polymers, and hence they can be viewed as spider-web organic polymers.

The mass spectra of polymer organic compounds considered here are the logical development of the spectra obtained with a purely carbon target. This marks the continuity of the results and proves them to be reliable and credible.

Given the novelty and importance of the results of the synthesis of carbon polymers, we consider the main achievement of these studies to be the first-ever abiogeneous synthesis of complex high-molecular organic compounds, which can be interpreted as modifications of polymer compounds. The fact that organic compounds interpretable only as organic polymers can be synthesized in the plasma torch identical to the plasma torch of a hypervelocity impact helps to remove the most complex discrepancies in the early scenarios of the prebiotic stage of evolution.

The results obtained also indicate that with a substantially increased size of the plasma-formation region the plasma torch of a hypervelocity impact can be viewed as the main source of polymer compounds that are required for the emergence of life and that could be synthesized at the early stage of the evolution of the Earth without solar energy input and irrespectively of the atmospheric composition.

The main and fundamental result of these studies is that they confirm the conclusions reported in Section 4.4 about the processes of self-organization of matter that occur in the torch plasma. However, this time these exceptional properties of the plasma torch are confirmed not just for an environment consisting of C and H atoms exclusively, but also for an environment that includes all the basic elements required for the synthesis of many key life-producing compounds – from amino acids to lipides.

4.7. Synthesis of Nucleotides and their Oligomers

The studies of the possibility of the synthesis of monomers of amino acids and their oligomers naturally evolved into the studies of the formation of nucleotides and their derivatives under similar conditions. The heightened interest in organic compounds of this class can be explained by the fact that many hypothetical models suggest RNA as the basic molecule for the emergence of the simplest form of living matter.

To experimentally investigate the possibility of the synthesis of nucleotides and their oligomers in the torch plasma, P – the key element required for the formation of nucleic acids – has been introduced into the target along with C, O, N, and H.

To this end, a target has been created to be subject to laser exposure, where fine-dispersed powder of the C^{13} isotope was mixed with ammonium hydrophosphate powder – $(NH_4)_2HPO_4$. Both ingredients were ultrapure substances.

The target was exposed to an IR laser operating in the Q-switched mode at the wavelength of 1.06 μm with a pulse duration of 0.3 ns and a power density of $W_L = 10^9$ W/cm^2. The focal-spot diameter was 50 μm.

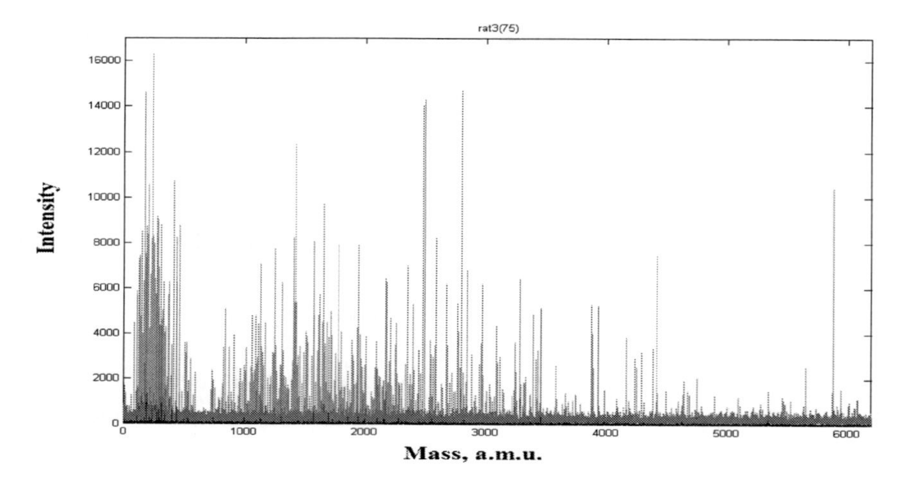

Figure 25. Typical mass spectrum of organic compounds synthesized in a laser-generated plasma torch with a target consisting of pure carbon mixed mechanically with $(NH_4)_2HPO_4$ for a spot diameter of ~50μm and pulse duration of ~1ns. According to a preliminary interpretation the observed mass peaks may contain nucleotides (left) and its oligomers (right).

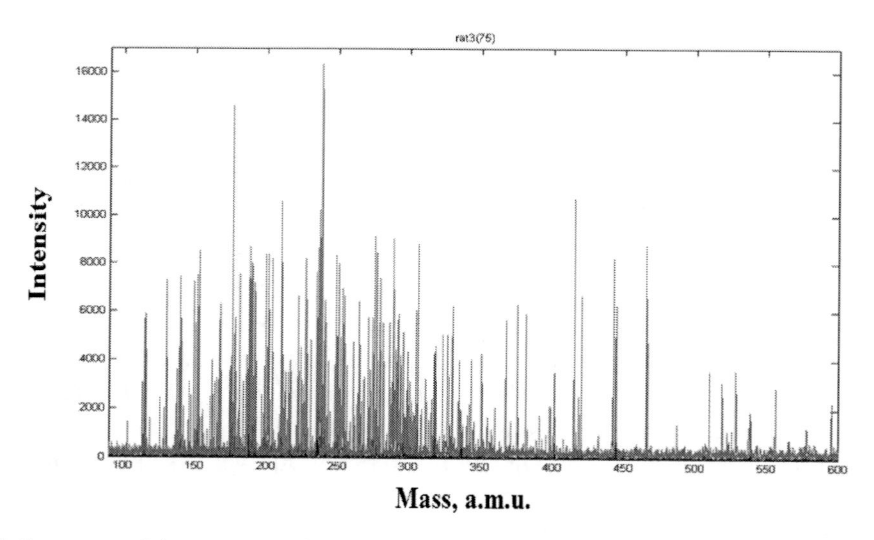

Figure 26. Fragment of the spectrum shown in Figure 25 in the mass interval from 100 to 600 a.m.u. «The carbon comb» with a step of 13 a.m.u. is immediately apparent.

The recorded mass spectra exhibited certain characteristic features: mass peaks broke conspicuously into two groups. The first group, as it is evident from Figure 25, was located in the mass interval from 100 to 400 a.m.u. with the maximum amplitude in the mass interval 200—300 a.m.u. The second group of peaks was located in the interval from 1000 to 4000 a.m.u. A minimum of peak intensities is observed in the mass interval from 450 to 960 a.m.u. As is evident from Figure 26, the so-called "carbon comb" with a step of 13 a.m.u. and with many interposed mass peaks located between the C_N peaks, where N varies from 9 to 32, shows up with increasing mass in the case of the first group. Such a periodicity of the mass peaks in the first group can be seen in almost any spectrum, sometimes with appreciable offsets of the distribution maximum.

The second group of mass peaks does not exhibit such a periodicity and the stability of the pattern is less pronounced. However, individual mass peaks of the second group may amount to 10000 to 20000 a.m.u., i.e., they may correspond to oligonucleotides (if such a compound was indeed synthesized) consisting of 30—60 monomers.

The interpretation of the observed mass spectra is therefore restricted to the analysis of the mass peaks of the first and second groups exclusively because their high reproducibility suggests that these peaks must be the most reliable ones.

A comparison of the distribution of the mass peaks of the first group with that of the peaks obtained in the synthesis of amino acids, i.e., in the absence of P, shows that the introduction of this element shifted rightward the maximum of the distribution of the mass peaks of the first group. This effect must be due to the increased mass of molecular compounds after P was introduced to the target. Hence synthesized substances included organic compounds, which contained P atoms along with C, O, N, and H. Because of the stochastic nature of the process, organic compounds synthesized in the torch may have an arbitrary, e.g., "nucleotide", structure. This hypothesis is based on the fact that the average mass of organic compounds of the first group coincides very accurately with the masses of nucleotides. This allows us to make an important conclusion that the first group may contain, along with other molecules, an indefinite amount of nucleotides. In any case, so far no chemical laws are known that would forbid such a process.

This hypothesis is indirectly confirmed by the absence of mass peaks in the mass interval from 450 to 900 a.m.u. and their appearance at larger masses corresponding to dimers, trimers, and oligomers of nucleotides, which are believed to be synthesized in the plasma torch.

We emphasize once again that the foregoing is nothing more than an interpretation of the experimental results obtained, and this interpretation should be followed by the identification of synthesized organic compounds. Do we really need to perform such extremely complex mass-spectrometric measurements in the case of spontaneous synthesis of organic compounds in the torch plasma, and what can we gain from an analysis of macromolecules?

Figure 27 shows the dependence of the masses of various organic compounds synthesized in the plasma torch on the effective diameter of the projectile, composition of the target, and duration of the laser exposure. This plot summarizes the results of the synthesis of organic compounds represented in Chapter 4. The identified compounds are marked by the asterisk (*). The coarse graphical extrapolation of the results obtained shown in Figure 27 implies that a 1-cm projectile is by a large margin capable of producing organic compounds with the masses as high as $5\ 10^5$ a.m.u., which are substantially higher than the supposed maximum mass of a hypothetical protoviroid – the primary form of living matter.

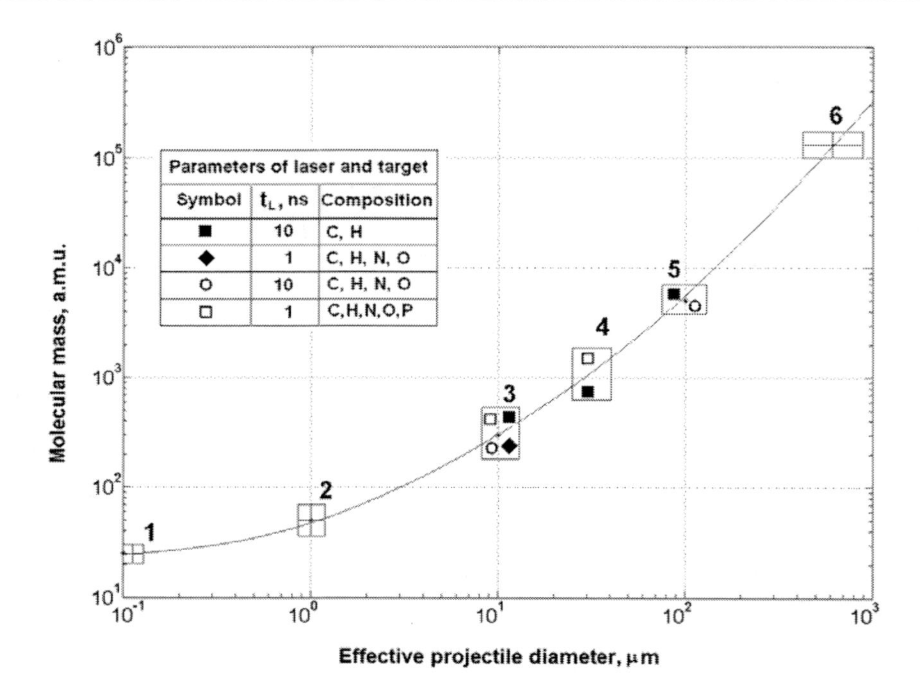

Figure 27. Dependence of the molecular mass of the products of synthesis on the effective size of the meteorite. The following typical domains of the synthesis of organic compounds are indicated: domains 1 and 2 correspond to the interstellar medium with the dust grain sizes of 0.1 and 1 μm and organic-molecule masses of 20 to 40 and 50 to 100 a.m.u., respectively. The experimental data points shown for domains 3—5 were obtained in laboratory simulations of impact processes. In the domain 3 the following symbols are used: ■ - carbines*, O – amino acids*, ◆- oligomers of amino acids, □ - nucleotides. In the domain 4: ■ - fullerenes *, □ - oligonucleotides. In the domain 5: ■ - hyperfullerenes *, O - highly branched acetylene carbines and dendrimers. Domain 6 corresponds to a macromolecule with a mass that is close to that of protoviroid and which was obtained by approximating the experimental curve. Asterisk (*) indicates identified compounds.

Note, however, that in this case the macromolecule so produced may resemble a protoviroid only in terms of molecular mass and bears no comparison neither in structure nor in functional capabilities.

***

The above results of the experimental simulations of the impact-produced plasma torch via laser exposure indicate that the conditions in the plasma make possible the efficient synthesis of organic compounds. These results can also be considered to apply to the impact-produced plasma torch.

The possibility of the synthesis of organic compounds in the torch should be considered as a natural property because such a plasma formation can be viewed as a special case of a plasma-chemical reactor. Natural reactor of this type, with its high catalytic activity, remains a poorly studied environment. It is, however, safe to suggest that the increase of the size of the plasma-forming region should result in a substantial complexification of synthesized organic compounds.

The new result consists in the possibility of the synthesis of simple and complex organic compounds in the impact-produced plasma torch, and in the idea of using this mechanism to explain abiogeneous synthesis of organic compounds on the primeval Earth during the prebiotic stage of the emergence of life.

The equipment used in the above experiments is capable of identifying only a part of the synthesized organic compounds detected by the time-of-flight mass spectrometer. The uncovered trend for the increasing complexification of synthesized compounds with increasing characteristic size of the plasma-formation region and decreasing duration of laser exposure can be reasonably explained based on the fundamental laws of plasma physics.

In this connection, the proposed mechanism of the synthesis of organic compounds, which is problem- and inconsistency-free, could very likely be realized under natural conditions, and thereby ensure highly efficient synthesis of simple and complex organic compounds, resulting in the formation and accumulation of organic compounds required for the emergence of living cell.

The main result of these experiments is especially valuable because a very simple laboratory simulation of an analog of the very common natural factor typical of the early Earth, and with no special tricks applied, made it possible to synthesize high-molecular organic compounds that proved to be practically identical to polymer organic compounds. Our analysis of organic compound structures synthesized in the torch unambiguously demonstrates that highly efficient processes of the self-organization of matter occur in the expanding torch plasma and result in the formation of carbines, fullerenes, hyperfullerenes, spider-web structures of organic compounds of acetylene carbons and dendrimers. The results obtained are indicative of the processes of self-organization of matter in the torch plasma, and this conclusion may be of crucial importance for the new concept of the prehistory of life.

Note that at this stage of research it is of little importance to which extent the polymer organic compounds synthesized in the torch are similar to certain biopolymers. Of much greater importance is the fact that the synthesis of extended polymer structures of organic compounds, which are indicative of the self-organization of matter, is possible in principle. This result may to a large measure be viewed as proving that polymer organic compounds whose functional and structural properties will eventually make possible the emergence of at least some characteristic signs of living matter may be synthesized in a plasma torch with a large characteristic size and other initial elemental ratios.

Chapter 5

OPTIMIZATION OF INTERACTION PARAMETERS AND IDENTIFICATION OF SYNTHESIS PRODUCTS

5.1. THE NECESSARY OPTIMIZATION

The experimental results reported in the previous chapter show that in a plasma torch carbon and hydrocarbon structures, as well as simple reactive organic compounds including the precursors of amino acids, are synthesized. These compounds have small molecular masses and are characterized by the periodicity of mass peaks – the features that can be used to identify them. However, with increasing mass, e.g., from 70 to 200 a.m.u., organic compounds synthesized in the torch become much more difficult to identify. The point is that at greater masses the same peak in the spectrum may correspond to an increasingly large number of different organic compounds. Therefore the ambiguity of the identification of such organic compounds increases with their mass. Given that the recorded mass peaks (1) contain only C, H, N, and O elements; (2) are not multiples of carbon clusters, and (3) their masses coincide with those of some amino acids, they can be with high probability interpreted as amino acids. It should, however, be born in mind, that a substance with the same molecular mass as an amino acid may actually not consist exclusively of amino acids. A single mass peak may include not only the corresponding amino acid, but also other organic compounds of the same molecular mass – the so-called amino acid isomers.

The analytical properties of the LASMA mass spectrometer used in these experiments do not allow it to separate isomers of the same compound and hence prevents the identification of amino acids. Dedicated techniques and instruments exist that should be used for such measurements. In such cases it is necessary to accumulate a certain amount of synthesis products to be further analyzed using alternative and more sensitive methods. In this case the first serious difficulties arise that are due to the low efficiency of the laser emitter of the LASMA instrument. In particular, the emitter allows a single mass spectrum to be obtained only once in 5 s. Such a slow accumulation of the synthesis products on the substrate makes the procedure very time consuming. Given that the mass of organic compounds synthesized in a single act of interaction in the plasma torch does not exceed 10^{-13} g, 10^4 to 10^5 laser exposures, or ~20 eight-hour work days, are needed to accumulate the amount of organic compounds required for a more in-depth analysis.

The only way to address this problem consisted in replacing the available laser emitter by a modern one, which would be capable of generating exposure pulses with a frequency of several dozen to several hundred Hz. In such emitters the energy comes to the active element of the main laser from a "pumping" pulsed semiconductor diode laser. This ensures highly efficient energy transfer to the active element and high emission pulse frequency.

This upgrade was a rather costly and time consuming process because of a number of non-standard optical units of the focusing system that had to be manufactured. As a result of the upgrade the refurbished LASMA instrument acquired unique analytical characteristics both for the study of the processes of the synthesis of organic compounds and for the elemental and isotopic analysis of various samples with the possibility of detecting ultralow concentrations of pollutant substances, which in some cases can be on the order of several ppt.

Another task that is at least equally important for the correct simulation of an impact plasma torch involved the optimization of laser pulse duration. It became clear, as a result of the progress in laser simulation of impact processes, that with the actual laser pulse duration of about 10 ns only the first nanosecond is spent on the formation of the torch, whereas the remaining 9 ns are spent on the generation of processes that are uncharacteristic of impact action. In particular, further plasma heating occurs after a dense plasma structure forms that is opaque to laser radiation. Such exposure changes fundamentally the final properties of the torch, destroys the organic compounds synthesized in it, and makes it impossible to correctly estimate the energy contribution to the plasma.

A comparison of the process of plasma formation as a result of laser exposure with the processes of the torch formation in the case of a hypervelocity impact shows that it takes about 1 ns for a 10—20 μm diameter projectile moving at a velocity of ~20 km/s to penetrate into the target. During this time a plasma torch forms at the impact site and this event is followed by adiabatic expansion-away of the plasma that is since then unaffected by other external factors. This means that the duration of laser exposure should be reduced to 1 ns in order to correctly simulate a hypervelocity impact of a 10—20 μm diameter meteorite.

Just a few years ago such an upgrade of the laser emitter would require considerable effort – e.g., a laser with unique parameters would have to be developed from scratch. However, in the case considered such a device has already been developed just in time – a next-generation diode-pumped laser emitter with a pulse duration that could be varied from 0.2 to 1 ns. Furthermore, the new emitter had highly stable pulse energy, which was a factor of greatest importance for the correct summation of the mass spectra obtained.

Therefore the replacement of the laser allowed us to address successfully a number of problems and brought about rapid progress. Thus the parameters of the new laser allowed us simultaneously:

- to increase the exposure rate at least by a factor of 100, in particular, from 0.2 to 30 Hz;
- to reduce the pulse duration by a factor of 10, from 10 to 1 ns;
- to reduce the scatter of pulse energy by a factor of 10, from 10 to 1%.

Consider now the most important consequences of the upgrade of the laser emitter in the LASMA instrument.

The increased exposure rate allowed us to increase substantially – by a factor of 100 – the productivity of the instrument. The new parameters resulted not only in a fast accumulation of data, but also in a much higher sensitivity over a finite measurement time.

The reduction of the pulse duration made it possible to preserve the organic compounds synthesized in the torch, raise the amplitude of mass peaks in the "amino acid synthesis domain", and increase the maximum mass of synthesized organic compounds to 300 a.m.u.

The high stability of the laser pulse ensured high reproducibility of the mass spectra, resulting in a higher sensitivity and mass resolution of the averaged spectra.

The upgraded instrument is capable of collecting enough mass spectra for the identification of amino acids when used together with other analytical instruments. Such identification was of special importance in the case considered. This "key" identification played an important part in the history of research.

In particular, amino acids have been synthesized over the past 50 years in numerous laboratory experiments simulating the effect of various natural factors onto a gaseous mix imitating the early Earth atmosphere. Rather efficient synthesis of amino acids was observed in a laboratory modeling of the plasma environment that forms in the ionosphere or in an electric discharge. In such a situation the synthesis of amino acids in dense high-temperature, high energy-density torch plasma was to be expected, however, the formation of these compounds had to be proved and identified without fail. These results could serve as a starting point for the studies of the torch plasma and the work on the new concept.

However, in the case considered these measurements were from the very beginning aimed at a different task – namely, at the study of the breaking of the mirror symmetry of amino acids generated in the plasma torch. Given the extremely high importance of the problem of mirror symmetry breaking in bioorganic world, the proposed upgrade of the laser emission was also aimed both at the proposed hypothesis of the possibility of such a breaking during the emergence of isomers in the plasma torch.

Thus, to measure the "sign" of the asymmetry of amino acids synthesized in the torch plasma, we had to learn how to accumulate them in sufficient quantities. The currently available instruments make it possible to collect the necessary quantity of synthesis products including amino acids to measure the "sign" of their asymmetry using new, highly sensitive techniques. Moreover, such measurements have already been made, and new experiments are being prepared to corroborate the correctness of these measurements. The results of these experiments will show how substantiated is the hypothesis that electromagnetic fields and radiation of the plasma torch are capable of generating local chiral physical fields, which may break the symmetry of amino acids synthesized in the torch.

Hence the upgrade of the LASMA mass spectrometer, which consisted in replacing the laser emitter by a next-generation radiation source, has substantially extended the functional capabilities of the available instrument. The resulting improvements made it possible to confirm the possibility of the synthesis of amino acids in the plasma torch and marked the beginning of experimental studies of the new idea about the possibility of the breaking of mirror symmetry in the torch plasma.

5.2. RESULTS THAT CONFIRM THE CORRECTNESS OF MODELING

The proposed new concept about the possibility of the synthesis of organic compounds in the plasma torch of a hypervelocity impact (Managadze, 2001, 2002a) was confirmed by direct impact experiments performed with the dust particle accelerator in Heidelberg (Stubig, 2002).

The facility in Heidelberg accelerated micron-sized aluminum, carbon, iron, and latex dust particles to 2—70 km/s. The masses and densities of these particles ranged from 10^{-15} to 10^{-9} g and from 1.1 to 7.9 g/cm^3, respectively. Impacts of particles moving at speeds 6 km/s and smaller resulted in the evaporation of the matter contaminating the surface of the target. Impacts with 18 km/s and higher velocities resulted in the predominant ejection of plasma ions of the matter contained in the accelerated microprojectiles or of the ions of the target material, and in the formation of a plasma torch.

The results reported here were obtained in the process of the calibration using the accelerators of CDA and CIDA onboard dust-impact tile-of-flight mass spectrometers (Stubig et al., 2001; Stubig and Grun, 1997) meant for the study of the stream of micrometeorites accelerated in the gravitational field of Jupiter. The results obtained with pure carbon and latex microparticles are of special interest for the concept proposed in this book.

A bombardment of a pure Rh target by carbon particles moving at a velocity of ~17 km/s produced intense peaks at 103 and 115 a.m.u. and typical carbon cluster peaks at equidistant mass positions with a characteristic step of 12 a.m.u. Given the results of laser simulation of impact processes, we proposed a novel interpretation of the data by identifying the mass peak at 115 a.m.u. with rhodium carbide (RhC) (12+103) and explaining its origin by the processes of synthesis occurring in the plasma torch produced by the impact.

Table 2. Possible molecular ions produced as a result of the bombardment of a silver target by latex micro particles (Stubig, 2002)

Mass (a.m.u.)	Possible molecules
27	CHN, C_2H_3
29	CH_3N, C_2H_5
31	CH_5N, H_3N_2
36	C_3
42	CH_2N_2, C_2H_4N, C_3H_6, N_3
44	CH_4N_2, C_2H_6N, C_3H_8, H_2N_3
46	CH_6N_2, H_4N_3
48	C_4
58	CH_4N_3, $C_2H_6N_2$, C_3H_8N, C_4H_{10}, H_2N_4
60	CH_6N_3, $C_2H_8N_2$, C_5
62	C_4N, C_5H_2, H_6N_4
63	C_4HN, C_5H_3
65	C_3HN_2, C_4H_3N, C_5H_5
67	C_2HN_3, $C_3H_3N_2$, C_4H_5N, C_5H_7
122	$C_5H_6N_4$, $C_6H_8N_3$, $C_7H_{10}N_2$, $C_8H_{12}N$, C_9H_{14}, C_9N, $C_{10}H_2$

We also bombarded a silver target by latex microparticles that moved at a velocity of ~16 km/s and consisted not only of carbon, but also of nitrogen and hydrogen atoms. The observed mass spectra showed the mass peaks of silver and the peaks that we interpreted as due to molecular ions containing H, C, and N.

The proposed interpretation of the mass peaks was based on choosing the optimum combination of C and N atoms supplemented by H atoms. Table 2 lists the organic compounds so obtained with the masses ranging from 27 to 122 a.m.u. The velocities of the impacts that produced the mass peaks mentioned above were higher than the critical velocity leading us to suggest, like in the previous case, that the molecular ions generated in the impact process were not fragments of target microparticle molecules, but rather compounds synthesized in the plasma torch. It is important that the mass spectra with the molecular-ion peaks found in the direct impact experiments mentioned above have not been previously observed, and that we were the first to explain their origin in terms of the new mechanism based on the processes that occur in the plasma torch.

To confirm the above hypotheses, we decided to reenact in a laser-produced torch the synthesis processes that occur in an impact plasma torch using the same substances, in particular, rhodium and carbon.

To this end, we prepared for laser simulation of impact processes a target consisting of a mechanical mix of Rh and C. The target was subject to laser pulse exposure at $\lambda \sim 1.06\mu$ with a duration of ~10 ns. The laser beam was focused into a ∅30-50μ diameter spot to achieve a power density of $W_L \sim 10^9$ W/cm^2.

Note that in this case the parameters of the simulation did not fully coincide with those of the impact experiment because, e.g., the plasma-formation regions had different volumes in the two cases. However, even without the appropriately chosen laser parameters the observed spectra were almost identical to those obtained in the impact experiments mentioned above.

Figure 28 shows two typical spectra obtained in direct impact experiments (a) and in experiments involving laser simulation of impact processes (b). The two spectra are practically identical in the composition and distribution of mass peaks. The difference in the peak width can be explained by the different mass resolutions of the two experiments, which (the peak halfwidth), in particular, is estimated at ~60 and ~300 for the CDA and LASMA instruments, respectively.

The experimental results obtained prove the validity of the newly proposed concept of the possible synthesis of polyatomic compounds including organic substances from inorganic matter in the plasma torch produced by a hypervelocity impact. This conclusion is demonstrated both by the experiments involving the simulation of a hypervelocity impact via laser exposure (Managadze, 2001, 2003) and by direct impact experiments using dust-particle accelerators (Stubig, 2002). The results obtained also confirm that laboratory simulation of a hypervelocity impact can be made with the optimum choice of simulation parameters.

Note that natural analogs of impact processes are not easy to observe under cosmic conditions because such events are rare. Moreover, the location of a future impact is impossible to predict. Laboratory experiments performed with dust-particle accelerators are subject to serious limitations in terms of both the mass and velocity of the projectile.

In view of the above, we consider the most appropriate strategy to be the one that combines extensive laser simulations of hypervelocity impacts using laser in a laboratory with occasional experiments involving direct impacts performed with microparticle accelerators with the aim to check some of the results of the simulations. Direct impact experiments cannot be performed as extensively as simulations because of their high cost.

5.3. Optimization of Laser Exposure Parameters

The results of laboratory simulations of hypervelocity impacts described above show that certain physical parameters of exposure factors affect the characteristics of the processes of the synthesis of new compounds in the plasma torch.

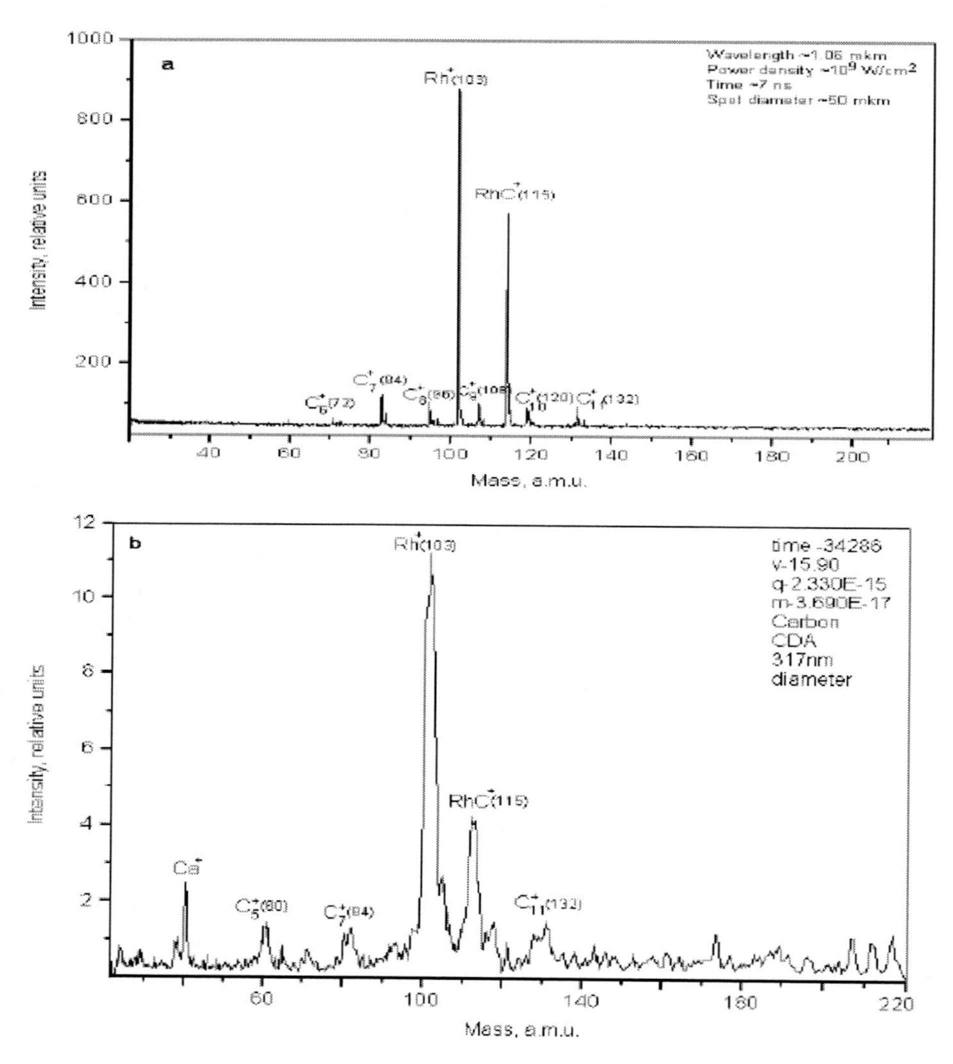

Figure 28. Comparison of the mass spectra obtained in the experiments on laser simulation of a hypervelocity impact (a) for a laser operating at λ~1.06 μm with a power density of 10^9W/cm^2 and in direct impact experiments using a dust particle accelerator and (b) in the case of an impact of a 317-nm 3.7 10^{-14} g particle moving at a velocity of 16 km/s colliding with a pure Rh target. This procedure ensured that the initial samples containing Rh and C had identical compositions in both experiments. It is important that rhodium carbide was synthesized along with hydrocarbons.

The effect of some of these parameters was easy to predict. Thus the increase of the spatial size of the plasma formation was expected to result in the synthesis of more complex and massive organic compounds. Experimental results confirm conclusively that is indeed the case. We described these results in Chapter 4.

However, the complexification of the structure and the increase of the yield of organic compounds in experiments with a factor of 30 shorter exposure times have not been predicted. Such a relation between the exposure parameters and the results of synthesis was nevertheless explained. This relation, in turn, indicated the necessity of the development of a simple and reliable method for determining the main parameters of the laser emitter to be used in experimental simulations of the impact process.

To answer this question, let us first analyze the effect of the laser pulse duration t_L on the formation time of laser plasma. The plasma-formation time is known to be the sum of the time scales of the following physical processes: maxwellization, ionization, and the establishment of local thermodynamic equilibrium (LTE).

For the plasma torch to be generated, the time scales of these processes, which we denote as t_{EE}, t_{ION}, and t_{EI}, respectively, must be much shorter than t_L. In Table 3 we compare the above time scales with the pulse durations for various laser emitters. It follows from these data that the necessary conditions for the generation of the torch are satisfied for $t_L \sim 10^{-8}$ to 10^{-9} s and are not satisfied for $t_L \sim 3\ 10^{-11}$ s. Experimental results are available for $t_L \sim 3\ 10^{-10}$ and for $3\ 10^{-8}$ s. These results, which we describe in Section 3.5, indicate that the formation of the plasma torch with increased yield of organic compounds and complexification of their structure was observed at $t_L \sim 3\ 10^{-10}$ s despite the fact that, according to Table 3, ionization did not reach 100% and early recombination occurred in this case. Hence when choosing the pulse duration for the emitter it would be reasonable to view $t_L \sim 3\ 10^{-10}$ s as the lower limit for this important parameter and consider t_L values from $5\ 10^{-10}$ to 10^{-9} s for actual operation.

Table 3. Effect of laser pulse duration on the basic properties of the plasma torch

No.	Name of the time scale of the physical process	Formula	Time scale	Laser pulse duration, sec			
				$\tau_L \sim 10^{-8}$	$\tau_L \sim 10^{-9}$	$\tau_L \sim 3 \cdot 10^{-10}$	$\tau_L \sim 3 \cdot 10^{-11}$
1	Time scale of plasma “maxwellization” (condition $\tau_{EE} << \tau_L$)	$\tau_{EE} = 0{,}26 T_E^{3/2} / / n_E \ln\Lambda)$	$\tau_{EE} \sim 10^{-14}$s, $\Lambda = 10$	$\tau_{EE} << \tau_L$	$\tau_{EE} << \tau_L$	$\tau_{EE} << \tau_L$	$\tau_{EE} << \tau_L$
2	Time scale of ionization development (condition of 100% ionization $\tau_{ION} << \tau_L$)	$\tau_{ION} \sim 0{,}3 \cdot 10^{-10} \times \times (\ln \cdot n(\tau_{ION} / / \ln(0))^{1/2}$	$\tau_{ION} \sim 10^{-10}$s	$\tau_{ION} << \tau_L$	$\tau_{ION} << \tau_L$	$\tau_{ION} < \tau_L$ 100% ionization is not achieved	$\tau_{ION} > \tau_L$ Conditions are not fulfilled
3	Time scale of local thermodynamic equilibration (LTE) (condition $\tau_{EI} << \tau_L$)	$\tau_{EI} \sim 252 A t_E^{3/2} \times \times (n_E z^2 \ln\Lambda)$	$\tau_{EI} \sim 10^{-10}$s, $\Lambda \sim 10$	$\tau_{EI} << \tau_L$	$\tau_{EI} << \tau_L$	$\tau_{EI} < \tau_L$ LTE is not achieved. Early recombination	$\tau_{EI} > \tau_L$ Conditions are not fulfilled.

Table 4. Laser emitter parameters for different pulse durations

№	Laser emitter properties	Pulse duration, sec.				
		$\sim 10^{-8}$	$\sim 10^{-8}$	10^{-9}	$\sim 3\cdot10^{-10}$	$\sim 3\cdot10^{-11}$
1	Pulse energy, mJ	20	$6\cdot10^{2}$	$20\cdot10^{3}$	0,3	1
2	Spot diameter, μm, at 10^9 W/cm^2	50	$1.5\cdot10^{3}$	$5\cdot10^{4}$	25	$1.5\cdot10^{3}$
3	Maximum power density for the 25 μm spot diameter, W/cm^2	$4\cdot10^{9}$	$3.6\cdot10^{12}$	$4\cdot10^{12}$	10^{9}	$3.6\cdot10^{12}$
4	Specific energy per ion, eV	10	10	1	0.4	0.04
5	Wavelength, μm	1.06	1.06	1.054	1.06	1.6
6	Mode composition	TEM_{02}	TEM_{00}	TEM_{00}	TEM_{00}	TEM_{00}
7	Laser pulse frequency, Hz	0.3	0.1	$3\cdot10^{-5}$	30	0,1

The main reason behind this choice is that with such emitters concrete results were obtained and the possibility of the synthesis of organic compounds was demonstrated with the masses in the interval corresponding to the monomers of a number of "key" organic compounds. Emitters of this class have become affordable that provide the required power density of $\sim 10^9$ W/cm^2. The required pulse energy is as low as 1—2 mJ. These emitters, which are pumped by diode lasers, can generate radiation pulses with a frequency of up to 100 Hz. As we already pointed out above, the high pulse frequency of the laser played the crucial part in the accumulation of the products of synthesis at the substrate in amounts sufficient for their further identification.

Table 4 lists the energy characteristics of various laser emitters used to obtain the main scientific results reported in this book. It follows from these parameters that the observed decrease of per ion energy with decreasing exposure duration results in the increase of the efficiency of the effect and complexification of organic compounds synthesized. It goes without saying that this qualitative result requires a more in-depth analysis. However, as a rough approximation, it can be explained as follows: we showed in Chapter 3 that complex organic compounds were synthesized with a pulse duration of 0.3 ns, whereas the per particle energy was only ~0.4 eV, which is about one order of magnitude lower than in the case of the data obtained with the first two emitters listed in Table 4, and the laser pulse duration of ~0.3 ns for this emitter is at the lower limit for the time scale of the establishment of local thermodynamic equilibrium (LTE).

However, the results of the experiments made using an emitter with a pulse duration of ~0.43 ns proved to be the most productive for the synthesis of complex organic compounds. These results can be explained if we assume that in the case of $t_L \sim 10$ ns and high specific per ion energy the plasma generated in the process becomes overheated and such state does not favor fast recombination and end of the synthesis.

A 30-fold reduction of the exposure time and the decrease of the energy contribution to the process of plasma formation produce the best initial conditions for synthesis processes. Note that in the comparison of the results obtained with different laser emitters one parameter remained unchanged – the power density, which was always kept at $\sim 10^9$ W/cm^2.

The above discussion facilitates the choice of the laser emitter with the optimum parameters to be used in experimental simulations of a hypervelocity impact. These parameters determine to a large extent whether the experiment as a whole will be successful and whether it will be possible to confirm the viability of the new concept. Therefore we paid special attention to the choice of the type and to the selection of the most important parameters of the laser emitter: after a long search and preliminary tests we chose a Nd:YAG laser emitter pumped by a diode laser, which provided the following parameters ranked by their importance:

Pulse duration	0.5—1 ns
Pulse energy	1—2 mJ
Pulse frequency	up to 100 Hz
Focal-spot diameter	25—50 μm

The radiation of a laser emitter with the above parameters was experimentally shown to be capable of ensuring the synthesis of organic compounds with the masses corresponding to monomers and, possibly, even dimers and trimers of amino acids, nitrogen bases, and other important organic compounds. It was, however, of interest to determine the properties of the micrometeorite whose impact could be simulated by such a laser emitter.

To this end, we must first estimate the mass M_L and volume V_L of the matter involved in the laser induced plasma formation. For the case considered with the laser crater depth of 2—3 μm and $W_L \sim 2\ 10^9$ W/cm^2 these quantities are equal to $M_L \sim 8\ 10^{-9}$ g and $V_L \sim 4\ 10^{-9}$ cm^3, respectively.

To determine the mass and size of the micrometeorite that is capable of involving such an amount of matter into the plasma-formation process, the inferred M_L and V_L should be decreased by a factor of ~4—5 because the plasma produced in hypervelocity impacts is known to consist of 80—90% target matter and only of 10—20% projectile matter. Hence in this case the mass and volume of the micrometeorite should be of about $M_M \sim 10^{-9}$ g and $V_L \sim 5\ 10^{-10}$ cm^3, respectively. The diameter of such a particle should be $d_M \sim 10^{-3}$ cm or ~10 μm.

Given that only 1—5% of the micrometeorite impact energy is spent for the formation of plasma, the initial energy E_M of the meteorite should be about 30 times greater than the energy E_L of the laser, or $E_M \sim 30E_L \sim 3\ 10^{-2}$ J. We can now use the inferred energy of the meteorite to estimate its velocity V_M, which is equal to $(2E_M/m_M)^2 \sim 10^7$ cm/s, implying an interaction time of $t_M = d_M/V_M \sim 10^{-10}$ s between the micrometeorite and the target. We can estimate the power density of such an impact only from the size of the micrometeorite, i.e., without the allowance for the crater. This quantity is equal to ~4 10^{12} W/cm^2, which is ~10^3 times greater than the density of laser radiation. The factor of 10^2 difference between W_M and W_L is due to the fact that the diameter of the crater produced by a hypervelocity impact of a micrometeorite is ~10 times greater than that of the micrometeorite. The volume of the crater may be 500 times greater than that of the micrometeorite. Hence the resulting power density for the impact is much smaller and equal to about ~10^{10} W/cm^2.

For clarity, we summarize the results in Table 5. The most important conclusions that can be drawn from these results consist in the following.

The parameters of laser exposure that results in the observed synthesis of complex organic compounds can be used to infer the parameters of the micrometeorite - e.g., velocity, mass, and diameter – whose impact produces the plasma torch similar to the laser torch.

Table 5. Comparison of the properties of laser pulses and meteorite impact

Characteristics	Laser pulse	Micrometeorite impact
Mass, involved in plasma formation, g	$\sim 10^{-9}$	10^{-9}
Energy used for plasma formation, J	$\sim 3\cdot10^{-4}$	$\sim 3\cdot10^{-4}$
Time of interaction, sec	$\sim 3\cdot10^{-10}$	$\sim 10^{-10}$
Power density, W/cm^2	$2\cdot10^{9}$	$4\cdot10^{10}$ *
Full energy of influence, J	$3\cdot10^{-4}$	10^{-4}
Crater volume, cm^3	$4\cdot10^{-9}$	10^{-7}
Crater diameter, μm	50	100

* Without the energy spent on crater formation.

In this case the most important processes including the synthesis of organic compounds should also be similar in the two cases considered. Thus, according to a preliminary estimate, such parameters as the energy contribution to the processes of plasma formation, the volume of the plasma-formation region, and the time of interaction during the impact of a micrometeorite with a diameter and mass of ∅10 μm and $\sim 10^{-9}$ g, respectively, moving at a velocity of $\sim 10^7$ cm/s should be of the same order of magnitude as the corresponding parameters of laser exposure with a focal diameter of ~50μm and a power density of $\sim 10^9$ W/cm^2. This means that the characteristic parameters of the plasma torch, the dynamics of its expansion-away, and the main physicochemical characteristics of the synthesis products should be the same in both cases.

An important conclusion from the above estimates is that the approximate micrometeorite parameters so determined are quite realistic and lots of such micrometeorites must be roaming in space. Thus the masses, sizes, and velocities of micrometeorites are known to be distributed over a wide range. The masses of these particles span from 10^{-15} to 10^{-3} g. The most numerous micrometeorites are those with masses ranging from 10^{-15} to 10^{-13} g. Micrometeorites with diameters smaller than 1 μm are accelerated by solar radiation pressure to velocities as high as 50 km/s. The bulk of the mass and energy flux are carried by the micrometeorites with the masses in the 10^{-8} to 10^{-3} g interval, their velocity in the interplanetary space amounts to 10 km/s. The velocity may become much higher when a micrometeorite enters the gravitational field of a planet.

It is therefore very important that the chosen laser emitter with $t_L \sim 1$ ns makes it possible to perform accurate simulation of a hypervelocity impact for micrometeorites with the diameters of ~10—20 μm and ensures the correct results for the most common particles. However, the possibility of accurate simulation of a micrometeorite impact in a laboratory becomes especially valuable from the scientific viewpoint given that these results, with certain unimportant restrictions, can be extended to the impact processes involving large meteorites. This is possible because in impact processes the increase of the characteristic size of the projectile does not change the main characteristics of the physical processes and the projectile-to-crater diameter ratio. This is true with the only exception that the synthesis products become more complex.

Thus the results obtained in experimental simulations show that intermediate-sized micrometeorite impacts may ensure the synthesis of monomers and, possibly, dimers and trimers. However, the observed tendency for organic compounds to become increasingly complex with increasing size of the plasma-formation region or laser focal-spot diameter in experimental simulations may also indicate the direction, which may lead to the synthesis of more complex organic compounds, e.g., short monomers.

To accomplish such a priority task, the size of the laser focal spot should be increased accordingly, which will make it possible to simulate an impact of a meteorite with a diameter of, e.g., 0.5 or 1 cm.

However, such an experiment is very difficult to perform in a laboratory. The serious problems to be addressed are due to the fact that the time of interaction with the target increases with the diameter of the meteorite. The longer interaction time, in turn, makes it necessary to increase the laser pulse duration, thereby breaking the balance between the basic parameters of the laser and impact exposure. It is therefore impossible for relatively large meteorites to maintain the similarity and at the same time preserve the parameters of the laser simulation of a micrometeorite impact. Thus, e.g., it takes $\sim 10^{-7}$ s or 100 ns for a 1-cm diameter meteorite moving at a velocity of 10^7 cm/s to "sink" into the target. To achieve the same interaction time, the laser pulse must have the same duration. This condition can be satisfied, however, the result will be a totally different dynamics of the formation of the plasma torch. In the case of an impact the hot spot forms in 1—2 ns and during the next 98 ns laser radiation heats the plasma of the newly produced torch and destroys the products of synthesis.

There are several ways out of this problem. The easiest solution consists in abandoning the requirement that all similarity parameters be identical in the two cases, and ensuring instead that all the basic parameters of the plasma of the generated torch are identical. However, even in this case some uncertainty remains that is due to the lack of experimental data about the properties of the plasma produced as a result of impacts of ~1-cm diameter meteorites. It is probable that it will be possible to obtain these data via computations.

There is another way of obtaining the necessary data – direct impact experiments performed in space and involving the measurement of the parameters of the torch plasma produced as a result of a collision between artificial bodies with known chemical composition. Such studies in space represent the so called class of active experiments, where analogs of natural physical processes are artificially reproduced. In the case considered we are dealing with impact processes, and the results of such experiments can be used to reproduce meteorite impact plasma in a laboratory for various combinations of the size, velocity and chemical composition of the projectile. Moreover, a dedicated experiment of this kind can also be used to study the synthesis of organic compounds as a result of a collision of inorganic compounds containing the constituent elements of organic compounds. In Section 8 we describe the original idea of the preparation and conducting of such an experiment in the near magnetosphere.

The results of cosmic experiments will make it possible to study the most important parameters of the "impact" plasma and then develop anew or find already available sources reproducing its analogs under laboratory conditions. The new sources required for the generation of a plasma torch to simulate impacts of centimeter-sized meteorites do not need to be laser based, because for the sizes considered the formation time of laser plasma would not coincide with that of impact plasma. However, this problem should be addressed and solved because overcoming this technical difficulty will make it possible to find fast answers to a number of key questions. We believe the most interesting and topical among these problems to be the one that is associated with understanding the degree of complexification and ordering of the synthesis products with increasing scale lengths of the plasma formation region and of the projectile.

5.4. Identification and Space Distribution of Carbines

The studies of the structure and of the space distribution of newly produced compounds may provide valuable information for understanding the synthesis processes in the plasma torch. For example, a detection of identical compounds, or, at least, compounds with a similar structure, in the plasma torches produced by laser exposure and by an impact would become an important proof of the identical nature of the processes that occur in these phenomena.

To this end, we analyzed the structure of the simplest hydrocarbon compounds, which had been commonly and with little effort synthesized in experiments and which have been detected as C_NH_M mass peaks making up the so-called "carbon comb". In these compounds there were at most four hydrogen atoms for several carbon atoms – a pattern that was indicative of a chain-linear structure of the synthesized substance, which resembled carbine.

Carbine (sp1) (Melnichenko et al., 1985) is, along with graphite, diamond, and fullerenes, yet another stable allotropic form of carbon. The structure of carbine is based on a linear carbon chain. In carbine carbon atoms are bound along the chain, whereas the interchain links are provided by the overlap of π orbitals of atoms.

Carbines were first synthesized in 1959 at the Institute of Organoelement Compounds of the Academy of Sciences of the USSR. The structure of crystalline samples was analyzed using methods of x-ray and electron diffraction. In nature such compounds were found in the meteorite impact crater Ries in Bavaria. They were identified (Goresy and Donnay, 1968) as a new allotropic form of carbon, which the above authors believed to be generated in the process of impact melting of graphite gneiss.

Two carbine modifications were identified, which were called α- and β-carbine. The hexagonal cells of these modifications have the following sizes:

$$a_\alpha = 5.08 \text{ E}, c_\alpha = 7.80 \text{ E},$$
$$a_\beta = 4.76 \text{ E}, c_\beta = 2.58 \text{ E}.$$

β- and α-carbine is believed to consist of polyene–$(C{\equiv}C\text{-}C{\equiv}C)_n$- and cumulene $=(C{=}C{=}C{=}C)_n=$ carbon chains, respectively.

Given that the possibility of the synthesis of carbines in the plasma torch was proven after their identification, the researchers decided to use them as "tracers" to study the distribution of substances synthesized in the torch both in the crater and in the adjacent regions. To this end, samples were prepared and measurements were performed using the methods of electron microscopy and spectrometry. Films made of compounds synthesized in the torch were analyzed after placing them onto a substrate. The substrate was placed inside the analyzer within the ion flux. Also analyzed were the surfaces of the carbon target before and after laser exposure. In particular, the bottom of the crater produced by laser exposure and the regions adjacent to the crater were studied.

The atomic structure of the films after their placement onto a substrate was analyzed using a JEM-100C transmission electron microscope operating in the diffraction and phase-contrast modes. The films for electron-microscopic analyses were applied to NaCl crystals, separated from the substrate by dissolving NaCl in water, and pressed out onto a microscopic net made of copper foil. Samples were also made by picking the matter from the surface in the region of laser spot and a thin layer of the adjacent regions.

The microstructural and electron diffraction patterns obtained using the electron microscope were then photographed with a digital camera. The high sensitivity and resolution of the camera made it possible to substantially reduce the intensity of the electron beam and prevent the destruction of the samples by the electron beam.

To obtain the background spectrum, the samples were analyzed using the method of Raman scattering, which allowed the characteristic frequencies of valence vibrations of bound atoms to be measured and the bond types in molecules and solid bodies to be determined.

We used the 4848 Å line of argon laser to record the Raman scattering spectra of the samples studied. We recorded spectra at room temperature in the inverse scattering geometry. The width of the spectral band of the instrument was 4 cm^{-1}. The shifts in the Raman scattering spectra were computed relative to the diamond-peak position at 1332 cm^{-1}.

When analyzing the spectra to interpret the signals we used the results of the computations of the spectra of various structural models of carbon chains.

To detect the structural changes due to laser exposure, we analyzed the initial composition of an ultrapure carbon target. The typical pattern of electron diffraction obtained for the initial material of the target includes a number of narrow diffraction rings. Table 6 summarizes the experimental interplanar spacings compared with the corresponding spacings for crystalline hexagonal graphite with the lattice parameters a = 2.46 Å and c = 6.71 Å. It is evident from these data that the target consisted of a single phase exclusively – of hexagonal graphite.

Table 6. Interplanar spacings in graphite crystal lattice

d_{exp} Å	d_{theor} Å	hkl
3.36	3.355	00.2
2.13	2.130	10,0
1.22	1.230	11,0
1.07	1.065	20,0

hkl are the Miller indices.

The studies of the crater floor showed a somewhat different pattern. Table 7 lists the diffraction-determined interplanar spacings for carbine with a = 5.34Å and c = 14Å and compares them with the results of theoretical computations. It follows conclusively from these data that new hydrocarbon structures that were identified as carbines formed after laser exposure.

An analysis of the phase-contrast pattern allowed us to estimate the sizes of little carbine crystals. We found them to be equal to ~10-20μm, which agrees with the maximum size of the crystals obtained earlier by other authors.

For samples subject to laser exposure in the atmosphere the characteristic pattern of electron diffraction that covers the crater and the adjacent regions revealed layered films of chain-linear carbon with the chains oriented normally to the surface (d = 4.02Å). Such a diffraction pattern is typical for condensed film systems with a small ordered region.

Table 7. Interplanar spacings of carbine crystal lattice

D_{exp} Å	d_{theor} Å	hkl
4,60	4,62	100
4,45	4,40	101
2,69	2,63	111
2,40	2,31	200
2,23	2,20	202
1,77	1,74	211
1,57	1,54	300
1,49	1,48	303
1,32	1,31	222

Thus the data obtained using an electron microscope lead us to conclude that hydrocarbon structures – carbines – that were not present there initially have been synthesized in a laser crater.

The results described above were confirmed by Raman scattering measurements. Two peaks were recorded for the initial target material – at 1580 and 1380 cm^{-1}, which correspond to the structure of polycrystalline graphite.

An analysis of the crater floor after laser exposure revealed an additional peak at 2100 cm^{-1}, which was indicative of the presence of polyene and polycumulene bonds, which are characteristic of carbine. About the same spectra were obtained for the regions adjacent to the crater. The main graphite peaks were found to be "washed-out" in the laser exposure region and this feature is indicative of the deordering of graphite, and fits well the overall pattern of the process.

An analysis of the angular distribution of the intensity of scattered electrons for applied films and a comparison of these data with a standard gold sample showed that the spectrum contains only one maximum at d=4.04E, which is typical of carbine. The size of the ordering region was 80 nm.

Thus the measurement results described above show that in a laser plasma torch with the parameters close to those found in the torch produced by a hypervelocity impact new hydrocarbon structures – carbines – are synthesized at the surface of a target that initially consisted of pure graphite exclusively. Carbines were found not only on the substrate, but also on the laser crater floor and in the adjacent regions.

Note that the synthesis of new stable hydrocarbon structures via laser exposure is a separate scientific and technical problem. The studies of the "graphite—carbine" phase transition in the processes of carbine synthesis under the action of laser radiation also fit into this field of research. The results obtained may prove to have a much wider application. Thus, according to Moscow State University experts in the synthesis of carbon structures, the above experiments were the first where polycrystalline carbine films had been obtained directly without ion stimulation. This may be a new, promising technological process of carbine production under the conditions with the most favorable thermodynamic parameters among those considered in the series of works on shock synthesis of carbine.

The most important results of the above studies also include experimental detection of ordered regions in the structural formations of carbines synthesized in the plasma torch. Note that these were the first results indicative of the processes of self-assembly and ordering of matter in the impact-produced plasma torch.

5.5. Synthesis and Identification of Amino Acids

Bona fide identification of organic compounds produced as a result of abiogenous synthesis is known to be the most important and labor-intensive task in the studies of the emergence of life under laboratory conditions. These difficulties are mostly due to high concentration of biogenic organic contaminants, and to the low efficiency of the laboratory synthesis of vital organic compounds with intense "background contamination" by various types of "junk". This situation makes it impossible to obtain a significantly higher-than-noise signal from the artificially synthesized organic compounds.

Special measures are usually undertaken to overcome these difficulties. We apply these same measures in our work. Appropriate redesign and adaptation of the analytical facilities and establishment of clean workplaces for the sample preparation allowed the organic compounds synthesized in the torch to be collected in amounts needed for correct chromatographic and chromato-mass-spectrometric analysis aimed at their identification.

To ensure fast accumulation of the synthesis products, we used the diode-pumped Nd:YAG laser described above, which is capable of operating with a pulse frequency of up to 100 Hz. The pulse energy and duration were equal to ~0.5 mJ and ~0.4ns, respectively, providing the required power density of $\sim 10^9$ W/cm^2 for a ~50—100 nm diameter focal spot.

Products synthesized in the plasma torch, which consisted of molecular ions and neutral molecules produced as a result of laser exposure, were deposited almost completely on the substrate with a central hole covering the upper hemisphere above the target. Laser radiation was delivered to the target through the central hole. A small part of plasma ions moved away from the target through the same hole inside a solid angle that did not exceed 1—2 square degrees. When reflected by the reflector, these ions reached the detector and produced mass peaks in the records. To control the efficiency of the synthesis of organic compounds, the spectra were recorded using a fast digital oscilloscope operating with a frequency of up to 30 Hz. The efficiency of the synthesis of organic compounds was also controlled visually on the monitor of an analog oscilloscope. If the characteristic mass peaks corresponding to the "carbon comb" disappeared, the target was shifted relative to the focal spot to make the peaks reappear.

The mechanical carbon and ammonium nitrate powder mix used at the initial stage was later replaced by a mix of carbon powder and ammonium nitrate melt. When it cools down, such a target proves to be more durable and more convenient for mass-spectrometric measurements.

According to approximate computations, with the laser emitter operated at a pulse frequency of ~100 Hz, synthesis products can be collected over 10^4 laser exposures for a total time of 15 minutes.

Figure 17 in Chapter 4 shows the mass spectrum obtained as a result of a single laser emitter pulse acting on a mechanical graphite and ammonium nitrate powder mix. The spectrum shows a conspicuous "carbon comb", i.e., a sequence of mass peaks corresponding to the masses of C_N, as well as mass peaks interposed between the C_N peaks. Some of these peaks are located after the C_N peaks and correspond to C_NH_M type compounds, i.e., to short carbon chains, where m usually does not exceed 4. However, the peaks located in the interval from C_N+4 to C_N+12 cannot be interpreted as hydrocarbon compounds and are of special interest. Some of them coincided with the masses of individual amino acids indicated in the spectrum; however, such mass peaks were not identified with amino acids. The fact that the mass spectra of the synthesized organic compounds contain peaks coincident with the masses of amino acids suggests that the observed mass peaks may also contain the signatures of amino acids. This is evidenced by the earlier published results of laboratory experiments performed by other authors. According to these results, amino acids are synthesized first and with high probability in laboratory analogs of various natural processes and especially in analogs of natural processes involving plasma. However, special attention was paid to the identification of amino acids synthesized among other organic compounds in the plasma torch.

To this end, synthesis products produced as a result of 10^4—10^6 laser exposures were brought, depending on the task considered, onto the hemispherical support described above. A series of preliminary experiments confirmed that the mass composition of the spectrum recorded from a single exposure was identical to the mass composition of the coadded spectra of repeated exposures. To this end, we recorded every 100th single spectrum obtained in a series of 4500 laser exposures of the target. We then coadded the 45 spectra so obtained and reduced them. The result is shown in Figure 29. The single spectrum shown in Figure 17 differs little from the result of coadding 45 of the above-mentioned spectra shown in Figure 29. This becomes evident given that the spectra were obtained in uncontrolled pulsed exposure processes with no special measures taken to prepare and homogenize the target surface.

To identify the amino acids in preliminary experiments, we collected on the support the synthesis products produced as a result of 10 to 40 thousand exposures.

Before placing the supports used for collecting the synthesis products into the vacuum chamber they were subjected to thorough treatment using various solvents and ultrasonic bath. The supports were then heated in a vacuum furnace for two to three hours at a temperature of 500—600 C.

The support with the synthesis products brought onto it was taken out of the vacuum chamber and placed into a special sealed container to be delivered to the place of its subsequent analysis. There, in the process of the preparation of the sample, the synthesis products were washed away from the support by purified water containing ~0.1% of hydrochloric acid.

To identify the amino acids at the initial stage, we decided to use Hitachi's (Japan) L-8800 ASM high-sensitivity chromatographic facility of A.N. Belozersky Institute of Physico-Chemical Biology of Moscow State University.

The instrument was tuned to measure the chemical composition and quantity of amino acids in the sample studied. To this end, we used the liquid column appropriate for the task and the technique of recording of chomatographic peaks. The substance from the support was washed away twice. After standard preparation, the sample obtained was introduced into the chromatographic column. After obtaining the first-wash chromatogram, a control experiment was performed with the wash water in order to detect the presence of amino acids. Then the second wash was analyzed.

Figure 30 shows the original printout with the results of the first wash with a chromatogram. Table 8 summarizes the results of two washes after data reduction. The positive conclusion about detecting amino acids was made if the ratio of the amplitudes of the first- and second-wash peaks to that of the peaks in the control measurements was greater than three. The last column of the table lists the peaks of the compounds, whose masses coincided with those of some amino acids, and which thereby could not be unambiguously identified in the spectra taken with LASMA instrument.

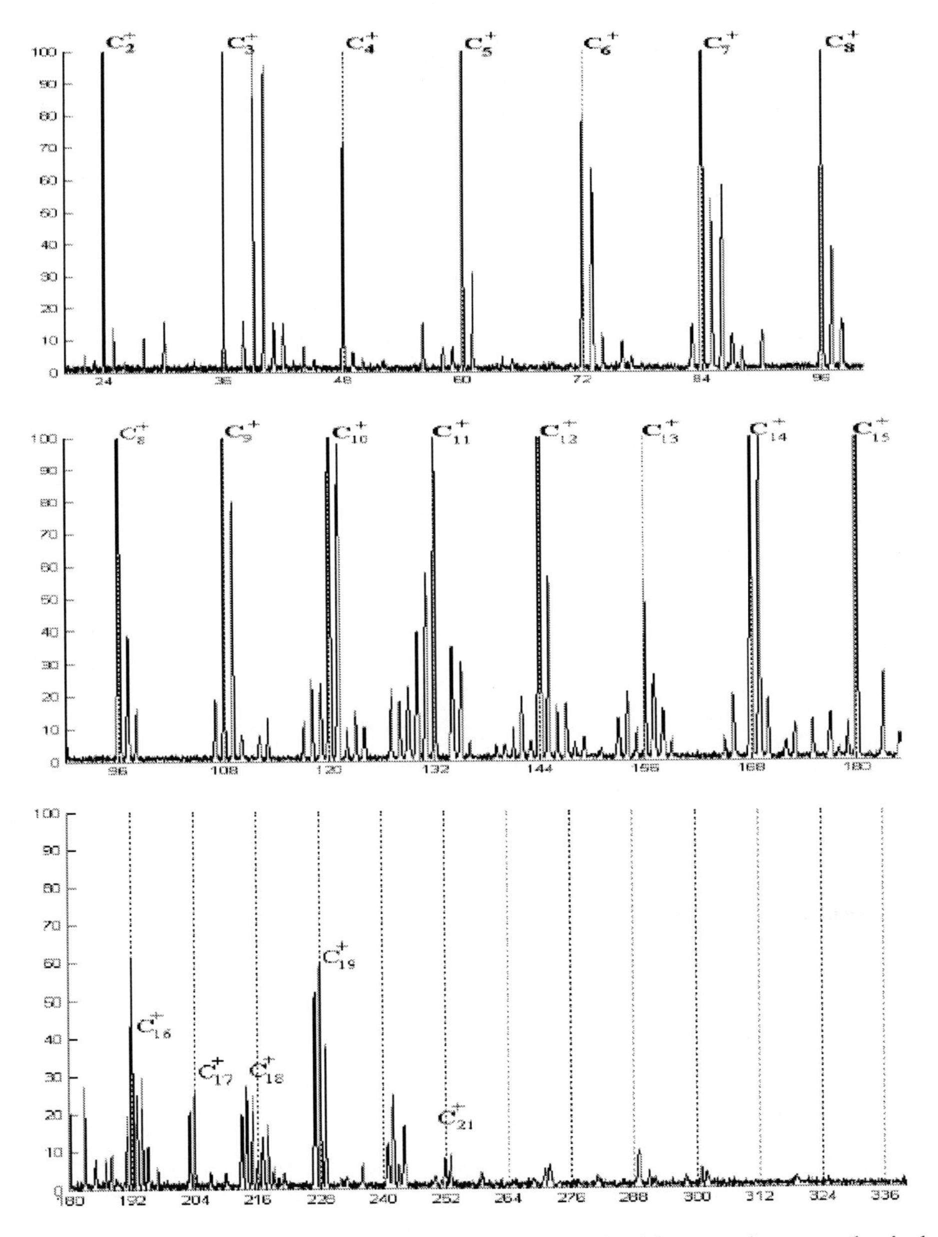

Figure 29. Result of co adding of 45 of the 4500 spectra obtained by exposing a mechanical mix of carbon and NH_4NO_3 to laser radiation. The single spectrum shown in Figure 17 differs little from the result of co adding of 45 spectra.

The results of chromatographic measurements show conclusively that amino acids are synthesized in the plasma torch. The "key" ones are: asparagine, serine, glutamic acid, methionine, lysine, histidine, and arginine. However, these data could not be used as the sole proof of such an important result because of their rather low signal-to-noise ratio, and we considered the potentialities of a number of other, more sensitive and selective, tools and techniques, which therefore provide more reliable information.

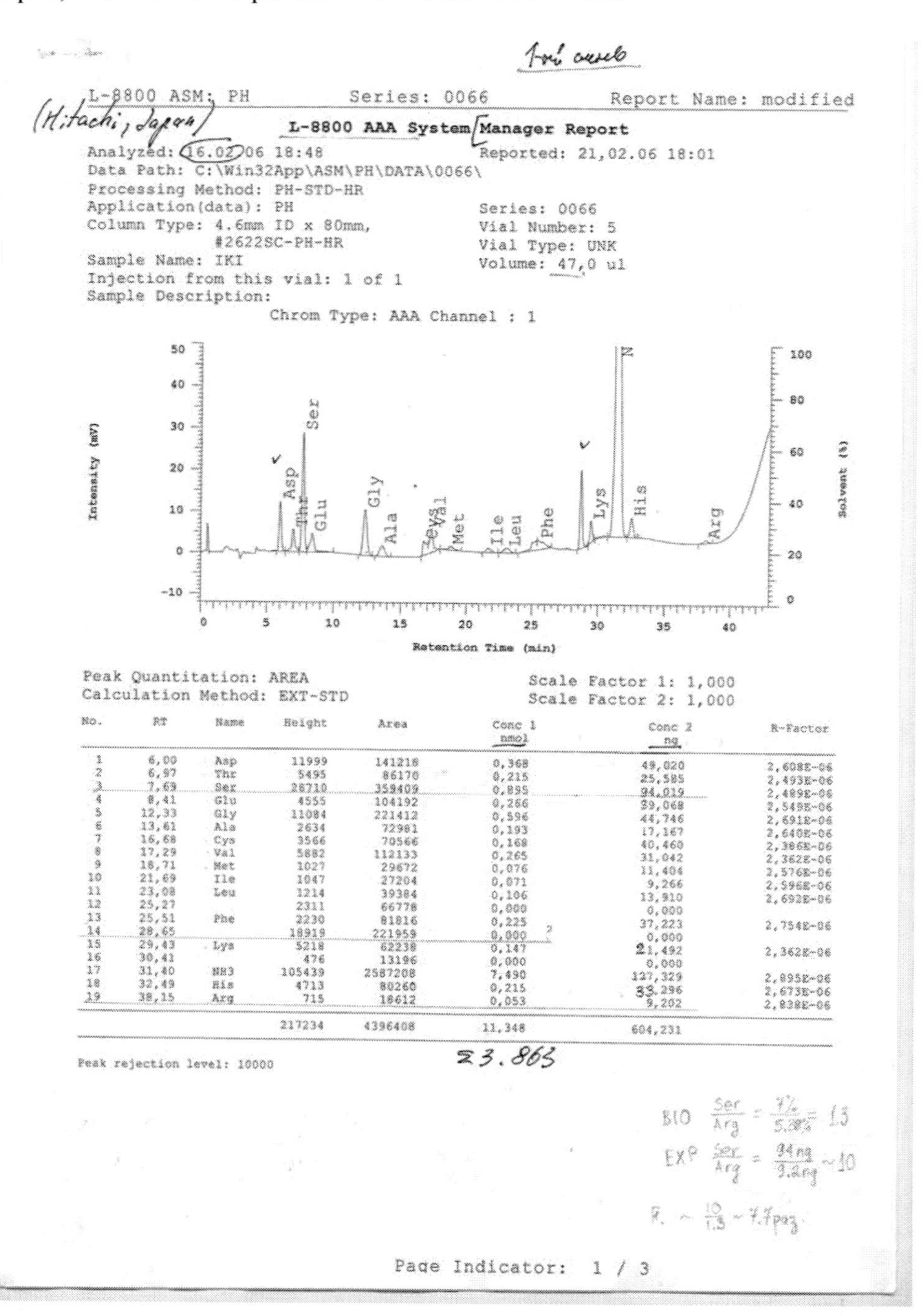

L-8800 ASM: PH Series: 0066 Report Name: modified

(Hitachi, Japan)

L-8800 AAA System Manager Report

Analyzed: 16.02.06 18:48 Reported: 21,02.06 18:01
Data Path: C:\Win32App\ASM\PH\DATA\0066\
Processing Method: PH-STD-HR
Application(data): PH Series: 0066
Column Type: 4.6mm ID x 80mm, #2622SC-PH-HR Vial Number: 5
Vial Type: UNK
Sample Name: IKI Volume: 47,0 ul
Injection from this vial: 1 of 1
Sample Description:

Chrom Type: AAA Channel : 1

Peak Quantitation: AREA Scale Factor 1: 1,000
Calculation Method: EXT-STD Scale Factor 2: 1,000

No.	RT	Name	Height	Area	Conc 1 nmol	Conc 2 ng	R-Factor
1	6,00	Asp	11999	141218	0,368	49,020	2,608E-06
2	6,97	Thr	5495	86170	0,215	25,585	2,493E-06
3	7,69	Ser	28710	359409	0,895	94,019	2,489E-06
4	8,41	Glu	4555	104192	0,266	39,068	2,549E-06
5	12,33	Gly	11084	221412	0,596	44,746	2,691E-06
6	13,61	Ala	2634	72981	0,193	17,167	2,640E-06
7	16,68	Cys	3566	70566	0,168	40,460	2,386E-06
8	17,29	Val	5882	112133	0,265	31,042	2,362E-06
9	18,71	Met	1027	29672	0,076	11,404	2,576E-06
10	21,69	Ile	1047	27204	0,071	9,266	2,596E-06
11	23,08	Leu	1214	39384	0,106	13,910	2,692E-06
12	25,27		2311	66778	0,000	0,000	
13	25,51	Phe	2230	81816	0,225	37,223	2,754E-06
14	28,65		18919	221959	0,000	0,000	
15	29,43	Lys	5218	62238	0,147	21,492	2,362E-06
16	30,41		476	13196	0,000	0,000	
17	31,40	NH3	105439	2587208	7,490	127,329	2,895E-06
18	32,49	His	4713	80260	0,215	33,296	2,673E-06
19	38,15	Arg	715	18612	0,053	9,202	2,838E-06
			217234	4396408	11,348	604,231	

Peak rejection level: 10000

23.865

BIO Ser/Arg = 7%/5.3% = 1.3

EXP Ser/Arg = 94 ng/9.2 ng ~ 10

Page Indicator: 1 / 3

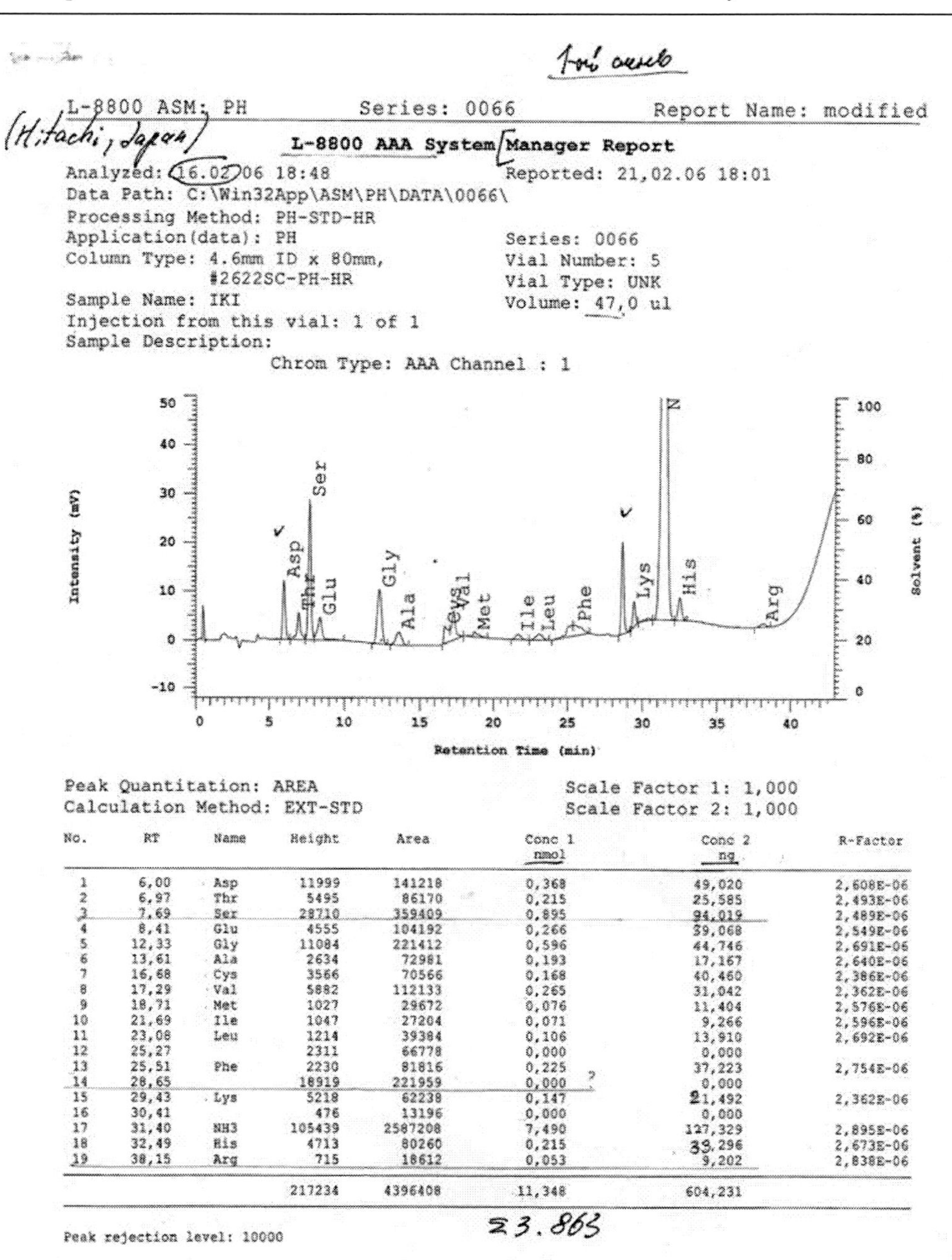

L-8800 ASM: PH Series: 0066 Report Name: modified
(Hitachi, Japan)

L-8800 AAA System Manager Report

Analyzed: 16.02.06 18:48 Reported: 21,02.06 18:01
Data Path: C:\Win32App\ASM\PH\DATA\0066\
Processing Method: PH-STD-HR
Application(data): PH Series: 0066
Column Type: 4.6mm ID x 80mm, #2622SC-PH-HR Vial Number: 5
Vial Type: UNK
Sample Name: IKI Volume: 47,0 ul
Injection from this vial: 1 of 1
Sample Description:

Chrom Type: AAA Channel : 1

Peak Quantitation: AREA Scale Factor 1: 1,000
Calculation Method: EXT-STD Scale Factor 2: 1,000

No.	RT	Name	Height	Area	Conc 1 nmol	Conc 2 ng	R-Factor
1	6,00	Asp	11999	141218	0,368	49,020	2,608E-06
2	6,97	Thr	5495	86170	0,215	25,585	2,493E-06
3	7,69	Ser	28710	359409	0,895	94,019	2,489E-06
4	8,41	Glu	4555	104192	0,266	39,068	2,549E-06
5	12,33	Gly	11084	221412	0,596	44,746	2,691E-06
6	13,61	Ala	2634	72981	0,193	17,167	2,640E-06
7	16,68	Cys	3566	70566	0,168	40,460	2,386E-06
8	17,29	Val	5882	112133	0,265	31,042	2,362E-06
9	18,71	Met	1027	29672	0,076	11,404	2,576E-06
10	21,69	Ile	1047	27204	0,071	9,266	2,596E-06
11	23,08	Leu	1214	39384	0,106	13,910	2,692E-06
12	25,27		2311	66778	0,000	0,000	
13	25,51	Phe	2230	81816	0,225	37,223	2,754E-06
14	28,65		18919	221959	0,000 ?	0,000	
15	29,43	Lys	5218	62238	0,147	21,492	2,362E-06
16	30,41		476	13196	0,000	0,000	
17	31,40	NH3	105439	2587208	7,490	127,329	2,895E-06
18	32,49	His	4713	80260	0,215	33,296	2,673E-06
19	38,15	Arg	715	18612	0,053	9,202	2,838E-06
			217234	4396408	11,348	604,231	

Σ 3.863

Peak rejection level: 10000

Figure 30. The original of listing with results of the analysis of the first wash with the chromatogram.

In accordance with the recommendations of the experts with much experience in these compounds we chose for further identification the reliable and well-developed technique of gas chromatography with subsequent introduction of the derivative amino acids in the gaseous state into the mass spectrometer, where gas was ionized by electron impacts. The amino acids separated in the chromatographic column were subject in the mass spectrometer simultaneously to ionization and fragmentation by an electron beam to ensure their highly reliable identification.

Table 8. Amino acids synthesized in a plasma torch and identified using liquid chromatography technique on Hitachi L 8800 instrument

№	Amino acid	Mass, a.m.u.	Signal to noise ratio of 1st wash-out S/N	Signal to noise ratio of 2nd wash-out S/N	Presence of amino acid provided S/N >3 in both wash-outs	Presence of mass peaks in TOF spectra recorded using LASMA instrument
1.	Asparagines acid, Asp	133	**5.4**	2	+	$C_{11}H^+$ peak
2.	Threonine, Thr	119	2.1	1.2	-	+
3.	Serine, Ser	105	**3.2**	1.5	+	+
4.	Glutamic acid, Glu	147	**3.6**	**7.5**	+	+
5.	Glycine, Gly	75	**3.5**	2.07	+	+
6.	Alanine, Ala	89	1.6	1.7	-	+
7.	Cysteine, Cys	121	0.9	0.9	-	$C_{10}H^+$peak
8.	Valine, Val	117	1.5	1.9	-	+
9.	Methionine, Met	149	1.3	**5.1**	+	+
10.	Isoleucine, □le	131	2.3	No data	-	+
11.	Leucine, Leu	131	2.0	1.2	-	+
12.	Phenylalanine, Phe	165	1.3	1.13	-	+
13.	Lysine, Lys	146	**3.6**	0.7	+	$C_{12}H_2^+$peak
14.	Histidine, His	155	**4.7**	1.5	+	+
15	Arginine, Arg	174	1.7	**7.1**	+	+

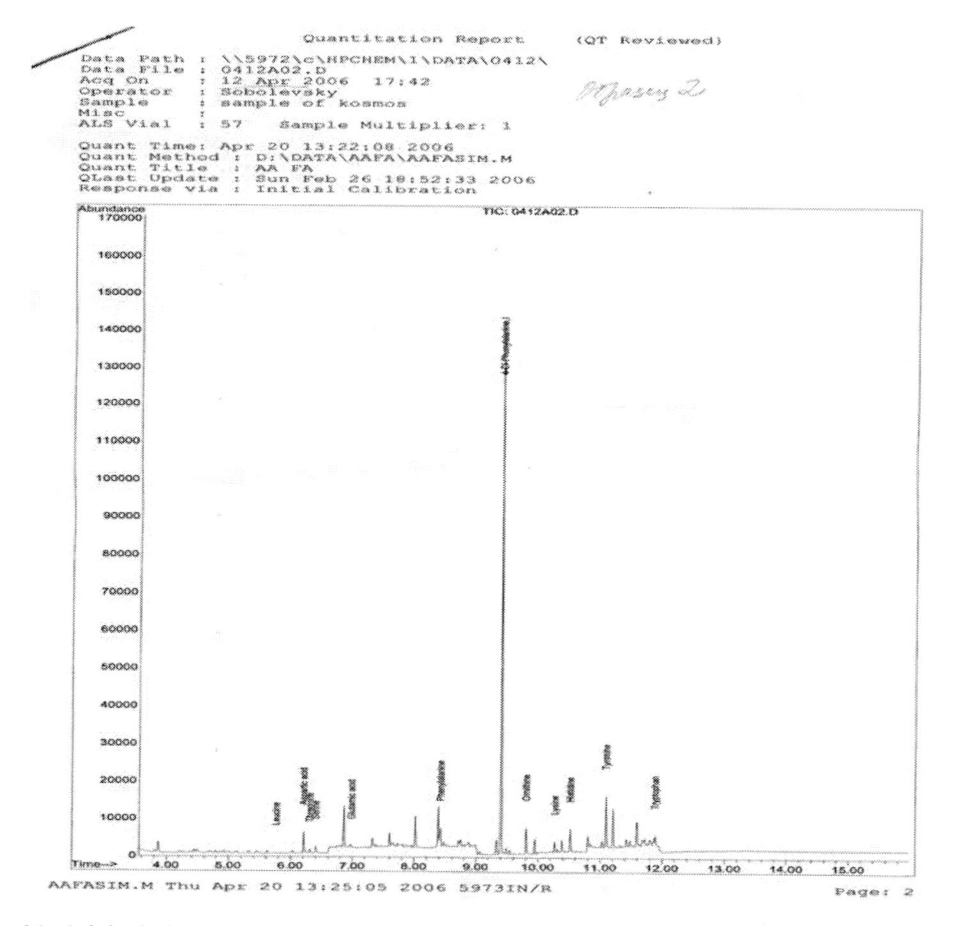

Figure 31. Original chromatogram used for the identification of amino acids using mass-spectrometry methods.

After washing the products from the support, a standard method of sample derivatization was applied, which allows the amino acid derivatives synthesized in the plasma torch to be transformed into the gaseous state.

Figure 31 shows the original chromatogram, which we then used for precise identification of amino acids using various mass-spectrometry methods. To perform these measurements, we collected the products of 4 10^4 laser exposures. The total concentration of synthesized amino acids was ~3.41 mg/l, which, given the small volume of the washing solvent, corresponded to ~1.7 10^{-6} g of synthesized amino acids. The corresponding quantity for 10^4 laser exposures must be about four times smaller and equal to ~4 10^{-7} g, or 400 ng. This estimate agrees by the order of magnitude with the independent estimate of the total mass of amino acids accumulated on the support inferred by analyzing the peak amplitudes in the mass spectrum recorded by LASMA instrument. According to these data, the total mass of amino acids was ~2 10^{-7} g.

An analysis of the sample revealed 11 amino acids: *lecithin, aspartic acid, threonine, serine, glutamic acid, phenylalaline, ornithine, lysine, histidine, tyrosine,* and *tryptophane*. The signal-to-noise ratio is S/N > 3.

Figure 32 presents the results of the analysis of the sample performed using the transition ion mode. The spectra were taken in the following order: water/reagent (blank) and the sample (sample 12.04.06). A total of 11 amino acids have been identified as a result of these measurements. Peaks were observed that corresponded to some fatty acids, which we interpreted as contaminants. Table 9 summarizes the data for the identified amino acids including the signal-to-noise ratios and signal amplitudes for individual peaks.

Table 9. Amino acids identified using gas chromatography and mass spectrometry performed in the transition ion mode with HP 5972 "Aglient" facility

Amino acid	Mass, a.m.u.	*Control*		*Measurements*		Detected
		Water/ reagent	Signal/ Noise	Aliquot	Signal/ Noise	
Alanine	89					
Glycine	75			360	0.6	-
Valine	117					
Gamma aminobutyric acid, GABA	103					
Leucine	131			1296	4.9	+
Isoleucine	131					
Asparagines acid	133			39018	318.4	+
Threonine	119			9138	13.7	+
Serine	105	420	0.5	9288	15.4	+
Proline	115					
Asparagine	132	1039	0.3	1591	0.3	-
Glutamic acid	147			1808	6.2	+
Methionine	149					
Phenylalanine	165	559	4.1	26577	172.8	+
Cysteine	121					
Glutamine	146					
Ornithine				62834	125.0	+
Lysine	146	469	7.6	22596	65.7	+
Histidine	155	386	3.7	20545	208.9	+
Tyrosine	181	4942	2.5	122675	100.4	+
Cystine	268					
Tryptophan	204			712	6,0	+

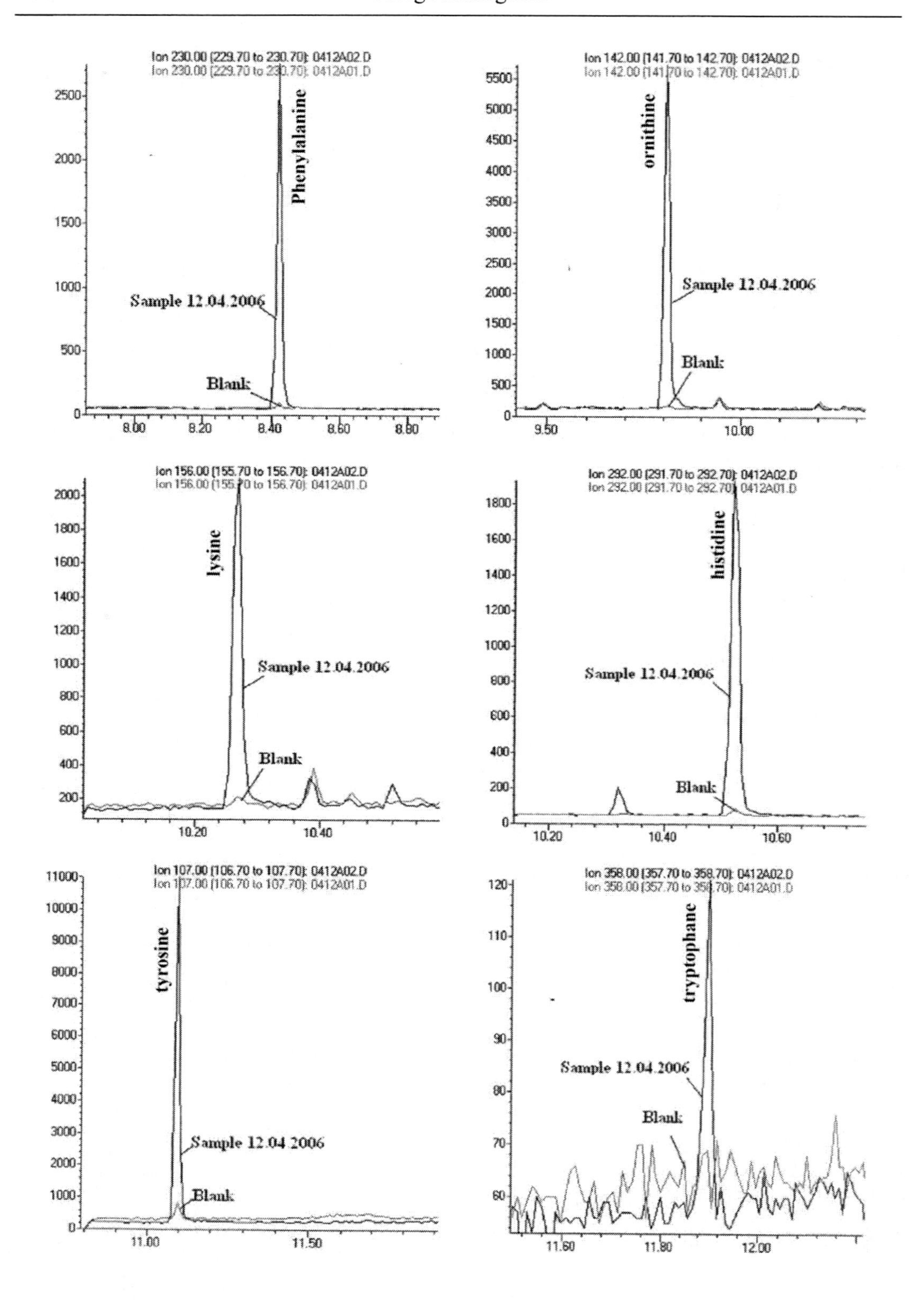
Ion 230.00 (229.70 to 230.70): 0412A02.D
Ion 230.00 (229.70 to 230.70): 0412A01.D
Phenylalanine
Sample 12.04.2006
Blank
Ion 142.00 (141.70 to 142.70): 0412A02.D
Ion 142.00 (141.70 to 142.70): 0412A01.D
ornithine
Sample 12.04.2006
Blank
Ion 156.00 (155.70 to 156.70): 0412A02.D
Ion 156.00 (155.70 to 156.70): 0412A01.D
lysine
Sample 12.04.2006
Blank
Ion 292.00 (291.70 to 292.70): 0412A02.D
Ion 292.00 (291.70 to 292.70): 0412A01.D
histidine
Sample 12.04.2006
Blank
Ion 107.00 (106.70 to 107.70): 0412A02.D
Ion 107.00 (106.70 to 107.70): 0412A01.D
tyrosine
Sample 12.04.2006
Blank
Ion 358.00 (357.70 to 358.70): 0412A02.D
Ion 358.00 (357.70 to 358.70): 0412A01.D
tryptophane
Sample 12.04.2006
Blank

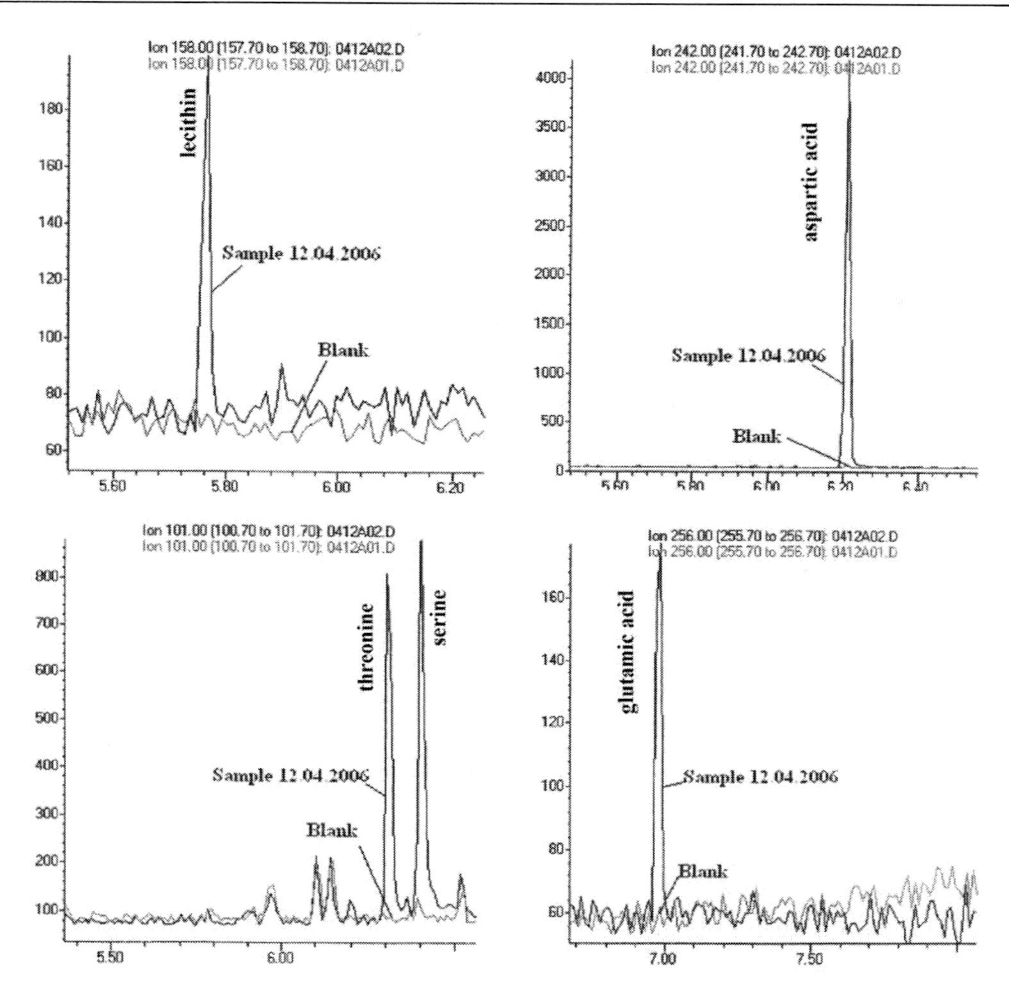

Figure 32. Results of a mass spectrometry/gas chromatography analysis of the sample in the transition ion mode using HP5972 "Aglient" instrument. The sample was redissolved in 500 μl 0.1 M HCl. The samples were recorded in the following order: - water /reagent (Blank), - sample (Sample 12.04.2006).

The results obtained, which are based on the measurements of individual-peak amplitudes, have a number of features that deserve a separate analysis and consist in the following:

(1) Anomalously high ornithine amino acid content equal to ~18%, which is untypical for biogenous amino acids.
(2) Anomalously low *leucine* content, which is equal to ~0.3%. Note that the average leucine content in biogenous amino acids is very high and amounts to 9.6%.
(3) The mass spectra exhibit no *alanine* and *glycine* signatures. According to the data obtained in earlier laboratory experiments simulating the synthesis of organic compounds in nature, these amino acids must be the first to be synthesized and must have high occurrence rate among biogenous amino acids amounting to 8 and 7%, respectively.

(4) The ratio of the maximum-to-minimum experimentally recorded content of amino acids purportedly synthesized in the plasma torch is ~200, which exceeds by more than one order of magnitude the ratio of the average occurrence rate of biogenous amino acids, which is equal to 10.
(5) The experiments show a factor of 2 to 8 (on the average, a factor of 3.5) increase of the number of molecules of various amino acids in response to a factor-of-four increase of the number of laser exposures: from 10^4 to $4 \cdot 10^4$.

In the case considered it is of interest to determine to what extent the above peculiarities of the results of chromato-mass-spectrometric analysis indicate that the identified amino acids are organic compounds synthesized in the plasma torch and are not biogenous amino acids due to the contamination of the sample, support, or reagents.

To answer this question, let us compare the results of the analysis. Thus the non-biogenous nature of the recorded amino acids is evidenced by the anomalously high content of ornithine, which is untypical of biogenous amino acids. The absence or extremely low content of glycine, alanine, and leucine – amino acids that are very abundant in bioorganic systems – is also indicative of the abiogenous nature of recorded amino acids. Extra supporting evidence is also provided by the anomalously large scatter of concentrations. And, finally, the increase of the concentration of amino acids with increasing exposure indicates that these amino acids are directly associated with the synthesis process because the contamination level must not increase with increasing number of laser pulses.

Each of the above facts by itself is indicative of possibly non-biological origin of the identified amino acids, however, a joint analysis of these facts leads us to conclude that the identified amino acids must have been synthesized in the plasma torch from the inorganic substance contained in the target.

When these works have been completed, we decided to perform similar studies aimed at the synthesis and identification of nucleotides and their oligomers. We also planned to continue the work on the synthesis of amino acids and their identification using various analytical tools and techniques. We have performed these works only partially because a more important task – that of addressing the problem of the possible breaking of symmetry in the torch - started to distract us.

The results presented above nevertheless played important part in the development of the instruments and techniques for a wide range of studies associated with the emergence of life. They proved to be of considerable interest among other reasons because they were the first ones to show that the key amino acids can be synthesized in the plasma torch of a hypervelocity impact. Moreover, it was possible to indentify such important organic compounds as aspartic acid, threonine, serine, glutamic acid, phenylalaline, ornithine, lysine, histidine, and tyrosine.

5.6. Studies that Confirm the Concept

Plasma torch, as a hitherto unknown medium, has been studied since the emergence of the first optical generators operating in the Q-switched mode. As we already pointed out in Section 3, at different stages of the development of these studies the torch plasma was viewed as a high-temperature medium with strong electromagnetic fields, which is suitable for the realization of thermonuclear reaction; as a pulsed source of plasma ions for time-of-flight analyzers; as a pulsed accelerator of micron-sized particles to high energies, and, finally, as a medium that is capable of raising ions of polyatomic clusters from the target.

The properties of the plasma torch as a medium capable of ensuring the synthesis of new substances did not attract particular interest. Research in this field was mostly of applied nature and aimed at the synthesis of various carbon structures with graphite subject to laser radiation.

Today the fact is beyond question that before the work began on the development of the new concept and before the first results about the peculiar properties of the plasma torch were published (Managadze, 2001, 2002a, 2003), no one viewed this medium as a natural phenomenon capable of ensuring the synthesis of new substances. And this especially applies to the synthesis of organic compounds associated with the emergence of life on the Earth and to the synthesis of simple organic compounds in interstellar gas-and-dust clouds (Managadze et al., 2003a, 2003b, 2003c). All these ideas were first suggested within the framework of the development of the concept proposed in this book. That is why we consider the plasma-chemical processes in the torch plasma to have the defining role in the proposed scenario.

However, the processes of the synthesis of new substances in the torch plasma appeared every now and then in various experiments and were observed by many researchers and have even been explained in some cases. How correct are these explanations?

In view of this, of great interest are the results obtained in the natural experiment VEGA carried out to study comet Halley, with PUMA instrument recording mass peaks corresponding to some organic compounds (Kissel and Krueger, 1987b). The above authors, with whom I have been closely cooperating for many years while working on VEGA and "Phobos" projects, interpreted the peaks of molecular ions of organic compounds as due to shock adsorption of organic dust particles frozen on the surface when they collided with the target. At the time when the paper was published (1987), nothing was known about the possibility of the synthesis of organic compounds in impact processes in space despite the comprehensive and in-depth studies that addressed such issues as protection against impacts in space and the formation of a plasma torch as a result of a micrometeorite impact. That is why the publication was received quietly, without criticism. Today we can quite soundly consider a different interpretation of the data. Thus, we can make the following conclusion based on the new ideas and on the results of the study of the torch given the initial conditions of the experiment. Earlier mass-spectrometric measurements can be reasonably explained in terms of the newly proposed mechanism of the synthesis of organic compounds in hypervelocity impacts. This happened in impacts of the dust particles of the tail of comet Halley with the target of PUMA instrument, which consisted of a pure silver plate.

Dust particles contained, with high probability, the constituent elements and compounds of organic matter and, in particular, carbon. If the impact velocity was equal to ~70—80 km/s, the dust matter may have, after its complete atomization and ionization, synthesized new organic compounds. These very compounds were detected by the onboard instrument. This conclusion is supported by the fact that the masses of organic compounds observed in space using PUMA onboard instrument coincide with those of organic compounds observed in the laboratory experiment performed to test an analog of the instrument. In this connection, it possibly makes sense to perform a joint analysis of old data in the light of new ideas and the results of laboratory simulations of impact processes.

Laska et al. (1996) and Zhang et al. (1999) used mass spectrometer to identify inorganic structures synthesized using laser radiation and, in particular, fullerenes and metallofullerenes. An important feature of this work is that the above authors used two modes of laser exposure of the carbon target: Q-switched mode with t_L~8 ns and free generation with t_L~230 ns at a wavelength of ~1.06 μm. The power density was equal to 10^5 (t_L~230 ns), 5 10^9, and 1.5 10^{10} W/cm^2 (t_L~8 ns) for the three typical spectra, respectively.

An analysis of the results obtained shows that after its complete atomization and ionization fullerenes were also synthesized in the gaseous state in the case of a relatively low power density, and with high efficiency in the plasma torch in the case of high power density. The spectra obtained clearly show that the nature of fullerenes differs in the cases considered. In the former case the model of low-temperature, tenuous plasma is realized, with low yield efficiency. In the latter case a high-density plasma torch is generated where intense synthesis of fullerenes and other complex compounds occurs.

The spectra obtained in the case of an exposure with a power density of $5 \cdot 10^9$ W/cm^2 agree well with the spectra obtained in this work whose results we report in Chapter 3.

The above comparison of experimental results leads us to conclude that in the case of low power density carbon cluster ions and other polyatomic ions evaporate from the surface and then combine into more complex structures. An increase of the power density up to ~10^9 W/cm^2 results in a qualitative leap in these processes because in this case atomic ions combine into polyatomic structures in the plasma torch during its expansion-away and cooling after complete atomization and ionization of the matter.

The most interesting result of Zhang et al. (1999) is that it demonstrates the process of the complexification of molecules with increasing power density. Thus, e.g., an increase of the power density from 5 to 15 10^9 W/cm^2 increases the number of carbon atoms in polymer structures from 30 to 300. A similar effect is observed in the case of the increase of the size of the plasma-formation region in experiments reported in this book.

A number of factors provide direct or indirect evidence for the synthesis of molecules in the plasma torch. In particular, the configuration of the experiment with "free expansion-away" prevents the detection of ions from the peripheral zone of laser exposure, where the power density may be substantially lower than needed for complete atomization, but at the same time sufficient for the desorption of a small quantity of clusters or molecular ions. We can therefore assume that in the mode of "free expansion-away" or in a mode with no artificially applied external electric field, the processes of ionization and further acceleration involve only the plasma ions from the central, high-temperature zone of the plasma formation to produce a torch.

Further important evidence for the synthesis of new compounds in the plasma torch is provided by the fact that experimentally observed multiply ionized hydrocarbons cannot be produced in thermal processes via successive ionization. They can be produced only by combining a multiply ionized carbon atom with, e.g., a neutral hydrogen atom. This is easy to explain: the energy of hydrocarbon bond $E_{CH} << E^{2+}$, where E^{2+} is the double ionization energy of carbon, which is equal to ~35 eV. Therefore ionization must begin after the dissociation processes.

The plasma nature of molecular ions produced is most clearly evidenced by the results of experiments with two-component powders. In this case, the plasma torch is the most likely region for the components to combine into more complex molecular ions.

Thus a joint analysis of the results obtained in this work and their comparison with those of other authors lead us to conclude that plasma torch is the optimum medium for the synthesis of new compounds. Laser exposure of the target makes it possible, like laser simulation of a hypervelocity impact, to reproduce a plasma torch where new substances including organic compounds can be synthesized during its expansion-away.

***

The experimental results of the original studies presented in this chapter contain data that are indicative of two facts of special importance.

The first fact demonstrates the high degree of similarity between the physicochemical processes in laser and impact produced plasma torches. In other words, the results of laser simulation reproduce completely the processes that occur during the impact, and these results can be trusted. i.e., the same compounds should be synthesized with high probability in impact interactions that are synthesized in laser plasma.

The second fact is that hydrocarbons and amino acids identified using various techniques can be viewed as products synthesized in the impact-produced plasma torch.

The main conclusion obtained by combining the above results is that the following compounds can be synthesized in the impact-produced plasma torch: carbines and amino acids, simple and complex carbon structures, and important intermediate compounds needed for the synthesis of amino acids. These compounds were identified using various analytical tools and techniques.

It was the right choice of the basic parameters of the laser emitter that made it possible to obtain bona fide experimental results that confirmed the new concept of abiogenous synthesis of organic compounds in hypervelocity impact processes and, especially, the increase of the mass of organic compounds synthesize din the torch. The adopted approach to the choice of the emitter demonstrated that the now available laboratory tools allow micrometeorite impacts to be simulated rather accurately.

The proposed approach also revealed that the simulation of impacts of larger meteorites with sizes from 1 mm and greater will require a revision of the simulation parameters including a possible transition to other (nonlaser) tools to produce an impact-like torch with relatively large projectiles.

The results presented in this chapter include important information that characterizes the new properties of the plasma torch. They are a logical extension of the results described in Chapter 4, which were obtained in the process of a laboratory simulation of the synthesis of organic compounds in the plasma torch of a hypervelocity impact, and add considerable support to both these data and their reliability.

Of the results presented in this chapter the similarity of impact- and laser-produced plasma torches is an issue that deserves particular attention. Only the high degree of similarity and identity between the processes that occur in these two phenomena would make it possible to obtain correct results in experiments on laboratory simulation of impact action. An independent experimental corroboration of the fact that impact- and laser-produced plasma torches are similar and the processes occurring in these media are identical as far as the synthesis of organic compounds is concerned, should be viewed as a result of special importance. These are new properties of the plasma torch and they have been found for the first time. Moreover, these results allowed us to develop special, fast experimental methods for investigating the proposed concept.

Other important issues for the study of the torch properties include the instrumental support of simulation experiments associated with a correct comparison of impact and laser experiments. The questions are:

- How to preserve synthesized organic compounds and how to protect them against destruction by a relatively long laser exposure?
- How to determine the effective diameter of the meteorite by the diameter of the focal spot of laser radiation?

These questions were answered experimentally and the adopted decisions ensured accurate simulation of hypervelocity impacts of micrometeorites with the effective diameters of 10—20 μm. This was achieved using laser exposure with a pulse duration of ~1 ns and an energy of ~1 mJ.

The use of a more powerful laser emitter with a pulse energy of up to 0.6 J and a pulse duration of ~7 ns allowed the effective diameter of the micrometeorite to be increased to ~100 μm and polymer organic compounds to be synthesized with molecular masses of ~4000 a.m.u. The available information about the possibility of the synthesis of key amino acids in the plasma torch combined with their identification played the crucial part in the interpretation of the resulting spectra of synthesized complex polymer organic compounds such as highly branched acetylene carbons, polypeptides or their dendrimers.

Of considerable interest were the results where the carbines that formed in the plasma torch have been used as “tracer atoms”. These experiments showed that organic compounds synthesized in the laser-produced torch are not localized in the plasma ejecta exclusively. They were also found both on the crater floor and in the adjacent regions in the area where melt and crunched matter are ejected.

The discovery of ordered fragments in the structure of carbines synthesized in the plasma torch should also be viewed as an important and promising result.

Thus the results presented in this paper allowed us to reconsider the tasks and potentialities of the experiments on laser simulation of the synthesis of organic compounds in the plasma torch of a hyper velocity impact. This, in turn, made it possible to introduce important corrections to the diagnostic equipment, the choice of the conditions for the experiments and to the process of data interpretation. This approach ensured fast progress and will make such progress possible in the future.

Chapter 6

SYMMETRY BREAKING IN THE PLASMA TORCH

6.1. MIRROR SYMMETRY BREAKING IN NATURE

The emergence and evolution of life on the Earth is a topic fraught with a number of complex issues including the problem of mirror symmetry breaking of bioorganic world, which has been very popular among the researchers for more than 150 years and still lacks a conclusive solution.

The core of the problem, if somewhat simply stated, is that in the process of their vital activity all forms of living matter on the Earth use only one of the possible mirror symmetric isomers of organic compounds, namely, L-isomers of amino acids and D-isomers of sugars, and never use the symmetric isomer. These compounds are said to be optically active because, when dissolved in water, they cause the polarization plane to rotate either left- (L) or rightward (D) depending on the spatial structure of a particular isomer. The peculiarity of the structure of these isomers is that they are mirror reflections of each other and have neither the symmetry center nor the symmetry plane. Such isomers are therefore impossible to superpose via translations and rotations. Insuperposable isomers of organic molecules are called chiral isomers. The term “chiral” originates from the Greek word “chira” that means “the hand”, because the right and left human hands are the conspicuous examples of such mirror symmetric, insuperposable antipodes (Figure 33).

If an optically active molecule has a single asymmetric center then it has only two optical isomers referred to as the L- or D-enantiomers. A medium containing equal numbers of L- and D-enantiomers is called a racemic mixture. Its solution is not optically active. A deviation from the equal enantiomer proportions makes the medium optically active. A solution containing only one enantiomer shows maximum optical activity and such a medium is referred to as optically pure. A polymer consisting of identical enantiomers is called homochiralic. Thus, e.g., all proteins are homochiralic because due to their biological origin they consist of L amino acids exclusively.

To better perceive the problem considered and to understand its complexity and relevance, let us refer to the original works (Kizel’, 1985; Goldanskii and Kuz'min, Kuz'min, 1989; Mason, 1991; Bonner, 1991 Keszthelyi, 1995; Avetisov and Goldanskii, 1996b). Each of these works gives an excellent presentation of the problem as a whole at different stages of its development, brings out its principal inconsistencies and the difficulties faced by the researchers trying to address it.

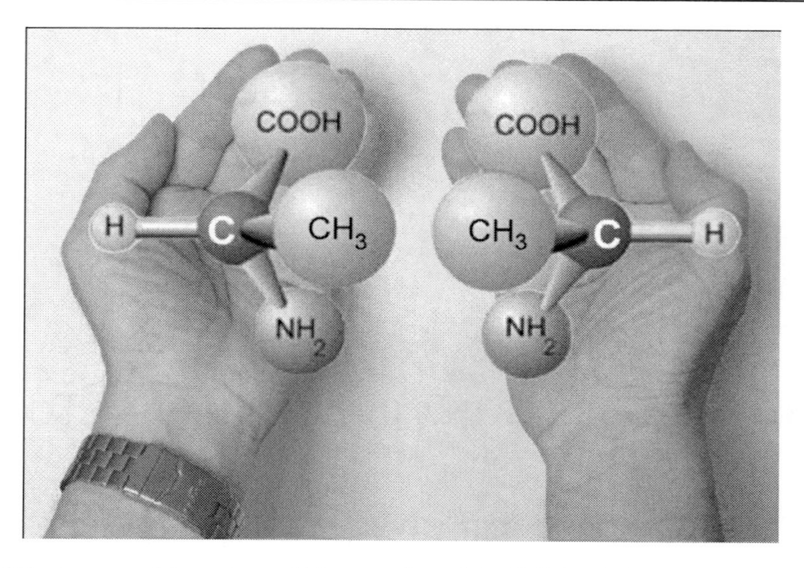

Figure 33. Of the two possible amino-acid isomers L and D all living beings use for their metabolism only those whose water solution rotates the polarization plane to the left, i.e., the so-called L-isomers. Like our left and right hands, L and D isomers cannot be superimposed by any rotations or translations. Hence the term "chirality" because "chira" in Greek means "hand". This figure shows the structure of two isomers of alamine amino acid.

In this connection, of considerable interest is the recent study of Davankov (2006), who developed a novel concept that elementary particles may constitute a homochiralic set of building blocks of matter, and advances new arguments supporting the idea that electrons and ions may possess a chirality of their own.

V. Davankov, who is an acknowledged expert in chirality issues, analyzes (Davankov, 2009) the new concept proposed in this book that the chirality of organic matter may develop in the plasma torch of a meteorite impact combined with his hypothesis about the chirality of elementary particles. He concludes that impact-produced torch plasma is the most suitable natural environment for the realization of this extremely important process.

However, the main questions arising in correction with this unique natural phenomenon – breaking of the mirror symmetry of bioorganic world – remain the same: when, why, and how did the symmetry breaking occur? Is such a breaking typical for the Earth's biosphere exclusively or is it a global phenomenon that accompanies the processes of the emergence of life?

In this chapter we analyze the hypothetical possibility of the symmetry breaking in the process of the synthesis of isomers in an impact-produced plasma torch. The results of laboratory analyses of the plasma torch produced by laser exposure showed that of the known «true» local physical fields that may produce the symmetry breaking during the emergence of enantiomers in the process of the expansion-away of the plasma torch we must first and foremost consider a combination of static magnetic field with constant orientation and linearly polarized radiation. It is important that the direction of the vectors of these factors is fixed and their orientation does not change from one act of influence to another. In the case considered the vectors should be collinear. The presence of these factors has been experimentally confirmed in the studies of the torch plasma and they are known to undoubtedly accompany the process of torch expansion-away.

The magnetic fields and plasma instabilities found experimentally in the plasma torch are very likely to make possible the generation of a circularly polarized wave – one of the most important and generally agreed «true» factors, which can also lead to the breaking of mirror symmetry. However, no such a wave from a torch has been so far recorded. This, hypothetically, may be due to this wave being «trapped» inside the plasma torch. The «trapped» wave cannot escape from the plasma-formation region. Therefore circularly polarized wave should have low intensity and can be detected only using hypersensitive polarimeters. Note in this connection that given the physical conditions in the plasma torch, the absence of circularly polarized radiation in this medium is much more difficult to explain than its presence. The causes of the presence of circularly polarized radiation are clear, understandable, and well founded.

The unidirectional electric and magnetic fields found (including via experimental evidence) in the plasma torch may be indicative of the «innate» asymmetry of the emerging medium. These fields can also be viewed as local chiral physical fields. They may probably result in the breaking of mirror symmetry of enantiomers in the process of their formation. However, this statement requires experimental verification.

The «true» chiral factors mentioned above, which arise during the expansion-away of the torch plasma, are referred to (for brevity) as «local chiral physical fields ». When combined, they may not only cause a weak breaking of symmetry, but also perform the important function of the «trapping field» for the fields of spontaneous processes.

These factors may result in isomer symmetry breaking not only for compounds that form in the «laser» torch. Given the complete identity of the physical processes in the plasma and impact torches, the result must be the same in the impact-produced torch plasma. This process may produce at least a minor initial symmetry breaking of organic compounds synthesized abiogenically under natural conditions.

In this chapter we also analyze other, equally interesting properties of the plasma torch that are associated with spontaneous symmetry breaking in a strongly nonequilibrium medium that is far from the thermodynamic branch of equilibrium. It is safe to suggest that the superposition of weak local chiral fields present in the torch onto spontaneous processes of symmetry breaking may have stimulated the formation of a homochiralic medium and structures of organic compounds synthesized in the plasma. Such structures are known to be needed for the formation of prebiological structures, and the «sign» of the asymmetry of local chiral physical fields of the torch, which do not reverse their «polarity» in the known media for the entire Universe, may have determined the «sign» of the asymmetry of bioorganic world.

To better understand the mechanism of spontaneous breaking of the symmetry of a physical system, it would be appropriate and useful to analyze a simplified model of this process. Let us consider a breaking of this kind for an initially symmetric system in the cases where the initial state is not a minimum-energy state and is therefore not energy advantageous. In this case, the lowest state with minimum energy is ambiguous or degenerate because it corresponds to a series of solutions. This means that in the case of spontaneous symmetry breaking one of these solutions should be realized.

A simple mechanical model may serve as an obvious case to illustrate spontaneous symmetry breaking in the general form. This model consists of an absolutely axisymmetric bottle with a convex bottom, as shown in Figure 34(1), with a ball falling into it exactly along the axis. However, the result is asymmetric: the ball ends up near the wall, off axis. The initial symmetry has broken spontaneously. It manifests itself only in the fact that the ball may roll to any side from the axis, i.e., the lowest state is degenerate with respect to rotations about the axis.

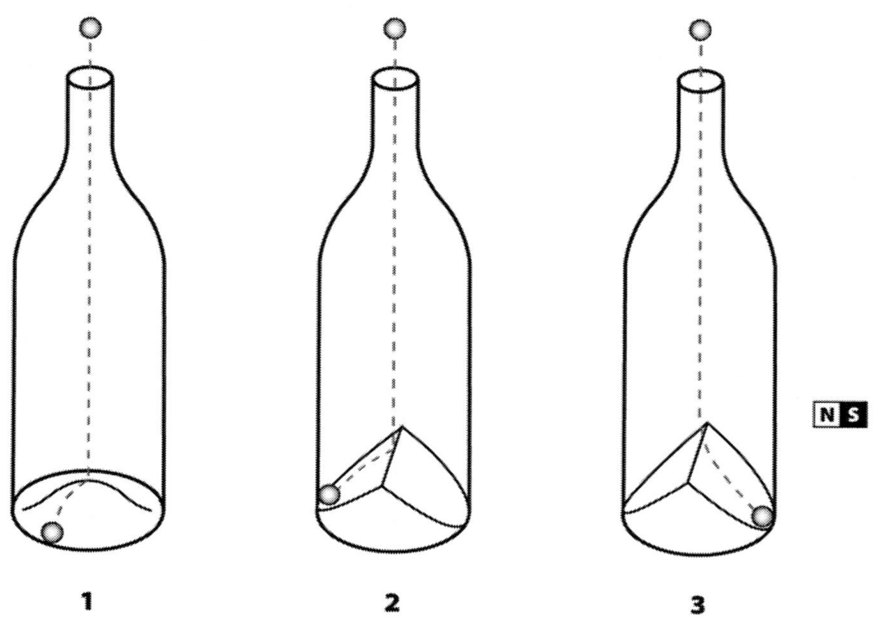

Figure 34. Mechanical model of the spontaneous symmetry breaking in a system without external influence (1 and 2) and in the presence (3) of a perturbing field eliminating the degeneration.

However, this mechanical model should be altered if we want to adapt it to spontaneous symmetry breaking in the case of enantiomorphism, i.e., in the presence of a process capable of producing mirror-identical modifications. In particular, for the case considered, where «left-» and «right-hand» isomers form, the bottom of the bottle should also have two equivalent minimum-energy domains as shown in Figure 34(2). In this case, spontaneous symmetry breaking corresponds to only two equiprobable solutions. This means that given the initial symmetry of the system the ball should end up in the «left-» or «right-hand» domain of the bottle, and the system remains degenerate, thereby ensuring the mirror symmetry or racemicity of synthesized enantiomers.

In the case of the mechanical model of spontaneous symmetry breaking considered here such a degeneracy can be «removed» by a permanent perturbation shown in Figure 34(3), e.g., a magnet located in the lower part of the bottle. Such a perturbation would ensure that the ball always ends up in one of the two possible cells. This means that spontaneous symmetry breaking results in the formation of only «left-hand» or only «right-hand» enantiomers, thereby ensuring their homochirality.

When comparing the mechanical model described above with the processes that occur in the plasma torch, one must take into account the fact that the perturbing magnetic field that acts on the iron ball is an analog of the trapping field or local chiral physical field that forms in the process of the expansion-away of the torch plasma. In this case, spontaneous breaking of mirror symmetry in the plasma torch occurs as a result of the formation of dissipative structures in a highly overheated and unstable plasma medium that is far from the thermodynamic branch of equilibrium. The combination of the above plasma processes –

presumably in some extremely rare cases – could have created the state of the medium meeting the conditions necessary for the synthesis of homochiralic macromolecular structures of organic compounds.

6.2. Status of the Symmetry Problem

The current status of the problem of the symmetry breaking of bioorganic world can be briefly formulated as follows.

Two fundamentally different scenarios of the breaking of mirror symmetry are considered (Goldanskii and Kuz'min, 1989). The first one is the scenario of asymmetric generation, according to which symmetry breaking occurred at the early chemical stage of evolution. The second scenario allows the possibility of biogenous breaking where symmetry breaking occurred at a later, biological stage. Biogenous scenario is unlikely because the above authors believe that the apparatus of self replication, which is needed for the emergence and development of life, could not form in a racemic medium.

The scenario of asymmetric origin also involved the analysis of two factors: symmetry breaking caused by physical fields and spontaneous symmetry breaking.

The effect of chiral physical fields – both local and global – came in handy first and foremost in the search for a mechanism of asymmetric formation and accumulation of chiral compounds on the Earth and in space. These factors could also be further enhanced via interaction with systems characterized by spontaneous breaking of mirror symmetry (Goldanskii and Kuz'min, 1989; Kizel', 1985).

Selective synthesis of enantiomers can be caused by local chiral fields, or by the so called chiral factors. Such factors include a combination of static magnetic field and linearly polarized radiation; circularly polarized light generated during solar flares; various combinations of electric and magnetic fields; a combination of gravitational and centrifugal, as well as mechanical forces, electromagnetic fields, and lightning discharges (Goldanskii and Kuz’min, 1989; Kizel’, 1985).

Global chiral influencing factors are produced by weak interactions. They include polarized products of beta decay and weak neutral currents. Their effect is always asymmetric, but too weak to cause appreciable asymmetry. (Zel'dovich and Saakyan, 1980; see also the reviews of Keszthelyi, 1995 and Bonner, 1984).

Selection criteria have been developed, which make it possible to determine which of the above local asymmetric factors are «true» and which are «false». Thus, according to Barron (1986; 1994), a combination of magnetic field with a linearly polarized wave and a circularly polarized wave are generally agreed «true» factors. On the other hand, a combination of polar and axial fields can, in principle, be viewed a necessary (but not a sufficient) condition for the formation of a chiral factor leading to asymmetric synthesis.

The scenario of symmetry breaking caused by physical fields also faces certain difficulties: the maximum achievable asymmetry in the currently known processes including the asymmetry achieved in laboratory experiments is evidently insufficient for the synthesis of homochiralic structures comparable to biopolymers (Avetisov and Goldanskii, 1996b). The complexity of such structures also makes it very difficult to describe their evolution toward increasing complexification. The point is that the number of evolving objects for homochiralic molecules consisting of about 50 or more links becomes so large that they cannot be produced by "exhausting" all possible combinations. Moreover, the presence of minor chiral defects in such molecules can result in specific phenomena and, in particular, in the «error catastrophe» (see, e.g., (Avetisov and Goldanskii, 1996b)), which prevents further assembly of homochiralic chains and replication. These and other constraints led the researchers to conclude that if symmetry breaking occurred at the chemical stage of evolution, then earlier known physical factors could hardly have played the crucial part in this process. This conclusion was also encouraged by the fact that before the publication of the paper by Managadze (2005a) no real local chiral factor has been proposed with a globally asymmetric distribution on the Earth.

While discussing this problem one should bear in mind that magnetic fields and polarized radiation of the plasma torch can be viewed as such true local chiral factors. Earlier it has been pointed out that the plasma environment of the torch is asymmetric on a planetary scale and, in the absence of antiworlds, even universally.

According to the scenario of spontaneous breaking of mirror symmetry, the formation of chirally pure forms of organic compounds is associated not with an external asymmetric influence, but rather with the self-organization of an initially racemic environment, where a sui generis nonequilibrium phase transition could have occurred. The researchers also considered the possibility of the subsequent enhancement of a weak chiral asymmetry triggered by spontaneous processes. According to the theory developed by Barron (1994), Morozov (1978, 1979), and Morozov and Goldanskii (1984), the process that leads to the self organization of chirally pure forms of organic compounds is based on the nonlinear properties of physicochemcial transformations that determined the formation of living systems of the protobiosphere. Soai et al. (1995) have experimentally confirmed the above concept by observing the enhancement of the enantiomer excess in autocatalytic synthesis reactions.

The aforesaid leads us to suggest that symmetry breaking can, in principle, be spontaneous, however, this conclusion cannot be considered to be sufficiently founded because of the high level of chiral purity required for the evolution of homochiralic structures of the biological level of complexity in catalytic processes of spontaneous symmetry breaking. This is the main inconsistency in the scenario of spontaneous symmetry breaking, which, however, does not rule out this scenario altogether (Avetisov and Goldanskii, 1996b).

The possibility of the symmetry breaking of macroscopic states of a chemical system that is far from the thermodynamic branch of equilibrium was analyzed by Morozov (1979), Morozov et al. (1982), Kondepudi and Nelson (1985), Goldanskii and Kuz'min (1989), Frank (1953), Nicolis and Prigogine (1981), Kondepudi and Nelson (1984), and Avetisov (1985). Prigogine and Kondepudi (1998) performed similar studies in terms of the concept of dissipative structures.

Dissipative structures are manifestations of nonequilibrium thermodynamic processes. However, one of the deepest consequences of these processes manifests itself in dualistic irreversibility: as a destructor of order in the vicinity of equilibrium and as a constructor of order far from equilibrium. Unlike equilibrium systems (Galimov, 2001), which pass to the state with minimum free energy, nonequilibrium systems may develop unpredictably: their state does is not always determined by a macroscopic equation. This happens because starting from identical initial conditions a nonequilibrium system may end up in different states. This may be due to fluctuations, minor irregularities, defects, and other accidental factors. The final state is impossible to predict, however, the system often reaches an «ordered state» characterized by spatiotemporal organization. The fundamental property of nonequilibrium systems shows up in their ability to pass into an ordered state as a result of a fluctuation – i.e. to exercise «order via fluctuations» (Prigogine, 1980).

The birth and maintenance of organized nonequilibrium states is due to dissipative processes, and therefore these states are referred to as dissipative structures (Prigogine, 1967).

When analyzing the problem of symmetry breaking the above authors point out, first, that such an asymmetry of the structure of molecules can show up only in systems that are far from equilibrium and that sustaining these asymmetries requires permanent, mostly catalytic, formation of one of the enantiomers that must occur despite their equiprobable birth. Second, according to the paradigm of order via fluctuations, in a system with the corresponding autocatalysis the thermodynamic branch containing L and D isomers in equal quantities may prove to be unstable. This may trigger a transition of the system into one of the possible asymmetric states.

It follows from the foregoing that a nonequilibrium chemical system is capable of producing and maintaining the asymmetry and gives only a general idea of where to seek the answer to the question of the mechanism of the selection of natural biochemical chiral molecules. Moreover, these results do not answer such questions as whether the chirality appeared before life or after the origin of life, in the process of the biological evolution of living matter, and what factors have determined the «sign» of its asymmetry.

However, the theory of symmetry breaking under nonequilibrium conditions provides an important method for estimating the likelihood of various hypothetical models. For example, concerning the initial conditions of the prebiological process of symmetry breaking, the system must be in a nonequilibrium state and subject to permanent influence from an external factor ensuring the predominant production or destruction of one of the enantiomers. It can be assumed that the parameters of such a permanently acting factor should also determine the "sign" of the asymmetry.

Avetisov and Goldanskii (1996b) analyzed the possibility of spontaneous breaking of mirror symmetry in nonequilibrium processes. The above authors showed that if conditions for the formation of homochiralic structures of the biochemical level of complexity are achieved in such systems then the same mechanism should also ensure the maintenance of the state of the medium before the development of enantiospecific functions, which are capable of maintaining on their own the evolution of homochiralic structures. Avetisov and Goldanskii also showed that this mechanism is the only one that is consistent with asymmetric contamination of the biosphere.

According to the above authors, the situation with the breaking of the mirror symmetry of bioorganic world allows us to make a number of assumptions suggesting that the breaking of symmetry could have occurred at the chemical stage of evolution. The currently known chiral physical factors could not have played the crucial role in the development of the asymmetry and, possibly, the evolutionary dynamics of the prebiological world may have consisted in the spontaneous breaking of mirror symmetry. However, each of these statements faces a number of difficulties and inconsistencies, which are impossible to circumvent given our current knowledge of nature. Thus, despite its 150-years long research history, the problem of the origin of mirror asymmetry currently offers more puzzles to be solved than answers to questions (Goldanskii and Kuz'min, 1989; Avetisov and Goldanskii, 1996b).

The above difficulties and inconsistencies may be due to the fact that the chemical system viewed as the main medium for the processes of symmetry breaking differs fundamentally from the plasma torch. Plasma torch is a different medium, which obeys different laws and which must therefore have properties different from those of the chemical medium. Otherwise there would be no such field of knowledge as plasma chemistry, and there would be no plasmachemical reactions. It is possible that the use of plasma chemistry laws in the analyses of the effects of dissipative structures and symmetry breaking in the torch plasma may eliminate the existing difficulties.

A comparison of the main properties of the plasma torch with the main ideas described above and the requirements of the concept of dissipative structures shows that the plasma medium generated in the process of a hypervelocity impact satisfied the most important requirements that are needed for the symmetry breaking in the process of the formation of isomers. Thus at the initial stage of adiabatic expansion-away nonequilibrium plasma is subject to permanent influence of "true" factors meeting all the necessary requirements to local chiral physical fields.

Naturally, symmetry must be broken, at least slightly, in organic compounds, e.g., in amino acids, synthesized under these conditions. Such a breaking of the "symmetry" sign must be observed in all impact processes, everywhere, throughout the entire planet, in the Universe, except for antiworlds. This will repeat itself always from one event to another. It is safe to assume that these processes should explain the predominance of L amino acids in meteorites where the chirality sign is coincident with that of terrestrial bioorganic world.

Impact plasma is a highly nonequilibrium medium with a superhigh catalytic activity and therefore it must easily develop a spontaneous symmetry breaking, which may be further maintained by the relatively weak local chiral physical field produced in the torch. Prigogine and Stingers (2005) analyze such a process or the chemical system and gravitational field.

In the plasma medium in question assembly and ordering of molecular structures occur with the realization of prebiological molecular selection in irreversible and fast plasmachemical reactions. It is probable that these molecules may prove to be homochiralic in some rare ejections of the torch plasma.

The proposed model is especially valuable given that the most important processes needed for its realization have been experimentally confirmed. Thus the results reported in Chapters 4 and 5 suggest that the synthesis of organic compounds in the torch is accompanied by their ordering.

The experimentally found breaking of isomer symmetry in organic compounds synthesized in the laser-produced plasma torch should be viewed as a result of greatest importance. These are preliminary findings and therefore it would be right to confine ourselves to the brief statement that we give below.

To support the experiment, we developed a special technique and made a facility that allows synthesis products to be accumulated at the support covering the upper hemisphere above the target. This facility made it possible to record the mass spectra of synthesized organic compounds simultaneously with their accumulation and thereby control the collection process of synthesis products.

The use of an external agent in the form of an IR laser operating in the diode pumping mode with a frequency of ~30 Hz made it possible to collect products for up to 1 million mass spectra in two work days. The preparation of the target, collection of organic compounds, and the retrieval of the support were carried out subject to the conditions of high sterility, and the support was transported in a sealed container. All the chemical compounds employed were ultrapure. The sterility of the medium and pureness of the samples have been continuously controlled by mass-spectrometric measurements.

Preliminary mass-spectrometric measurements showed that the sample contained mostly amino acids. Therefore measurements were aimed primarily on the search for the symmetry breaking in amino acids.

At the initial stage of the work the sample was sent to the research laboratory of V.Schurig at the Institute of Organic Chemistry at the University of Tubingen (Germany). Measurements were made with the chromato-mass-spectrometer equipped with a chiral column meant for the analysis of amino acids. The measurements showed the sample to be uncontaminated, but the small amount of the sample prevented obtaining reliable results.

For further measurements we chose a high-sensitivity technique and, at the same time, increased tenfold the total number of spectra collected on the support.

These measurements yielded the first positive results; however, they were not sufficiently reliable because of the small amount of the sample. We therefore decided to repeat the measurements.

To address this problem – i.e., how to collect a greater amount of amino acid monomers, – we have developed and made a vacuum facility with a high-frequency laser exposure, which made it possible to increase the number of synthesized molecules by a factor of 1000.

The new plasma properties that have been discovered, which make possible the occurrence of processes in the torch, eliminate a number of difficulties concerning the problem of the breaking of mirror symmetry under the conditions of «cold» cosmic scenario. These difficulties are due to the specific features of chemical processes at ultralow temperatures, which prevent the full use of some scenarios of mirror symmetry breaking. Thus low temperature in the interstellar medium prevents the use of the model of spontaneous breaking of mirror symmetry because organic compounds frozen into the icy envelopes of dust particles become too immobile for the realization of autocatalysis (Goldanskii and Kuz'min, 1989). For this reason, the existence of a single physical mechanism of the synthesis of organic compounds on the Earth and in space and associated with impact processes could ensure the similarity of chemical processes and remove the arising restriction.

When we analyze the most important inconsistencies of the problem, there is an impression that the processes of the breaking of mirror symmetry in nature involved hitherto unknown mechanisms, and that by finding these mechanisms and taking them into account we will be able to achieve progress and overcome the difficulties. We therefore consider finding in nature of a new mechanism of possible symmetry breaking irrespectively of the scenario it can be associated with - the influence of physical fields or spontaneous symmetry breaking - to be a result of greatest importance, which deserves the most thorough analysis.

The mechanism of abiogenous synthesis of organic compounds in plasma outbursts that we propose in this book belongs to the type of mechanisms just described. It is universal because it can operate both in interstellar gas and dust clouds and in planets during the early stages of their formation. This mechanism can deliver the local chiral factor to the entire planet and to interstellar clouds, and it may result in the observed breaking of mirror symmetry of isomers on the Earth and in space.

The possibility of abiogeneous synthesis has been until recently viewed as the first principal component of the new concept. However, with time, we began to consider the most important component to be the capability of the plasma torch to generate a medium possessing all the properties of a «true» local chiral physical field and to asymmetric formation of enantiomers during the process of their synthesis (Barron, 1986; 1994). And this result can be viewed as very important despite its preliminary nature.

In view of this, it becomes necessary to more thoroughly analyze the mechanisms and conditions of the formation of the electromagnetic fields of the plasma torch and the generation of polarized radiation. The results of these experiments will show rather conclusively how realistic is the possibility of the formation of "true" local chiral physical fields in the torch plasma resulting in symmetry breaking.

6.3. Results of the Measurement of Electromagnetic Fields in the Plasma Torch

According to the results of direct impact and model experiments, the processes of the synthesis of organic compounds in the plasma torch of a hypervelocity impact can be viewed as real rather than hypothetical processes. Until recently, this conclusion was difficult to make when analyzing the possibility of the generation of local chiral physical fields in the plasma torch of a hypervelocity impact. The fact that physical fields close or similar to chiral fields exist in the plasma torch can be viewed as an experimental fact. In this case, the hypothesis consisted in the assumption that physical fields produced under certain initial conditions are capable of ensuring asymmetric synthesis of enantiomers.

Let us now consider the results of experimental studies concerning the possibility of the generation of chiral fields in the plasma torch, and the possibility of their artificial reproduction for further experimental investigation. Studies of local physical characteristics of the plasma torch generated as a result of a hypervelocity impact cannot be performed in dust-impact experiments in open space because the location of the impact is impossible to predict. The impact domain is also difficult to locate and correct measurements difficult to perform in laboratory experiments with dust-particle accelerators because of the small characteristic size of the plasma outburst. These difficulties can be circumvented by creating via a laser exposure a plasma torch identical to the plasma torch of a hypervelocity impact. This approach will make it possible to locate the plasma torch and thoroughly analyze its physical properties including the local electric and magnetic fields, using fast equipment meant for plasma diagnostics. Despite the problems due to the high rate of the process considered (Korobkin et al., 1977; Bychenkov et al., 1993; Stamper, 1991), extensive experimental data (Stamper, 1991) related to the studies of plasma formations generated by laser radiation are available.

During the preparation to these experiments, special tasks were formulated that were associated with laser fusion, whereas no tasks have been set involving the simulation of plasma processes that occur in the case of an impact. However, many of these experiments were made with such laser exposure parameters that provided the initial conditions identical to those that ensure a qualitative simulation of the impact process. Therefore the results obtained can be used to reproduce and study the physical processes and fields generated in the process of a hypervelocity impact.

Ensuring more or less equal (by the order of magnitude) power densities of laser and impact effects can be viewed as the most simple and reliable simulation parameter for achieving a qualitative identity of the two processes. The power density for an impact effect can be determined from the formula W_{IMP}~$10^{-7}\rho/3\cdot v^3$ W/cm^2, where ρ is the density of matter in the projectile (in g/cm^3) and v is the velocity of the projectile (in cm/s). The power density is ~10^{14} W/cm^2 for a density of ρ~3 and impact velocity of 100 km/s. Therefore a power density from 10^{13} to 10^{15} W/cm^2 (Bychenkov et al., 1993) can be considered to be acceptable as a similarity parameter for modeling the impact process.

For an analysis of the possibility of the generation of local chiral physcial fields in the plasma torch we need, first and foremost, the data about the magnitudes of electric and magnetic fields and their mutual layout.

It is believed (Bykovskii and Nevolin, 1985) that in expansion-away processes in both laser- and impact-generated plasma ions are accelerated via pressure gradient during hydrodynamic expansion-away of the plasma and in the self-consistent electrostatic field generated at the periphery as a result of the escape of hot electrons. According to the results of Mendel and Olsen (1975), the magnitude of this field at a 5 mm distance from a carbon target is ~1900 W/cm in the case of a power density of 10^{11} W/cm^2 and an exposure diameter of ~60 μm. This magnitude increases with increasing power density. The direction of electric field is constant judging by the direction of the motion of low-energy ions, and is mostly isotropic, however, a certain part of energetic and multiply charged ions are accelerated inside a cone with an opening angle ranging from 20 to 40° (Phipps and Dreyfus, 1993).

The measurements of the magnitude and configuration of magnetic field were initially made using small magnetic probes (Korobkin et al., 1977), and later, employing complex equipment using the Faraday Effect, which is based on the rotation of the polarization plane of probe radiation transmitting through the plasma torch and its neighborhood (Bychenkov et al., 1993; Stamper, 1991).

Experiments performed with a laser exposure with a power density of W_L~10^{13}–10^{15} W/cm^2 showed that a ring-shaped or toroidal magnetic field with a magnitude of 100 to 200 kG is generated above the target in the zone surrounding the plasma torch. Subsequent experimental studies of the plasma torch also revealed the presence of axial field with a magnitude of 0.5 to 1 MG (Stamper, 1991).

The direction of the ring-shaped magnetic field, like that of the axial field, is reproduced for each laser exposure. According to Bychenkov et al. (1993) and Stamper et al, (1991), the generation of the ring-shaped field is due to the excitation of the thermal e.m.f. and ponderomotive e.m.f. Briand et al. (1985) explain the presence of axial magnetic field by the dynamo effect that arises in the plasma torch with a strong anisotropy. Also observed are high-frequency vortex electrical fields generated by plasma instabilities (Stamper,1991).

The configuration of electromagnetic fields produced during the generation of the torch meets all the criteria imposed on chiral fields because, as is well known (Barron, 1986; 1994), a combination of the vectors of magnetic and electric fields that are not perpendicular to each other (Kizel', 1985) for nonequilibrium processes can be considered to be a sufficient condition for the generation of a chiral field, and its effect may result in asymmetric synthesis of enantiomers. According to Gol'dsnskii and Kuz'min (1989) and Kizel' (19875), the chirality factor can be substantially higher given that plasma radiation may be circularly polarized in the case considered. No publications reporting the results of such measurements have been found despite the fact that provided the presence of strong magnetic fields in the plasma torch there are all conditions for the generation of radiation of this kind.

The highly anisotropic, high-temperature, and very dense plasma is the main source that generates electromagnetic fields and plasma instabilities in the plasma torch. Therefore the physical parameters of the "laser" and "impact" plasma should be close to each other if the power density of exposure is the same in the two cases. This should be sufficient to achieve a qualitative agreement in the two processes. Accurate selection of the main parameters of modeling is necessary for performing quantitative measurements and dedicated experiments of laser simulation of chiral fields generated in impact processes (Managadze and Podgornyi, 1968, 1969).

It follows from the above that in the plasma outburst of a hypervelocity impact illustrated in Figure 35. organic compounds are synthesized and this process can be accompanied by the generation of a field meeting the requirements to local chiral physical fields, which results in the breaking of the symmetry of newly synthesized enantiomers.

The generation of a magnetic field in the torch plasma combined with linearly polarized radiation makes it more likely for "true" chiral fields to form. And this scenario may be the most exacting because, as we show in the next section, both factors – field and radiation – are generated simultaneously.

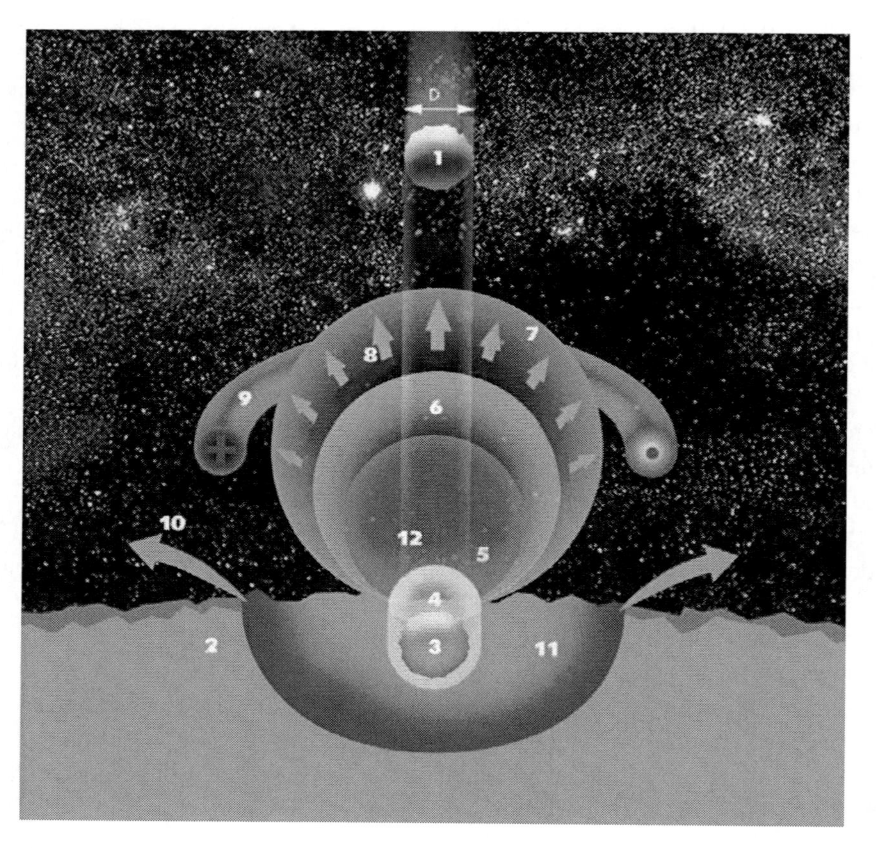

Figure 35. Schematic sketch of the main physical domains, electromagnetic fields, and characteristic sizes of the plasma torch during its expansion-away in vacuum. Plasma is generated by a hypervelocity impact of a body of diameter D. The formation and expansion-away of plasma precedes the ejection of ground material of the target from the intermediate crater of diameter 5D and depth 2—3D. Meteorite 1 approaches target 2. Here 3 is the plasma-formation domain; 4, the hot spot; 5, the zone of the beginning of the expansion-away; 6, the zone of hot electrons; 7, the zone of ion-beam formation; 8, the electric-field vectors; 9, the ring-shaped magnetic field; 10, the direction of the ejection of target matter; 11, the transitional crater, and 12, the vector of axial magnetic field.

6.4. MEASUREMENTS OF TORCH PLASMA POLARIZATION

We pointed out above that circularly polarized radiation is one of the most efficient factors to trigger symmetry breaking when acting on a racemic mix of enantiomers (Goldanskii and Kuz'min, 1989). In such radiation, depending on its symmetry "sign", the vector of the electric field of the light wave rotates in one of two different directions: clockwise or counterclockwise. As a result of its interaction with various cosmic irregularities, unpolarized light may produce circularly polarized radiation with different "signs" and different degrees of polarization. Some researchers try to explain the asymmetry of bioorganic world as due to the eventual asymmetry of circularly polarized radiation in space. Therefore the discovery of even partially circularly polarized radiation in the torch plasma could be viewed as further important evidence for the asymmetry of the plasma outburst. We found no positive results of such measurements and, possibly, such studies have never been made. However, we already pointed out above that the absence of such radiation in the plasma torch cannot be explained. Hence it must be present.

The special status of polarization measurements of plasma emission can be explained by the large amount of information that these measurements provide. This is due to the physical processes that ensure the generation of polarized radiation, which are of great importance for the problem considered. Thus, for polarized light to appear in the plasma torch, unpolarized radiation of the plasma must pass through a natural or induced anisotropic medium. This means that in the process of plasma expansion-away asymmetric and ordered anisotropic media must develop in it. In such a medium symmetry breaking is to be expected when monomers form in it that could later have produced homochiralic polymer structures. It is also important that all this information is obtained from a hard-to-reach zone, namely, from the recombination region of the torch plasma, where monomers of organic compounds are purportedly synthesized, and where other plasma diagnostics tricks do not work at all or work incorrectly.

It follows from the aforesaid that studies of symmetry breaking in the most interesting inner regions of the plasma torch do not need to focus on circularly polarized radiation exclusively. There is no doubt that valuable data should be provided both by the results of the measurements of linearly polarized and elliptically polarized radiation.

In this connection, of great interest are papers and reviews (Baronova et al., 2007 a, b, c, 2008), which summarize the data of polarization measurements of various artificial plasma formations from discharges and pinches to plasma torches. Even a cursory glance at these publications allows us to make the important conclusion that linearly polarized radiation is observed almost in all known artificial plasma formations, many of which have natural analogs. These results indicate that the breaking of isotropy and the emergence of order is one of the most characteristic features of plasma formations. Their origin is to be linked, like in the case of the torch plasma, to the presence of unidirectional electric and magnetic fields in hese formations.

Kieffer et al. (1992, 1993.) report the results of their investigation of linearly polarized radiation from the region of a plasma torch generated by the radiation of two lasers in turn in the Q-switched mode. Similar results were obtained by Inubushi at al. (2006).

In the experiments of G.Keffier the main 1-ps pulse at a wavelength of 1.053 μm, which provided a power density of $W_L \sim 8\ 10^{14}$ W/cm^2, was preceded by a 60-ps pulse providing a power density of $W_L \sim 8\ 10^{11}$ W/cm^2. The combined effect of laser emitters resulted in the heating the plasma in the torch to 500 eV or $5{\cdot}10^6$ C. At such a high temperature linearly polarized radiation of the plasma was recorded in the wavelength interval from 7 to 8 Å. To record the result, more than 30 individual-exposure spectra have been averaged. The recorded polarized radiation was generated in the torch plasma and was unrelated to the polarized radiation of the lasers. These results are presented in Figure 36.

For each laser exposure two peaks, on the average, were recorded with the amplitudes (in arbitrary units) equal to 8 and 2 for polarized and unpolarized radiation, respectively. The peaks were 0.05 Å apart.

The results obtained, which are based on the ratios of average peak amplitudes, are indicative of high degree of linear polarization, which amounts to 75 % and more. However, the polarization degree could have been substantially higher and may have reached 100% in some cases, but it was possibly "washed out" as a result of averaging of individual observations.

Note that the observed linearly polarized radiation generated at x-ray frequencies is difficult to directly associate with the process of formation of molecular ions of polypeptide monomers or nucleic acids. However, if the physical processes of the breaking of the isotropy of the plasma medium are viewed as processes characteristic of the torch, the effect should also be observed in cooler plasma and at wavelengths corresponding to formation of monomers.

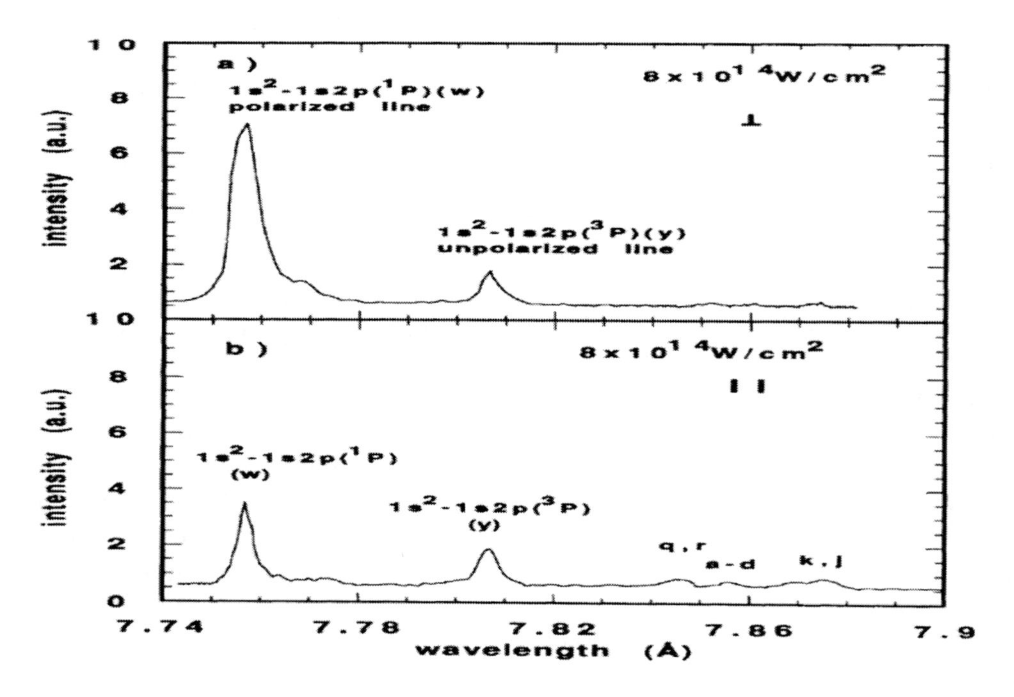

Figure 36. Helium and lithium-like aluminum emission lines of laser plasma at the wavelength of ~0.8 nm recorded by Kieffer (1992, 1993). The simultaneous presence of linearly polarized emission and magnetic field in the medium is one of the most important "true" physical advantage factors resulting in the symmetry breaking. The figure shows the recorded spectra accumulated over 30 shots: panel (a) shows is the polarized component in the perpendicular direction and panel (b), the component in the parallel direction.

This is evidenced by the effect of the «loss» of mass spectra, which shows up conspicuously for the torch plasma in the process of the synthesis of complex organic compounds. Thus the formation of complex polymer structures in the torch for the given exposure parameters and for the laboratory time scale has a very low probability of 10^{-3}-10^{-4}. This means that in the process coadding there may be one "informative" spectrum for one thousand "uniformative" spectra. This rate may prove to be too low for laboratory conditions. However, such a rate should be viewed as sufficiently high and quite acceptable for cosmic time scales, which are measured in hundreds of millions of years.

The degree of the symmetry breaking in the processes of the synthesis of organic compounds in the torch may be correlated with the amplitude of polarized radiation in the plasma and with the formation of irregularities. This hypothesis can be tested experimentally by recording simultaneously the degree of symmetry breaking of organic compounds and the amplitude of polarized radiation from the same torch. In this case, it will be possible to detect the correlation between the two processes. However, experiments of this kind are currently impossible to perform because of the problems with the determination of the symmetry breaking in organic compounds in the case of single exposures.

When performed with different exposures and subsequent averaging of the results obtained, such experiments may provide very important data. Thus, based on the available results, the probabilities of various events occurring in the torch do not differ very much in different experiments with similar initial conditions. It is this approach that was mostly used in experiments performed to study the properties of the torch plasma.

The results obtained in experiments with data averaging suggest that in some cases polarization degree in the plasma torch may reach 100%. In such cases, the result obtained can be hypothetically associated with the processes of spontaneous ordering of the plasma medium of the torch. This may cause a strong symmetry breaking in the process of the synthesis of enantiomers and followed by the assembly of homochiralic molecules of organic compounds in the torch.

Note that the presence of "unidirectional" linearly polarized radiation emerging from the plasma torch may be indicative of a substantial anisotropy of the medium where organic compounds are synthesized. These conditions affect appropriately the degree of the symmetry breaking of enantiomers that form in the plasma medium of the torch.

However, of greater importance for the symmetry breaking is the effect of the combination of linearly polarized radiation with magnetic field, and this fact should not be neglected. In this connection, linearly polarized radiation can be viewed as a bona fide factor that ensures, together with magnetic field, the formation of "true" local chiral physical fields, and as a factor that bears information about symmetry breaking in spontaneous processes. It is also important that linearly polarized radiation escapes rather easily from the plasma medium of the torch and «delivers» information about the processes that occur inside the torch to the «consumer».

As we mentioned above, the possibility of the detection of circularly polarized radiation in the plasma torch may become crucial evidence for understanding the symmetry breaking in this medium. In this connection, of special interest is the hypothesis of G.Sholin concerning this problem. G.Sholin is an expert in plasma polarimetric spectroscopy who works at the Institute of Hydrogen Energy and Plasma Technology Institute of the Russian Research Center "Kurchatov Institute" and has years-long experience in such problems (Sholin, 1968; Sholin and Oks, 1973; Baronova et al. 1999, 2003).

Thus, according to G.Sholin, an np-1s type resonance transition may occur in the plasma torch and the emission produced by this transition can be circularly polarized if observed from a certain direction.

This is resonance emission and it may be generated in the transition region between the torch plasma and neutral gas or unperturbed atmosphere.

For these reasons, emission may to be «trapped», and in this case it becomes very difficult for it to escape from the formation region.

The strong anisotropy that develops in the torch as a result of the combined effect of magnetic field and linearly polarized radiation in the presence of circularly polarized radiation in the recombining plasma of the torch may result in the formation of molecular compounds with strong symmetry breaking beginning with the synthesis of monomers.

G.Sholin believes that observations of circularly polarized emission using modern diagnostic techniques should be very difficult to perform for a variety of reasons. The most important ones are the very low yield of this radiation from the plasma and the fact that it can be generated at wavelengths that do not exceed 100 nm.

To detect and record radiation with such a low intensity, new techniques for recording of circularly polarized radiation should be developed or the already available techniques should be improved. Recording the aforementioned circularly polarized resonance radiation, which, in addition, is in the "trapped" state, would also require a substantial increase of the sensitivity of the existing diagnostic equipment.

6.5. The «Scheme of Genesis»

In the previous chapters we analyzed the natural processes that could accompany a hypervelocity meteorite impact and ensure the synthesis of simple and complex organic compounds that led to the emergence of primary forms of living matter on the Earth. We also showed in these chapters that without various natural mechanisms ensuring the breaking of the mirror symmetry impact processes cannot ensure the synthesis of extended homochiralic polymer molecules needed for the emergence of life in its most primitive form.

The breaking of mirror symmetry under natural conditions is the subject if this (sixth) chapter. We show here that the torch plasma generated as a result of an impact may possess important properties, which could ensure the formation of homochiralic polymer structures.

If the arguments presented in this book to support the realizability of various processes and mechanisms are considered separately, it becomes clear that they are not equally reliable. Thus, along with highly reliable results based on experimental evidence for the synthesis of organic compounds in the torch plasma, we also consider the possibility of the formation of local chiral physical fields, which has the status of a hypothesis, albeit well founded. Note that the data concerning the possibility of the formation of local chiral fields were also obtained experimentally, but as a result of other works performed by other authors. It would therefore be appropriate to verify and confirm the correctness of their applicability.

We show below that such processes are numerous and that they all are different links of the same event of extreme importance – the generation of the first living being. Therefore the justification of the use of these processes for the reconstruction of such an important event may further clarify the event as a whole and the correctness of the applicability of these processes.

That is why the idea of the so-called « scheme of genesis» started up. Such a scheme consisting of a sequence of events in the chain leading to the emergence of living matter could help find solutions explaining how the crucial processes operate in nature. These include natural processes, which facilitate the realization of individual stages of the proposed concept.

In this connection, of special interest is the asymmetry of amino acids found in Murchison meteorite. If the proposed mechanism of the synthesis of organic compounds in the plasma torch of a hypervelocity impact is realized in nature, it must ensure that, e.g., amino acids both in space and in the bioorganic world in the Earth have the same asymmetry "sign". Such an asymmetry must also be observed in laboratory simulations of the processes of the synthesis of amino acids in the torch plasma.

The predominance of L amino acids in the bioorganic world is a well established fact. At the same time, the analysis of the "sign" of the asymmetry of amino acids contained in Murchison meteorite showed the predominance of L-enentiomer of alanine. In particular, the D/L ratio for alanine was found to be 0.85±0.05 (Engel et al., 1990). The D/L ratio for another, glutamic amino acid is equal to 0.54. According to the above authors, the experimentally observed symmetry breaking cannot be explained by terrestrial contamination of the sample. The result obtained provides the first evidence suggesting that organic compounds both on the Earth and in space, including those found in interstellar gas and dust clouds, have possibly been synthesized in identical physical processes, namely, in plasma torches of hypervelocity impacts. The significance of this result is beyond question, however, it is yet unclear how it could be used in composing the "scheme".

A preliminary analysis of the «genesis scheme» discussed here assumes that it must be a comprehensive scheme. Or a scheme of the interaction of physical and plasmachemical processes that under natural conditions could result in a chain of events starting from the synthesis of organic compounds from nonliving substance to the emergence of matter possessing some signs of vivication, and that would provide a medium for the survival and evolutionary development of such systems.

Hence the «scheme of genesis» must include all stages from the meteorite impact to the completion of the formation and heating of the crater. In this book we describe, as far as possible, how we reproduce these processes under laboratory conditions or discuss the results of other authors concerning these events. It is pertinent to note that we do this for all processes except for the «animation of matter». It is a very important prerequisite because the main task of the book consists in creating the conditions for the generation, and after the generation, the conditions for survival, whereas the process of integration proper of the primary being can be arbitrarily chosen among numerous hypothetical models proposed by other authors.

This also means that I have not got my own model of the primary living being and do not propose such a model, and have no intention to develop such a model. I therefore assume the responsibility solely for the medium and conditions. The other aspect of the issue consists in that among all the existing models I choose protoviroid (Altstein, 1987) as my working model because it is the easiest to understand, "palpable", and promising.

Figure 37 shows the approximate scheme of the physical and plasmachemical processes that develop in the torch plasma and the possible ways of their interaction that ensure the creation of conditions necessary for the generation of living matter. We assume that the emergence of living matter could have occurred in both the plasma torch and in the impact crater at a substantial depth under the surface. All nonbiological processes on the scheme are enclosed into rectangles and only the process of the «animation» of matter is shown by a circle.

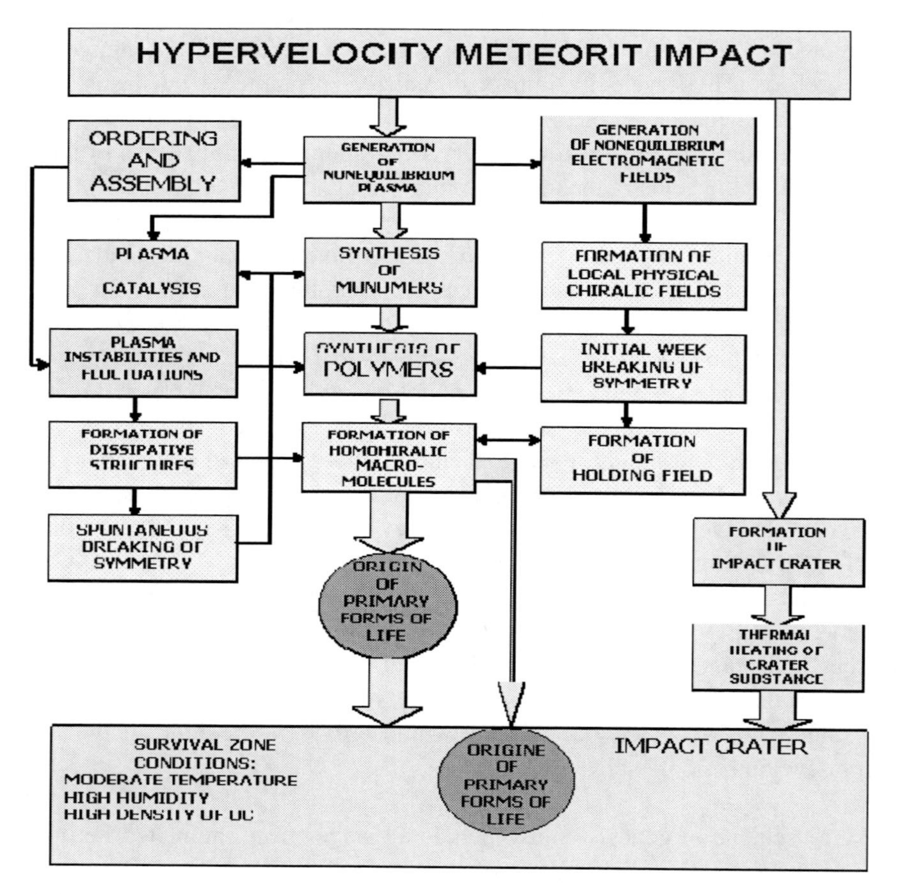

Figure 37. Scheme of the interaction of processes which could ensure the conditions for the emergence of primary forms of living matter in plasmachemical reactions in a torch or in an impact crater in the case if homochiralic macromolecules get to the crater. The green background means that the realizability of the corresponding processes in nature has been experimentally demonstrated or that there are experimental indications of such a realizability. The yellow and red background indicates that the corresponding processes are hypothetical, however some of them are supported by certain experimental background.

A joint analysis of the processes represented on the « scheme of genesis» and of the possible ways of their interaction indicates that the proposed model of generation is based on real physical phenomena that occur in nature and is therefore realizable and viable.

The probability of the emergence of living matter in the order proposed by the scheme will probably be determined not by the preparation of conditions for the emergence, but by the probability of the formation of a viable structure. This in the hypothetical model of a protoviroid this probability is estimated as a rare event and is mostly a result of the stochastic process of the combination of progenes. The minimum probability of such an event, which is approximately equal to ~ 10^{-35} (Altstein 1987), is limited by the number or progenes that formed on the Earth during several hundred million years preceding the emergence of the first living being – a protoviroid. Hence we should not count on the possibility of reproducing primary forms living matter under laboratory conditions in the foreseeable future. Today we cannot compete with nature neither in terms of time scales, charactersitic sizes, or the number of experiments. It will become possible to circumvent this problem only in the future provided that a possibility in principle are found and appropriate techniques are developed that will make it possible to perform dedicated experiments aimed at laboratory modeling of the processes of the emergence, under the conditions with specially chosen, accurate characteristics of the initial matters and parameters of the medium.

We are currently facing certain difficulties associated with the study of the processes of symmetry breaking in the plasma torch. In particular, we know nothing about the size of the «synthesis zone» in the plasma torch and its location. No theoretical principles have been developed for «dissipative plasma» structures because the authors of such hypotheses studied nonequilibrium processes based chemical reactions exclusively. At the same time, they did not consider at all the plasmachemical reactions and plasma instabilities observed in the torch plasma. Thus explosive phenomena, which occur in the case of the separation of free radicals and which I.Prigogine considers to be a subject of a new field of research – «fluctuation chemistry» – are practically undescribable in theoretical terms. Therefore under the circumstances plasmachemical methods will remain the only possible way – at least in the years to come – to study the laws of «fluctuation plasmachemistry» that can be reproduced in the plasma torch. Studies of this kind should planned with the allowance for the fact that the possibilities of performing a full-scale magnetic experiment involving the reproduction of impact processes in head-on collision type space experiments may facilitate substantially the solution of many complex tasks associated with the specifics of the expansion-away of nonequilibrium torch plasma.

Note that some of the results and conclusions reported in this chapter are of preliminary nature and need bona fide experimental verification. The point is that we are now at the initial stage of new and promising studies and many interesting results and scientific discoveries await us.

***

The hypothesis proposed in this chapter concerning the generation of local chiral physical fields in the plasma torch of a hypervelocity impact, which has got a preliminary experimental confirmation, is a result of great importance for understanding the cause of the breaking of the mirror symmetry of bioorganic world. This result may, at least in the future, clarify the role the processes of the synthesis of organic compounds in the plasma torch could have played in the plasma torch. These results may also indicate that the breaking could have occurred at the initial stage of the chemical evolution.

Experimental verification of the idea that plasma torch possesses a hitherto unknown property that makes possible the synthesis of organic compounds transforms this hypothetical process into real. The same conclusion applies, with some restrictions, to the processes of mirror symmetry breaking in the torch.

The possibility of the synthesis of organic compounds in the impact-produced plasma torch is now not in question because the physical and plasmachemical processes of this phenomenon and mechanisms are well-understood. Moreover, they have sound experimental confirmation.

The presence in the plasma torch of physical factors meeting the conditions to local chiral physical fields capable of producing the symmetry breaking could be viewed as important, albeit indirect evidence for the nonaccidental nature of the resulting combination of these processes. The simultaneous occurrence of these processes is efficient in energy terms and typical for many natural phenomena. The combination of the synthesis of organic compounds with the process of symmetry breaking can be viewed as interrelated and synchronous natural processes that have determined the character of bioorganic world.

For the plasma torch to be generated as an inevitable process accompanying a hypervelocity impact, a minimum number of initial conditions should be met, which are automatically realized at the stage of the birth of planetary systems. Thus the velocity of the impact is provided by the gravitational field, whereas the matter for the synthesis including carbon is contained in the meteorite. It follows from this that the process should be both very likely and highly efficient.

The formation in the plasma torch of «truly» chiral factors in the form of a composition of magnetic fields and linearly polarized radiation, or the generation of circularly polarized radiation, which result in the formation of local chiral physical fields can be viewed as inevitable events. It is therefore unlikely for the initial symmetry breaking in the torch to have no effect in the form of a minor breaking of enantiomer symmetry.

The discovery in the plasma torch of initial conditions that can result in the formation of a kind of dissipative structures and produce a strong breaking of mirror symmetry via spontaneous processes should be viewed as a result of great importance.

Hopefully, further in-depth analysis of the specific features of the proposed mechanism of the synthesis of organic compounds in the plasma torch with symmetry breaking may clarify the true viability and value of the new concept.

A joint analysis of the results of theoretical and experimental studies described in this book will make it possible to formulate some preliminary conclusions about the processes that occur in the torch plasma and favor the breaking of mirror symmetry. It is important that these natural processes occur simultaneously with the synthesis of organic compounds in the same medium. The natural asymmetry of these processes is indirectly evidenced by the results of some experimental studies. Some authors report observing such properties as features. Therefore such information should not be ignored, especially because these conclusions, albeit hypothetical, appear to be quite reasonable.

We currently do not know how strong could be the breaking of mirror symmetry in the plasma due to local chiral physical fields of the torch. This breaking could be weak; however, the symmetry breaking of this kind could play the crucial part in the formation of the asymmetry «sign» of bioorganic world.

This mechanism could trigger the symmetry breaking observed in nature and could have caused this breaking because the physical fields of the torch possess innate asymmetry, whose polarity is the same for the entire Earth, Solar System, and the Universe.

Because of their unidirectional nature, the local chiral physical fields that arise in the torch may have played the role of «holding field» for spontaneous processes and determine the unique asymmetry "sign" anyplace where life could have emerged.

Spontaneous symmetry breaking in the torch plasma could, in turn, facilitate the formation of homochiralic molecular structures.

Today, a number of mechanisms have been proposed that are capable to produce spontaneous breaking of symmetry. Some of these mechanisms have been experimentally justified. Such processes can also occur in strongly nonequilibrium plasma that is far from the thermodynamic branch of equilibrium and that represents a kind of Prigogine's dissipative structures. These processes could have facilitated strong symmetry breaking during the synthesis of enantiomers and ensure the assembly of homochiralic molecular structures of organic compounds with the asymmetry "sign" possibly determined by the local physical field.

If the hypothetical prerequisites described above become experimentally confirmed in the future, this would mean that the main idea of the new concept is true and that plasma torch can indeed be identified with the medium where the primary form of living matter could have emerged. However, this will become possible only after the main «bastion» of the eternal problem falls – i.e., after a natural mechanism of the asymmetry of living matter is found. This has become the chief task.

Chapter 7

EXTRATERRESTRIAL LIFE AND HYPOTHETICAL SCENARIOS OF ITS ORIGIN

7.1. "EXTREME" BEINGS AND LIFE BEYOND THE EARTH

In the previous chapters we discussed the results of early works, analyzed them and compared them with the new concept of the emergence of the primary forms of living matter in the processes of a hypervelocity meteorite impact. The experimental evidence presented there pointed to the possibility of the realization of such a scenario on the early Earth. We also showed that to ensure the viability of the proposed "impact" mechanism, a minimum set of initial conditions is required and these conditions proved to be rather easy to provide. These circumstances naturally focused the researchers' attention on the problem of the applicability of the processes of a hypervelocity impact as a medium conducive to the emergence of extraterrestrial life, in particular, on other planets and satellites of planets in the Solar System. This chapter is dedicated to this important problem.

It follows from the currently available material evidence about the situation on the early Earth that all the conditions necessary for the emergence of life were ready several hundred million years after the formation of our planet. At that time the temperature on the Earth was moderate, there was water and primordial atmosphere to keep it. Under such conditions, life on the Earth could emerge in different natural processes including the processes of a hypervelocity impact. Extra conditions necessary for the generation of a plasma torch were in this case provided due to the high velocity of the meteorite impact, which amounted to $v_{IMP} > v_{CR}$.

During the formation of the Earth carbon could be delivered as a constituent of meteorite bodies, which mostly consisted of planetesimals and cometary nuclei. The carbon content in these bodies amounted to 3 to 15%, but they also contained all the other elements needed for the formation of organic compounds and used to sustain the vital functions of any organisms. The presence of water on the early Earth indicated that the moderate surface temperature of the planet ensured the preservation and accumulation of organic compounds synthesized in various natural processes.

When we analyze the possibility of the emergence of extraterrestrial life, it is of special interest to estimate the probability of the development on other planets and planetary satellites of the Solar System of initial conditions necessary for the realization of a process similar to the one that occurred on the Earth. It is also of interest, in this connection, to find to what extend the use of the opportunities provided by the hypervelocity impact mechanisms were needed to this end.

As is well known, the initial conditions on neither of the Solar System planets of planetary satellites can match the "comfortable" conditions on the Earth. Because of the low surface temperatures of these bodies neither of them contains liquid water. The lack of water excludes most of the numerous known terrestrial scenarios of the emergence of life, i.e., practically all scenarios except the one proposed in this book, which is based on the mechanism of a hypervelocity meteorite impact.

There are reasons to believe that on many planets and planetary satellites the peculiarities of the processes that accompany the hypervelocity impact and the properties of the plasma torch could have made possible both fast emergence of the most primitive forms of living matter and the development of conditions necessary for the onset of the terrestrial-type stage of chemical evolution. The extraterrestrial life that originated on other cosmic objects could have survived as a result of the surprising capability – known for terrestrial microorganisms – of preserving their viability under extreme conditions. Terrestrial-like life supposedly could not emerge under low-temperature conditions at the surfaces of many cosmic objects, e.g., under close to present-day Martian conditions. However, life could develop in deep layers of celestial bodies, in the water lakes located in these layers and created by the tidal forces, or in the deep damp geological rock layers heated to moderate temperatures and rich in organic compounds synthesized as a result of impact processes. Such a combination of initial conditions could evidently also be realized on the early Earth and make possible the emergence of life – first and foremost, under the Earth surface.

Thus we should better start the study of the possibility of the existence of extraterrestrial life by analyzing the remarkable capability of terrestrial microorganisms to survive under extreme conditions.

In recent years, many studies have been made of microorganism communities that maintain their vitality under various, seemingly lethal conditions. These results suggest that, in particular, microorganisms are capable of existing not only at the surface and at the subsurface Earth layers, but also in deep sedimentary rocks and deep in the bottom ground of the ocean. The lack of light and oxygen in these habitats is often combined with superhigh ambient pressure and high temperature. At the same time, it is possible that the microorganisms found could have originated under these very extreme conditions. Microorganisms have been studied that prefer harsh Arctic and Antarctic conditions, inhabit ancient ices and permafrost that have not thawed for millions of years. Such microorganisms, which can exist under extreme conditions of the "death zone", have been called "extremophiles".

The attempts undertaken so far in order to find mechanisms that could ensure the emergence of life in deep water bodies or in the Earth crust interiors were hindered by the lack of efficient mechanisms that could, via synthesis processes, lead to the enrichment of these domains with organic compounds or ensure the transportation of these compounds to such difficult-to-reach zones. This implies that the emergence of life under the surface of the planet should be based on hitherto unknown processes and mechanisms. Therefore the hypervelocity impact configuration considered in this book – the configuration, which could make possible the introduction into the subterranean water bodies or into deep geological layers of a part of the organic compounds synthesized in the torch - is of special interest. It is also important that the penetration depth of organic compounds in impact processes may be quite large and amount to 15—30 km for 5 to 10-km large meteorite bodies.

The new approach to the origin-of-life problem extends substantially the application domain of the proposed mechanism of the synthesis of organic compounds in hypervelocity impact processes and ensures the delivery of the final products of the synthesis produced in the plasma torch of the impact to a place where they can be preserved. Such places may be located in deep, warm, and moist layers of the celestial body. However, the synthesis of organic compounds in other, "nonimpact" processes should be rather inefficient at low and ultralow surface temperatures on planets and their satellites.

The heating of deep water bodies or geological rocks on celestial bodies can be driven by tidal forces or a meteorite impact. Tidal forces, which act permanently, prevent the freezing of water. However, if water has frozen, the impact energy can thaw it and, depending on the mass and velocity of the projectile, this energy would maintain the temperature at the level needed to produce liquid water over a long time interval. A simple formula for estimating the crater cooling time as a function of the meteorite diameter can be found in Chapter 2. According to these estimates, the crater cooling time scale should be equal to 10 million years for a 10-km diameter meteorite impact. When this time elapses, water freezes again and the medium becomes "preserved" along with the organic compounds contained in it until the next impact. Therefore some of the "tideless" bodies of the Solar System must contain ice "record" displaying the process of progressive complexification of primary forms of living matter and organic compounds synthesized in the torch. We will hardly be able to "read" such unique records on Jovian icy satellites or in cometary nuclei to see what were the first living beings, but perhaps our offspring will.

7.2. Concentration of Synthesis Products in a Penetrating Impact

The processes of the formation and expansion-away of the plasma torch discussed in Chapter 3 referred to the case of a half-infinite target. In this case the depth of the craters produced both by a laser exposure (H_L) and by impact (H_{IMP}) was much less than the thickness H_D of the target.

If H_L or $H_{IMP} > H_D$, the target is destroyed or punched through and two plasma torches form: one at the face side (i.e., at the side facing the external action) and one on the back side of the thin target.

Data are available for the parameters of the ion flows escaping from the face and back sides of the thin target. Thus Oporew (1967) found that in the case of a 5-μm thick punched-through target the energy and number of ions escaping the target from its back side amounted to 70 and 25% of the energy and number of ions escaping from the face side, respectively. For a thicker (30-μm thick) target the corresponding fractions proved to be equal to 40 and 25%, respectively. No ion flux from the back side was recorded for a 60-μm thick target.

This effect was used in various laser technologies and, in particular, in LASMA-500 time-of-flight mass-spectrometer, to perform elemental and isotopic analysis of biological objects with the sample location selection accuracy to within 1-μm. In the instrument just mentioned the plasma ions of the sample studied escaping from the back side of a thin target penetrated by laser radiation are introduced into a time-of-flight mass-analyzer.

It is important that in the configuration considered the mass compositions of the plasma ions of both torches are identical and the diameter of the hole in the target coincides with the diameter of the laser focal spot.

A number of authors studying the punch-through of a metallic foil by a high-power laser found, along with the plasma torch, submicron-sized particles of molten matter on the back side of the target. These particles may have above-critical velocities and were therefore used as projectiles in laboratory experiments performed to study the physics of hypervelocity impacts.

Consider now the interaction of the meteorite with a finite-thickness target in hypervelocity impact processes. The results of numerical simulations and observations of the existing impact craters on the Earth surface can be used to determine, to a coarse approximation, the dependence of the crater depth on the diameter of the projectile.

The diameter and depth of the intermediate crater are known to exceed the diameter of the projectile by a factor of five and two-to-three, respectively. Hence we find, for characteristic sizes of impacts, that a ~0.5-km diameter meteorite moving at a velocity of 15-20 km/s should punch a ~1.5-km diameter hole in the (rock or ice) target in the case of a 2.5-km initially thick layer. Note that it is a very coarse estimate because the characteristic size of the result of an impact should depend on many factors including the hardness of the projectile and target material, the incidence angle and size of the meteorite. However, in the process of the penetration of the target by the projectile the meteorite matter converted into hot and dense plasma produces two torches – on the face and back sides of the target - moving in opposite directions.

The processes of the expansion-away of these two torches are very likely to be similar and, provided certain proportions between the size of the projectile and the thickness of the target, about half of organic compounds synthesized the torches should end up on the back side of the target, i.e., they should move perpendicularly away from it in the direction of the projectile velocity vector.

The fact that the torch expansion-away processes are identical also implies the complete symmetry of their electric and magnetic fields. In this case the chirality "sign" of the isomers that form in the torches should be preserved whatever the direction of the motion of the torch plasma relative to the target.

Thus in the case of a punch-through by a meteorite of rock or ice layer the expansion-away of the plasma torch and organic compounds synthesized in it should occur perpendicularly away from the face and back sides of the target. In this case the breaking of the symmetry of synthesized isomers should have the same sign of "chirality" for both torches.

To understand the scale of the meteorite impact, let us consider the characteristic sizes of known craters produced by impacts of large meteorites.

The thickness of the continental Earth crust, where the energy of impacts is mostly "damped", ranges from 30 to 50 km. Experts estimate that the Yucatan meteorite that fell on the Earth about 65 million years ago in the Gulf of Mexico area had a velocity of 12—15 km/s and a diameter of 14 to 20 km. The Chicxulub crater produced by this meteorite has a diameter of 180 km (the diameter of the intermediate crater is ~60 km) and a depth of at least 30 km. It is known that during the meteorite bombardment, during the first 100—200 million years, the Earth was hit by a total of 10^7 meteorites of 10-km diameter and each of them punched at least a 30-km thick layer.

The high intensity of the meteorite flux onto the early Earth and the resulting penetrating impact processes could have made possible the accumulation, preservation, and concentration of organic compounds under the surface. This process could have been facilitated by underground lakes and ice-covered water bodies, which are not uncommon on modern Earth. Underground and subglacial lakes on the Earth are usually located in the crust and on ice-covered continents, respectively, like, e.g., Lake Vostok in Antarctic. It is safe to suggest that such local formations must have also been very common on the early Earth, where they have played the crucial part in the preparation of regions capable of ensuring the survival of the first organisms.

As we pointed out above, all the necessary conditions for the emergence of life at the Earth's surface were also already in place from the first several hundred million years since the formation of our planet. Thus the synthesis of complex organic compounds was made possible by hypervelocity impact processes and the appropriate temperature and water at the Earth surface. Life simply could not but emerge under such favorable conditions. However, can the hypothesis about the emergence of life at the Earth's surface be considered to the only plausible one or the chemical stage of evolution proceeded in parallel at different places, under different conditions and via different processes?

According to G.Joyce (1989), the answer to the question about the origin of life "if it ever known, will not be a single statement…". In this connection it would be of interest to answer the question of whether life could emerge on the Earth in confined water bodies, under ice and rock layers via the mechanism of the synthesis of organic compounds in penetrating meteorite impact processes. The answer can be found at the end of this chapter in the paragraph where we consider a hypothetical scenario of such an event.

One of the most important properties of ice is known to be that its density is lower than that of water. Therefore when water freezes and transforms into ice, the latter emerges and produces a protective layer above the water surface preventing the complete in-depth freezing of the water body. Therefore the presence of water under ice should be viewed as a characteristic feature of water basins. The formation of subglacial water bodies is favored by longitudinal and seasonal temperature variations. That is why there are many places on the Earth with under-ice areas filled with water. There are also many underground lakes. Water temperature in these formations may vary from -3 to +15 C, which is quite suitable for newly formed microorganisms to survive.

Therefore of crucial importance may have been a combination of intense meteorite bombardment on the early Earth with the possibility of the introduction of the synthesis products originating in the plasma torch of a penetrating meteorite impact into deep-seated water bodies or into damp rock layers. In particular, such external factors could have provided real conditions for the formation of the most primitive forms of living matter and ensure that these structures end up in an environment that makes possible their survival and evolution.

During the penetrating impact through a target located above the water body the high pressure in the nascent torch at the initial stage of the formation of the hot spot could have pushed organic compounds synthesized in the plasma directly into water. A photo of such an event reproduced via a laser pulse was used to produce the hypothetical image of a penetrating meteorite impact onto Europa shown in Figure 39.

Such a configuration made it possible to preserve in water practically the entire mix of simple and complex organic compounds synthesized in the impact to be further integrated into more complex compounds. Intense impact heating of the medium for a substantial time period could serve as the main source of energy for chemical and biological processes. When the temperature in the impact region was negative, the upper water layer, when it froze to produce a protective ice cover, allowed water to preserve the heat produced during the impact. This provided the necessary conditions for further evolution.

The experimental results presented in this book allow us to support the proposed concept stating that the first primitive organisms could have originated just in such impact processes. These organisms, which were initially located at large depths and under harsh conditions, could become the most ancient progenitors of modern "extreme" beings or "extremophiles". Curiously, the "impact" concept "fits" rather harmoniously into the scenario of the origin of both terrestrial and extraterrestrial life.

Hence the physical and plasmachemical processes of a meteorite impact could ensure the injection of organic compounds synthesized in the plasma torch into deep-seated water bodies or moist rock layers. Under certain circumstances, such external action could have created the conditions for the onset of the chemical stage of evolution, and under other circumstances, e.g., in the case of the formation of the most primitive forms of life in the plasma torch, favor their survival and evolutionary development.

It is evident that the same or similar process could be realized on other planets or planetary satellites in the Solar System provided that they have in their deep layers water bodies covered by a rock or ice layer. For this process to be realized, the thickness of the covers above such structures should be "punchable" for hypervelocity meteorite impact processes. The thickness of the layer that is punchable for impacts with above-critical velocities could be rather large and exceed the diameter of the meteorite by a factor of two to three. Therefore during the formation of the Solar Systems considerable depths with water bodies located below them in the form of near-surface lakes and seas possibly heated to moderate temperatures by tidal forces were accessible on most of the planets and planetary satellites for major meteorites with diameters ranging from 5 to 10 km.

7.3. Mars and Microbial Life

Mars is traditionally believed to be the most suitable planet for the emergence of life second only to the Earth. This statement can now be called into question and attributed to the historical specifics and current conditions at the surface of this planet. Reliable information about the conditions on Mars and on other planets and planetary satellites of the Solar System can be found in many early and current publications (Moroz, 1978; Ksanfomaliti, 1997).

The low density of the Martian atmosphere holds out no hope of the existence of terrestrial like life on its surface. Hard solar UV radiation not only can kill microorganisms or primitive plants, but also destroy any life-related organic compounds. This conclusion is based on the data provided by the onboard instruments of the Viking mission landing module mounted there in order to find at least a single experimental fact that could be explained by microbial activity or traces of such activity in the past (Goldsmith and Owen, 1980).

Subsequent missions also proved to be disappointing in this respect. However, we cannot say conclusively that there is no life on Mars.

Our hope that microbial life may nevertheless exist on Mars stems from the discovery of methane in the atmosphere of the planet. This gas must be unstable and have a very short lifetime in the Martian atmosphere. Its rather high abundance in the atmosphere implies that it must be permanently resupplied from somewhere. There are only two possible sources of methane resupply: volcanic activity, which was not found to currently exist on Mars, and microbial life in deep-seated layers of the planet. Martian microbes, like their terrestrial brethren, could produce methane from hydrogen and carbon dioxide in the absence of oxygen.

The results of the discoveries of methane in the Martian atmosphere appear reliable because they were made via independent measurements with major ground-based telescopes and onboard measurements of the ESA "Mars Express" spacecraft.

These results may be directly related to the above processes of the injection of complex organic compounds into deep-seated water bodies, ice deposits, or rock layers during hypervelocity meteorite impacts. The point is that, according to the results of the most recent studies of the planet, underneath the arid and lifeless desert at its surfacet there may be underground lakes within the reach of a meteorite impact. Organic compounds injected into such water bodies during heavy meteorite bombardment at the early stage of the formation of planets could have created the conditions for the development of methane-producing microorganisms, which may have survived until now. This hypothesis is not entirely devoid of sense because it is based on natural mechanisms, which actually operated during the stage of heavy meteorite bombardment and not only on the Earth and Mars, but also on other planets and planetary satellites of the Solar System.

There is another hypothesis that is also capable of explaining the existence of microbial life on Mars, and this hypothesis is associated with the Earth.

The analyses of the initial conditions required for the emergence of life on the Earth in meteorite impact processes showed that the presence of liquid water at the surface of a planet is sufficient for the emergence of living matter. The presence of water at the surface automatically determined the range of temperatures on the planet and implied the existence of a weak atmosphere with a density of ~0.01 of its current value. Hence the presence of liquid water at the surface of a planet could be viewed as a criterion of its, at least microbial, habitability. This hypothesis has a direct bearing on the early Mars, because until the age of 1—1.5 billion years it may have had both a dense atmosphere and water at its surface.

Indeed, according to Martian record (Carr, 2004), there may have been liquid water at the surface of the red planet at the time considered. If these conditions were satisfied then for at least 1 billion years Mars must have also had the necessary conditions for the emergence of microbial life. Life may have emerged both at the surface of the planet and in its deep underground layers. In the former case, when the condition of Mars changed, life could have later "resettled" underground. In this case life on Mars would be hard to exterminate after that.

Hence owing to the processes that may occur in meteorite impacts, the young Mars, like the Earth, possessed all the necessary conditions for the emergence of life. Therefore the probability of finding microbial life on Mars today is by no means equal to zero. In the case of success, irrespectively of whether the beings found are alive or fossilized, the discovery of extraterrestrial life may become the most important event of the third millennium.

The quest for microbial life or traces thereof on another planet using onboard instruments operating in automatic mode has so far been unsuccessful. To succeed, we have to find cells or metabolites and identify the microoragisms. In most of the cases this requires extremely difficult and multistage preparation of the sample, and centrifugation of soil or ice samples with repeated replacement of liquid ingredients and solvents. These studies should include complex chromato-mass-spectrometric analysis with repeated calibrations, which are usually difficult to perform even in ground-based laboratories. However, there are simpler and more easily available methods for the analysis of the samples, which make it possible to first detect signs of life and determine the state of its carriers. These methods are based on mass-spectrometry measurements of the elemental composition of the microbial mass and we discuss them in detail in Chapter 8.

The measurement techniques and onboard instruments for identification of microbial life have been developed for more than 20 years at the Laboratory of active diagnostic of the Space Research Institute of the Russian Academy of Sciences (Managadze, 1994, 2002b). These works are carried out at the same department where the new concept of the prehistory of life has been developed and where the main results described in this book have been obtained. These relatively simple techniques are aimed at finding the signs of life.

Searches for fossil life, which could exist on Mars 2 to 3.5 billion years ago, face extra and serious problems. In this case, to correctly determine the sample intake with high probability, the age of the rocks should be determined where fossil microorganisms can potentially be found. However, such measurements were until recently impossible to perform under space conditions because of the lack of appropriate onboard facilities.

The simplest laboratory technique of the measurement of rock age, which is based on the determination of Pb isotope ratios, is too complex and difficult to perform even under ground-based conditions. A common laboratory measurement facility of such an instrument needs an entire room to accommodate it and weights at least one ton. To be able to perform such measurements on other planets, onboard facility must weight no more than 3—4 kg. Besides a highly sensitive mass-spectrometer, an onboard measurement facility should also incorporate a state-of-the-art chemical reactor capable of preparing the sample onboard the landing module.

The analytical characteristics of the instrument must be equally sophisticated. Thus the enrichment ratio of the geological samples collected should be as high as $\sim 10^4$. Such enrichment ratios are needed to concentrate lead whose content in Martian rocks does not exceed 10^{-6} g/g.

I proposed a laboratory prototype of the onboard instrument capable of determining the age of Martian rocks, which was then developed by the Laboratory of active diagnostics of the Space Research Institute of the Russian Academy of Sciences. The required enrichment of the sample was provided by a tiny chemical reactor capable of completely dissolving the rock sample. In the prototype onboard reactor this was achieved by subjecting the sample to strong 300 C hot hydrochloric acid at a pressure of ~40 atm. We found, in the process of the work, that it is difficult to correctly choose the material for coating the inner walls of the reactor cell: this material should not dissolve when in hot hydrochloric acid at high pressure. Practically all acid-resistant metal alloys dissolved in such an environment, whereas strong plastics began to melt. To address this problem, a new carbon-based compound was developed, which has the form of paste and withstands such exposures.

A complete cycle of onboard measurements made to detect signs of life should include two determinations. These, in particular, include the determination of the age of the rock sample from the Pb isotopic ratio and the determination of the elemental composition of the suspected microbial biomass after its isolation. These measurements can be made using the same mass-spectrometer after the sample is enriched and the required amount of the biomass is selected. Such a procedure reduces substantially the mass of the onboard facility and simplifies the control software. The flight prototype of LASMA instrument can be used as a mass-spectrometer.

In this case, the first to be performed may be the manipulations aimed at extracting unfossilized microbial mass from the ice lattice followed by the measurements of the elemental and isotopic composition of this mass. The instrument can then be used to estimate the age of the rocks located in the immediate vicinity.

The results of measurements will make it possible to determine with high reliability, based on the C, H, N, and O mass-peak rations, whether the sample contains terrestrial-type biomass.

If the answer is positive, the measured P, S, Ca, and K peak ratios (Mulyukin et al., 2002) may provide extra information about the presence of cells and their condition (anabiotic, fossilized). In Chapter 8 we describe in more detail the potentialities of the mass-spectroscopic methods in detecting micrioorganisms and determining their properties based on the elemental-composition measurements.

7.4. What does Europa's Ice Shell Protect?

Europa is the most interesting cosmic object where the new mechanisms of the synthesis of organic compounds in hypervelocity impact processes can be used to their full extent for the emergence of life. This primarily concerns the processes of the introduction of organic compounds synthesized in the torch plasma into the water-filled inner cavity of the giant planet's satellite during a penetrating meteorite impact.

Whereas many scenarios of the emergence of life from inorganic substance under the action of various natural factors have been proposed for the Earth over the past half century, only a few such scenarios based on simple and realistic processes have been proposed for Europa. This is because of the unusual configuration of Europa, which is supposedly covered by a 10 to 30 km thick ice shell. The diameter of the outer ice sphere is equal to ~3122 km. The ice surface hides the solid part of the planet with a diameter of ~2900 km consisting of rock and a small metallic core at the center. The region between the the icy and solid surfaces of the planet is supposedly filled by an ocean.

The surface of Europa is smooth with hills no higher than several hundred meters. It is made up of "young" and therefore clean ice. The age of the surface does not exceed 30 million years and is therefore indicative of its high geological activity. Therefore the traces of intense early meteorite bombardment, which abound at the surfaces of such Jovian satellites as Ganymede or Callisto, have been almost completely obliterated on Europa.

The structure of the biggest surface crater Tyre (see Figure 38) is attributed, in accordance with one of the two possible interpretations, to a penetrating meteorite impact. According to this version, a massive meteorite punched the ice shell, and the icy mess raised by water froze to produce a crater with a wall diameter of ~100 km. The estimates presented in Section 7.2 allow us, if we assume that the diameter of the wall coincides with that of the hole punched in the surface ice layer, to coarsely estimate the maximum thickness of the ice shell. Given that in the case considered the diameter of the meteorite should be equal to ~10 km, it should be capable of punching a ~30-km thick layer if moving at a realistic velocity. The inferred thickness of the ice shell agrees well with other published estimates, possibly because the same initial condition criteria were used in all cases.

Another young crater with a diameter of ~25 km, presumably produced by the impact of a 2.5-km diameter meteorite could not cross the entire ice shell thickness because it penetrated only to a depth of no more than 8 km. In this case the traces of the impact-produced ejecta can be seen at a distance of several hundred km.

The surface of Europa, which undergoes permanent change and is furrowed by numerous intersecting lines, cracks, and fractures, is very cold. Its temperature ranges from -150 to -190 C. Some ice structures are indicative of the possible breakthrough of water through the cracks onto the surface and of cryovolcanic water eruptions. The relief of some surface portions suggests that ice must have been completely molten there at some time and icebergs floated in water. These features of the processes in Europa's icy shell are very important because they may ensure the penetration of organic compounds synthesized at the surface as a result of hypervelocity impact processes into inner cavities and into the ocean.

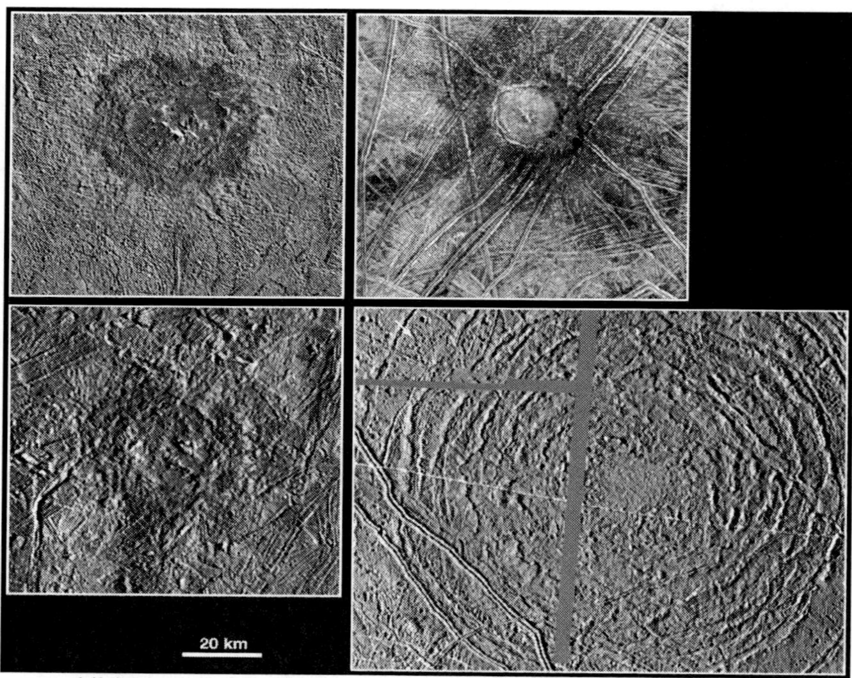

The photos are published at NASA website www.nasa.gov.

Figure 38. Impact craters on Jovian satellite Europa (clockwise from the top left corner) - Pwill, Cilix, Tyre, and Mannann'an – recorded by Galileo interplanetary probe.

Europa lacks atmosphere and the composition of water in its ocean supposedly resembles that of water found in terrestrial deep-seated geothermal or under-ice lakes, where life may exist. There are other, more pessimistic versions of the composition of water in Europa's ocean, which suggest that it may contain acids and may be unsuitable for microbial life. However, the discovery of evidence indicative of penetrating meteorite impacts should be considered to be the most important result of the studies of this satellite.

During an impact onto Europa's icy surface the matter contained in the meteorite becomes completely atomized and 100% ionized. As a result, a high-temperature and ultrahigh pressure plasma torch is created and a through-hole is punched in the ice with a diameter five times that of the meteorite.

This is followed, as we showed in Figure 39, by explosive expansion-away of the plasma cloud and formation of a torch, which expands into vacuum (i.e., into space). The synthesis products partly end up in the ocean after the meteorite punches a hole in the ice shell. This is evidenced by the results of laboratory experiments described in Chapter 5 of this book and published by Managadze (2009). According to these results, a considerable fraction of organic compounds synthesized in the torch and, in particular, carbines, was found at the bottom of the laser-produced crater.

The concentration of organic compounds in Europa's subglacial ocean has until recently been very difficult to estimate. The difficulties were primarily due to the lack of any material evidence or realistic hypotheses concerning the initial conditions needed for such an estimate. Many researchers tried to address these issues. The results of some of these estimates can be found in the papers of Chyba (2000) and Borucki et al. (2002).

The data needed for estimating the concentration of organic compounds in Europa's ocean became available only after laboratory simulations of the processes of the generation of an underwater plasma torch produced by a hypervelocity impact and the experimental determination of the efficiency of the synthesis of organic compounds.

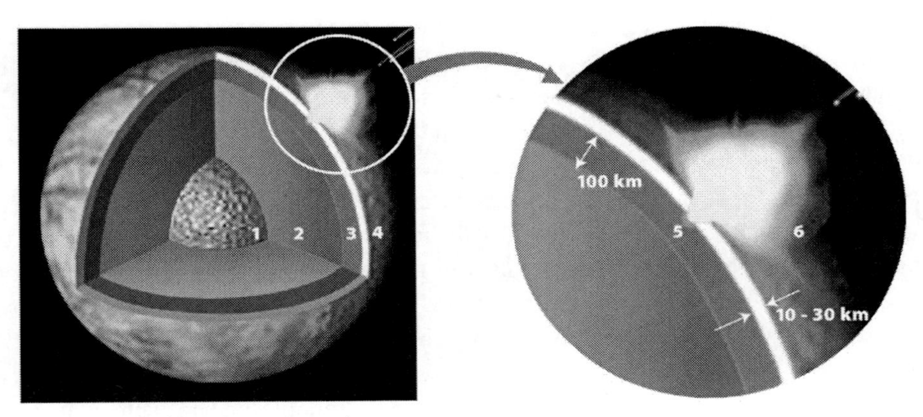

(Source - NASA website www.nasa.gov).

Figure 39. Schematic view of the penetrative impact of a major meteorite onto the Jovian satellite Europa. This schematic sketch uses photos of plasma torches produced in laboratory experiments simulating a hypervelocity impact. 1. Metallic core; 2. Rocky formation; 3. Ocean; 4. Ice cover; 5. Plasma torch as viewed from the back side of the ice cover; 6. Plasma torch as viewed from the front side of the surface of Europa.

Let us now estimate the concentration of organic compounds in Europa's ocean that could have been injected at the early stage of the formation of the satellite in the process of a penetrating meteorite impact. We assume that only 10% of all synthesis products could have reached the ocean and that the thickness of the ice shell above the ocean does not exceed ~10 km and that it can be punched through by a ~3—5 km diameter meteorite.

The Earth is known to have been hit by $\sim 10^{10}$ meteorites of similar sizes during the first 200 million years of its existence. According to published estimates (Pechernikova and Vityazev, 2008), 90% of these bodies were common chondrites, 10% were carbonaceous chondrites, and 1% - icy cometary nuclei. The pattern for Jupiter and its satellites was somewhat different. In particular, because of the strong gravitational field of the giant planet the meteorite flux could have been much more intense. Given that the planet is within the reach of comets, the infall may have consisted mostly of cometary nuclei where the concentration of carbon could have been as high as ~15—25% (Chyba et al., 1990).

For a coarse estimate, we can further assume that the total number of cometary nuclei with the diameters of 3—5 km during the two periods of early and late bombardment of Europa's surface – i.e., during 500 million years – did not exceed the number of meteorites fallen onto the Earth during the first 200 million years of its history. Given the areas of the surfaces of the Earth and Europa, the number of cometary nuclei fallen onto the Jovian satellite could be as large as $5 \cdot 10^8$ and the total volume of the meteorites, with the volume of cometary nuclei taken into account, could amount to $\sim 5 \cdot 10^9$ km^3.

According to another, independent, estimate based on the count of craters on the Jovian satellites Ganymede and Callisto, the total number of cometary nuclei fallen onto Europa's surface during the same 500 million years was equal to $5 \cdot 10^7$. The $5 \cdot 10^7$ - $5 \cdot 10^8$ interval for the number of 3—5-km diameter meteorite impacts appears quite realistic. The total average volume of cometary nuclei could in this case be equal to 10^9 km^3.

If we assume that Europa's ocean is 100-km deep then its volume should be equal to 10^9 km^3. This means that there must be 1 cm^3 meteorite matter for every 1 cm^3 of water. Note that the average concentration of carbon in a cometary core is equal to 10%. Let us assume that the effective yield of one molecule of organic compound, which, on the average, contains five carbon atoms, is equal to 1%. In view of the above quantities, the total concentration of organic compound molecules with molecular masses of 100 a.m.u. in Europa's ocean should be equal to ~0.1%. One must also bear in mind that the concentration of these organic compounds may vary substantially - e.g., from 0.01 to 1% - depending on the initial conditions. Note also that if the lower limit is optimal for chemical reactions, the upper limit appears to be too saturated. One should also bear in mind that this estimate was obtained assuming that organic compounds synthesized in the torch plasma are not destroyed and do not evolve during the period of their accumulation, i.e., during 500 million years. If taken into account, these factors may change considerably the overall pattern of the processes and the resulting estimate.

The result obtained is very important because it is reliable and proven: it is based on experimental data. Hence we can view the results obtained as reliable and credible because they are based on the operability of each individual stage of the proposed process and each individual episode. Their combination produces the final effect, which, in turn, is severely confined by the initial conditions, which can be varied only slightly. Thus the credibility of the results obtained is ensured by known laws of nature, initial conditions based on material evidence, and the data obtained in experimental laboratory simulations of impacts. Furthermore, the use of such data prevents large errors, which usually plague the computations that are made "from scratch". Relative estimates are known to be the most reliable and free of large errors.

The structure and configuration of Europa has a number of exceptional positive features that are worth noting. First, the high credibility of the conclusion that organic compounds are introduced into the ocean provided that the energy and size of the meteorite allow it to punch through the ice shell. Thus in the case of the Earth or Mars such a process may be realized only if the meteorite "happens to hit" an empty cavity or a subsurface lake. In the case of Europa this happens for every impact. Furthermore, unlike what we have on the Earth, in the case of the emergence of any form living matter in Europa's ocean and at any stage of its development it would always be protected against a meteorite attack from space. Impact catastrophes on Europa should always be local in nature because the destructive shocks produced by the impact are rapidly damped in the liquid medium. Therefore no mass extinctions should occur on Europa as a result of meteorite impacts, whereas many such events are believed to have occurred in the past on our Earth.

However, all the above does not mean that Europa's biosphere, if it exists, should not face other serious problems. Thus, e.g., the lack of solar energy may limit substantially the evolutionary processes at the later stages of the development of the biosphere.

The fact that Europa resides in the gravitational field of the giant planet Jupiter should have, to a first approximation, increased the probability of meteorite impacts onto its surface and this could have played a positive role for the synthesis and accumulation of organic compounds. Today, with all the data obtained onboard interplanetary probes compared to the results of laboratory studies, while developing a hypothetical scenario of the emergence of life on Europa we can hardly find a more efficient, reliable, and physically clear mechanism than the one proposed in this book, with such inexhaustible energy capabilities as a meteorite impact. In this connection, it would be justifiable to view Europa as a secure oasis of life residing in an ice fortress and aimed at protecting and preserving the unique subglacial biosphere populated, at least, by microorganisms.

7.5. Could Life have Originated on Enceladus or Titan?

The scenario considered above where in the process of a hypervelocity impact the resulting synthesis products are introduced into geological cavities and subsurface water bodies on planets or planetary satellites with very low surface temperatures may fundamentally alter the deeply rooted concepts about what kinds of cosmic bodies should be searched for primitive forms of microbial life.

According to traditional scenarios, the prehistory of the emergence of life included many stages distributed in time: from the formation of "raw materials" necessary for the synthesis of simple organic compounds from the components of an atmosphere with a strictly defined composition to the combination of monomers of organic compounds into complex polymer chains – precursors of biological macromolecules (Dickerson, 1978). A realization of such a complex scenario is hardly to be expected at the surfaces of Jovian and Saturnian satellites because of their extremely low temperatures.

A number of authors view the discovery of subsurface water bodies on some planetary satellites of the Solar System and, in particularly, on Enceladus (Porco et al., 2005) (see below) as an indicator of the presence of life on these objects without trying to find a mechanism of its emergence. However, the scarce possibilities for the synthesis of complex, and especially, polymer organic compounds at the cold surfaces of such cosmic objects appear to be totally unpromising in the astrobiological context discussed here. For life to emerge in the case considered, a simple and reliable mechanism is needed, which must ensure highly efficient synthesis of organic compounds and, at the same time, the delivery of the necessary elements and inorganic compounds to the domain of origination and survival, which appears to be impossible to achieve without a meteorite impact.

In the presence of water in the subsurface water bodies of cosmic objects the combination of the proposed impact mechanism of the synthesis of organic compounds with the possibility of delivering the substances needed for the emergence of living matter to the subsurface cavities via the energy of meteorite bodies radically changes the situation. Such processes provide virtually everything that is needed: (1) the initial conditions for the formation of components in the form of complex organic compounds that are necessary for the emergence of the primitive forms of living matter and (2) their delivery to their destination.

The proposed concept of the prehistory of life allows us to see in a new light the possibility of the emergence of life on any object in the Solar System including the Saturnian satellites Titan and Enceladus. Therefore of great interest are the most recent results obtrained for these cosmic objects, which point to the presence of indicators of life under the surfaces of these bodies.

The successful "Cassini" and "Huygens" space missions aimed at the study of the "Saturnian world" reignited the interest in the "eternal problem" – the search for extraterrestrial life, at least in its most primitive forms. Both space probes succeeded in this task: "Cassini" while remote sensing of the Saturnian satellites Titan and Enceladus (Porco et al., 2005), and "Huygens" on its mission to Titan involving the successful landing of a probe onto the surface of the satellite (Ksanfomaliti, 2005).

Titan is the only planetary satellite with a diameter 1.5 times greater than that of the Moon. It has a rather dense oxygen-free atmosphere, which consists of 95% nitrogen and 5% methane and has a pressure of ~1.6 atm. Because of its surface temperature of -178 C, it has rivers and lakes of liquid methane with admixture of ethane at its surface. Titan's composition is about half water ice and half rocky material. Such a structure does not rule out the existence of water lakes under Titan's surface.

The presence of methane in Titan's atmosphere leads many researchers to suggest that this satellite may host microbial life. This hypothesis is supported by the fact that if Titan had no methane resupply sources this gas could have been destroyed by solar ultraviolet radiation in 10 million years. However, this is not the case and there is hope that life in appropriate forms may be hidden on Titan at a depth of ~30 km in the hypothetical water ocean.

Given that no form of terrestrial life is expected to be able to originate at the cold surface of Titan, the researchers suggest that it may have emerged in the depths of the hypothetical ocean. This process may have occurred, like on the Earth, during the heavy meteorite bombardment of planets and planetary satellites of the Solar System. Ten to fifteen km diameter meteorite bodies were capable, with a large margin, to punch the 30-km thick rock and ice layer and ensure the introduction of a large amount of organic compounds into these cavities.

According to coarse estimates, the number of 10-km diameter meteorites that fell onto the surface of Titan during the very first 200 million years of its history could be quite substantial and amount to $\sim 3 \cdot 10^5$. The temperature of water in lakes situated a 10-30 km depths could have been above 0 C. The heating at the meteorite impact site could have increased substantially the temperature for a period of ~10 million years so that the chemical processes of the complexification of organic compounds could occur under reasonable conditions and at moderate rates.

The development of a hypothetical scenario of the chemical stage of evolution for the emergence of life on Titan involved the earlier assumption about the existence of a subsurface water body or ocean on this satellite. This assumption appears quite solid given that about half of Titan's volume consists of water ice, which can, via internal sources of energy, yield liquid water without resorting to external impact actions.

The presence of water under the icy surface of Europa and deep in Titan's interior so far remains only a hypothesis. Moreover, we have no bona fide evidence for the existence of liquid water on other cosmic objects. And this is true for all objects of the Solar System with the exception of the Saturnian satellite Enceladus. The most recent results were obtained by the scientific instruments onboard "Cassini" interplanetary probe.

The surface of Enceladus, which consists of water ice, is the cleanest planetary satellite in the Solar System and one of the most interesting Saturnian satellites. With a diameter of 500 km it has a spherical shape. It has an average density of ~1.1 g/cm3 and anomalously light surface whose temperature is of about -200 C. Earlier studies showed that the surface of Enceladus evolves: it exhibits traces of global liquid flows, which destroy the old crater relief that they encounter on their way. The age of craterless regions is estimated at ~100 million years. A few years ago a hypothesis was suggested (Ksanfomaliti, 1997) that Enceladus may host water erupting ice volcanoes with water possibly containing ammonia and methane dissolved in it. Such warm flows supposedly caused melting of old relief features.

"Cassini" space probe solved the puzzle of Enceladus to a considerable extent. Moreover, onboard TV cameras recorded powerful water fountains erupting from subglacial reservoirs in the South Pole region (Figure40).

Water basins on Enceladus are supposed to be located under a ~10 m thick ice layer. The surface of Enceladus is covered by an even layer of pure water ice, which gives the satellite its white color.

After the analysis of the data obtained from "Cassini" some researchers and, in particular, C.Porco, who heads the team that studied Enceladus, put it at the top of the list of candidate life-bearing cosmic bodies.

Enceladus is indeed an ideal cosmic body, where highly efficient synthesis of complex organic compounds could occur in the processes of penetrating impacts of relatively small meteorites with diameters of 4 m and greater. This increased substantially the likelihood of the emergence of life in impact processes on Enceladus. The signs observed at the surface of Enceladus are indicative of at least five stages of geological evolution. They are hypothetically associated with eruptions of warm water flows, which could have been partially due to cataclysms at the surface of the satellite caused by impacts of large meteorites.

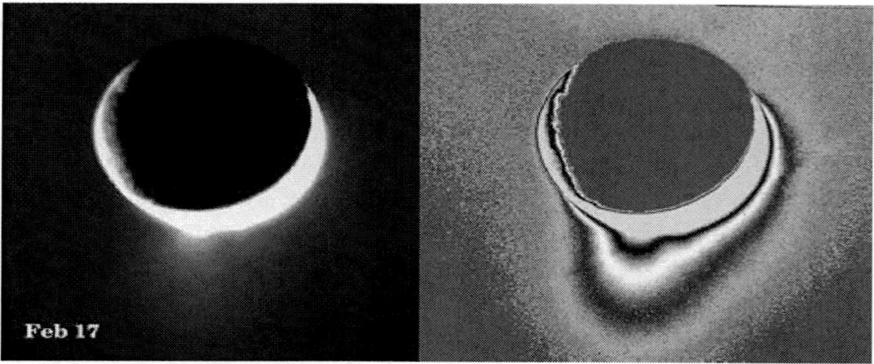

(Source - NASA website www.nasa.gov).

Figure 40. Water emissions from Saturnian satellite Enceladus recorded onboard the Cassini interplanetary probe.

The foregoing leads us to conclude that the initial conditions, which nature succeeded to provide on Enceladus, are ideally suitable for the realization of the synthesis of complex organic compounds in penetrating meteorite impact processes because the resulting synthesis products are very likely to end up in a subglacial water body. This is especially true because another mechanism is hardly to be discovered in the future that would be capable of equally efficiently and easily ensuring the realization of the process of the emergence of life or the preparation of the environment to this process under severe conditions of space.

7.6. Model of Europa and the Synthesis of Organic Compounds in an Underwater Torch

The realization of the configuration of a penetrating meteorite impact is a phenomenon of extreme importance. It is the only natural phenomenon that is capable of ensuring direct introduction of organic compounds synthesized in the torch into deep rock layers or deep-seated subsurface water bodies on cosmic objects.

Until now the researchers did not consider such processes as capable of causing the emergence of life on the Earth, other planets or their satellites, because they had not been studied in sufficient detail. However, this natural phenomenon deserves a detailed investigation because it could have occurred repeatedly both on the early Earth and on planetary satellites with ultracold surfaces provided that they had an ocean or other water resources under their surface.

The experimental confirmation of the possibility of the introduction and accumulation of complex organic compounds, e.g., in the Europa's ocean, in the process of a penetrating meteorite impact could reverse the aversion for the idea of the habitability of this ocean.

As we already pointed out above, of great interest in the analyses of a meteorite impact onto Europa's icy surface is the configuration of the target, which consists of a finite-thickness ice layer propped up by half-infinite water layer. Before its complete destruction a 5 to 15-km diameter projectile would punch through Europa's supposedly 10 to 30-km thick ice crust. This means that hot plasma should be "squeezed" into water by the undestroyed part of the projectile piston, expand in this medium and produce a high-temperature plasma cavity.

This possibility is, at least, supported by the fundamental differences between many important properties of water and ice.

Thus, unlike water, ice at low temperatures is rather viscous and explosion resistant solid medium. The sound speed in water is equal to 1400 m/s, whereas in ice it is equal to 4000 m/s at 0 C and amounts to 5000 m/s at lower temperatures. The thermal conductivity of water is 15 times higher than that of ice. Such a difference between the physical properties of the two juxtaposed media ensures that the processes of the formation and propagation of shocks should also differ. The penetration of the "hot spot" into water under the conditions considered here should result in the development of an underwater plasma formation where organic compounds can be synthesized.

The effect of the breaking of the lower boundary of the ice layer by high-pressure hot plasma in the presence of a projectile piston resembles, in a certain sense, an underground explosion and is also similar to the effect of laser radiation on a target covered by a water layer. Such a configuration can be viewed as a satisfactory, and, possibly, the only possible model of a penetrating impact under terrestrial conditions in the case of a combined target consisting of ice crust adjacent to the surface of a deep water body.

To confirm the possibility of the synthesis of organic compounds in the plasma torch generated under water surface, laboratory experiments were performed to simulate the configuration of a penetrating meteorite impact through Europa's icy crust. To reproduce this process under laboratory conditions, the only possible configuration of a model experiment of this kind was proposed that can be implemented on the Earth.

To this end, 40-mm long 15-mm diameter sealed cylindrical containers were made with antireflection optical windows for 1.06-μm laser radiation. As shown in Figure 41, an ultrapure carbon tablet is placed at the bottom of such a container. The region above the tablet was filled with ultrapure water solution of ammonium nitrate. A ~0.6 J 7-ns long laser radiation pulse was focussed into a 2-mm diameter spot at the surface of the carbon tablet. The surface of the water solution highly saturated with ammonium nitrate was located ~20 mm above that of the carbon tablet. This ensured the concentration of the torch plasma inside the solution and the participation of the matter of the thin surface layer of the liquid located above the tablet in the formation of the torch plasma.

Two identical containers were prepared subject to all the requirements ensuring high purity of the experiment as a whole. Carbon tablets were then mounted in and solutions introduced into these containers. One of the containers was subject to laser exposure, whereas the other container served as a control.

The target has been subject to laser radiation for three hours with a total of 1000 laser exposures made.

After the completion of the experiment both containers were sent to Shemyakin and Ovchinnikov Institute of Bioorganic Chemistry of the Russian Academy of Sciences for mass-spectrometric analysis.

The analysis was made using a Brooker MALDI-TOF-TOF type facility at the Laboratory of proteonics.

The resulting spectra exhibited mass peaks of various intensities up to the masses of ~700—800 a.m.u. associated with noise due to the presence of the matrix material in both samples. Such noise peaks in the mass spectra are common and inevitable for instruments that operated based on MALDI principle.

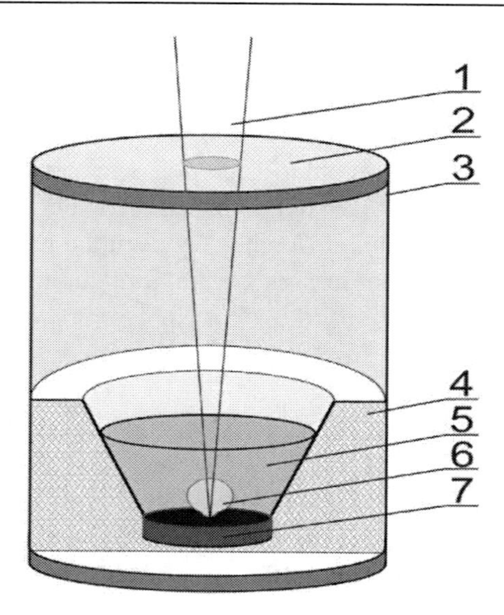

Figure 41. Sealed container for the generation of an underwater torch with the accumulation in aqueous solutions of organic compounds synthesized in the process of plasma expansion-away. (1) Configuration of laser exposure; (2) optical window; (3) case; (4) tablet holder; (5) ammonium nitrate water solution; (6) plasma torch; (7) carbon tablet.

Beyond the mass peaks of the matrix the spectra of the exposed and control samples differed radically. In particular, mass peaks almost disappeared in the control sample, whereas they were observed out to 2000 a.m.u. in the exposed sample.

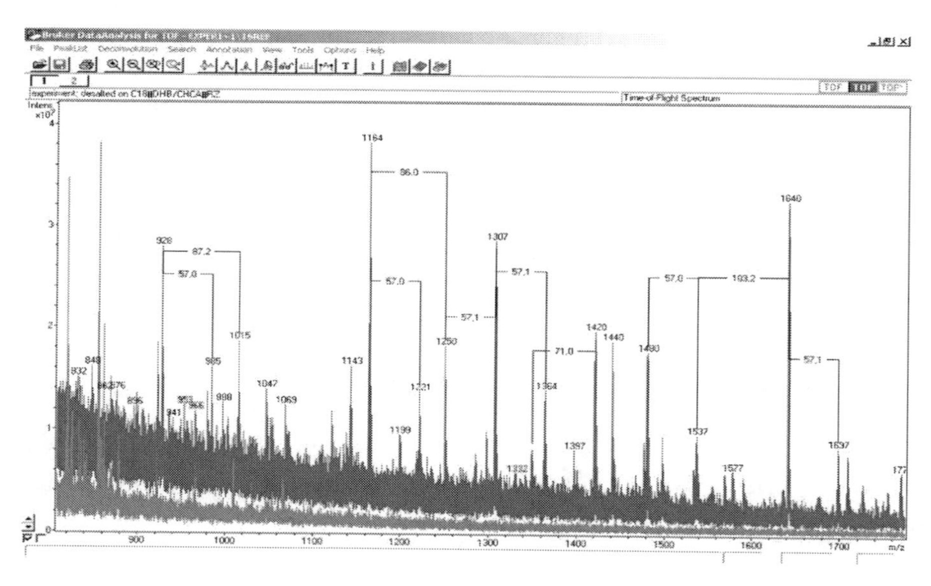

Figure 42. Mass peaks of organic compounds synthesized in the underwater plasma torch as a result of laser exposure of a carbon target and water solution of NH_4NO_3. Some interpeak intervals are interpreted as Gly, Ala, Ser, and Cys. The control measurements shown in this plot exhibit no mass peaks in the interval from 1100 to 1700 a.m.u.

Figure 42 shows the mass spectra corresponding to the working and control samples. These spectra clearly show that the observed mass peaks that arise as a result of laser exposure of the working sample have a characteristic interval with the mass difference coincident with the masses of the bases of some amino acids.

Thus first experimental evidence was obtained that proved the possibility of the synthesis of organic compounds in such a configuration, which can be realized in nature in the process of a hypervelocity meteorite impact.

Despite the great importance of the results obtained, they were to be expected. The point is that the extra pressure amounting to 10^5—10^6 torr produced in the process of the formation of the hot spot can produce a plasma cavern inside which part of the observed organic compounds are synthesized. Later, when the cavern "collapses", these organic compounds may get into the solution, where they remain preserved.

Note that a configuration resembling that of a penetrating impact could have arisen on the early Earth as a result of a direct meteorite impact into water. In this case the diameter of the meteorite should be $D_{IMP} > H_W$, where H_W is the thickness of the water layer (Shuvalov et al., 2008).

In the case of a meteorite impact such that $D_{IMP} < H_W$ a considerable part of the meteorite may remain unfused, thereby ensuring "soft" landing of various organic structures produced in space irrespectively of their complexity. Such a result, if combined with the new possibilities of hypothetical penetration of life, makes it necessary to more thoroughly analyze the possibility of panspermia.

7.7. Where Could have Originated the Primary Form of Living Matter?

The analyses of the problem of the emergence of primary forms of living matter should especially focus on searching answers to the crucial questions: when and how could the first forms of living matter have originated in the processes that accompany a meteorite impact. A hypothetical scenario for this event should be developed so as to provide the best demonstration of the strengths and weaknesses of the new concept and its most vulnerable points.

We currently lack even approximate knowledge of when the event occurred that we should identify with the emergence of life. Can we identify this event of crucial importance with the emergence of the first forms of living matter? When could this event happen? These questions have no unambiguous answers; the answers to these questions usually depend on the particular hypothetical model considered. For the impact mechanism of the origin of life proposed this book we substantiated in Chapter 1 the viewpoint that the emergence of the primary form of living matter marked the time of the emergence of life on the Earth. Two questions remain open: when and where? Below we answer these questions trying, as far as possible, to minimize dealing with biological processes.

According to the results of measurements of carbon isotopic ratios (Schidlowski, 1998), we know for a certainty that a living cell existed on the Earth as early as 3.8 billion years ago. A number of researchers believe that it must have taken about 300 million years for living cell to evolve from its most primitive form to a mature being. This means that life on the Earth could not have originated later than 4.1 billion years ago.

However, a number of experts have a different view and believe that life originated almost contemporaneously with the Earth or perhaps several million years later, i.e., 4.5 billion years ago. Thus the gap between the two hypotheses amounts to 400 million years.

The problem of the time of the emergence of the first living organism embodying the most primitive form of life is linked to a number of initial conditions that have no direct bearing on meteorite impact processes. Thus the volume and surface temperatures of the early Earth could nullify all the advantages of impact mechanisms. Hence the important role of impacts in the processes of the emergence of life should be discussed only if reliable data become available suggesting that local and planetary-scale conditions, on the whole, favored the development of needed links between the events.

To illustrate the potential of the new concept, we must choose among various hypothetical scenarios the one that is most difficult to realize. Currently, this is the variant with early emergence of life. Recall that the experimental results presented above prove the possibility in principle of the synthesis of complex organic compounds in the plasma torch of a hypervelocity impact.

According to modern views (Vityazev et al., 1990), the core and mantle of the Earth formed at the final stage of the growth of the planet. This process developed almost simultaneously with the formation of the primitive crust, hydrosphere, and atmosphere and supposedly occurred from 30 to 100 million years after the formation of the Sun, i.e., less than 25% of the 400-million year "gap" between the two scenarios. It is now beyond question that 75 million years after its formation the Earth had already accumulated ~90% of its mass. This, given the scatter of the initial velocities of the meteorites that bombarded the Earth's surface, provided sufficient gravitational acceleration for the generation of a plasma torch. Thus as early as 30 million years after the birth of the planet the Earth could have already possessed a primitive hydrosphere held in place by the primary atmosphere with a density equal to ~0.01 of the current value and therefore sufficient to this end. The processes of the formation of the planet were accompanied by the processes of the synthesis of polymer organic compounds produced in the torch plasma. Hence the two key components – water and complex organic compounds could have been available on the Earth as early as 30 million years after the formation of the planet. Moreover, dense dust clouds that formed in the process of impact cratering ensured the preservation and accumulation of organic compounds. Therefore the conditions for productive operation of impact mechanisms were in place on the early Earth almost from the very beginning of its existence.

However, the formation of a plasma torch and complex organic compounds did not immediately imply the emergence of life. Could a primitive organism develop in the torch plasma produced by the impact of a large meteorite? Yes, probably.

This was made possible by the properties of the torch plasma and its exceptionally high catalytic activity, which could ensure maximum rates of plasmachemical reactions. The estimates reported in this book, which are based on experimental results, prove conclusively enough that during the adiabatic expansion-away of the torch plasma the rates of plasmachemical reactions could have increased by a factor of 10^8—10^9 compared to those of the chemical reactions that occurred under natural conditions of the early Earth via natural catalysts. Therefore the formation time of organic compounds in plasmachemical reactions in the plasma torch of a meteorite impact would be mush shorter than the time required for the synthesis of such organic compounds in natural chemical reactions.

According to Altstein's estimates (Altstein, 1987), the time needed for the assembly of a ~100 000 a.m.u. protoviroid under favorable natural conditions could be of about ~10 minutes. The results of the extrapolation of experimentally observed impact processes shown in Figure 27 suggest that molecular structures of such a mass can form as a result of an impact of a 1-cm diameter meteorite, and that this would take, given the catalytic activity of the plasma torch, less than 6 μs.

The results presented in Figure 27 may be interpreted in a different way, but the final answer would be the same. Thus it is evident from the experimental plot that a meteorite with an effective diameter of 0.1 mm can ensure the synthesis of molecules with masses up to 5000 a.m.u. Given that the mass of the protoviroid could be of about 10^5 a.m.u., the extrapolation considered implies that the diameter of a meteorite ensuring the formation of such a molecule would not exceed 1 cm.

If estimated using the results of Pechernikova and Vityazev (2008), the effective influx of centimeter-sized meteorites onto the Earth's surface during the first 200 million years of the planet's history could amount to 10^{25}. If we assume that the probability of assembly of ordered molecular structures with the mass of 10^5 a.m.u. was equal to 10^{-9} and that each impact could have produced 10^{19} such structures, the total number of such structures produced during the time period considered would amount to 10^{40}.

Therefore given the exceptional properties of the torch plasma described above, the development of protoviroid-like structures as a result of countless trials with a wide range of initial conditions cannot be ruled out completely.

Experimental studies of the self-organization of matter in the processes considered show that the probability of the emergence of the primary reproductive structure may be very low if the effective yield of complex molecular structures is high. If the attempt of the formation of life would fail because of the small meteorite diameter of for other reasons, other mechanisms could be brought into play automatically. In particular, molecules synthesized in the plasma torch could be injected into the warm and moist fine-dispersed subsurface medium of the impact crater produced as a result of impact processing of rocks and making possible further complexification of the resulting molecular structures.

The energy of an impact of a 10-km diameter meteorite could heat a crater region with a diameter and depth of 100 and 23-30 km, respectively. The concentration of organic compounds with an average mass of 1000 a.m.u. in some parts of the crater could amount to 10^{20}-10^{21} molecules/cm^3 and the cooling time could be of about 10 million years. One would therefore expect that given such a high concentration of organic compounds in the medium, a primary protoviroid-like living system could have formed, survived, and triggered the processes of evolutionary development during the crater volume cooling time.

Hence the emergence of the primary forms of living matter in the processes that accompany a hypervelocity meteorite impact cannot be ruled out even at the initial stage of the formation of the planet. No factors have been found to be able to prevent this event of crucial importance in the history of the Earth. Therefore, given the initial conditions and properties of the plasma torch as the ambient environment, this event appears to be quite realizable even under the extreme conditions of the early Earth. Hence life should have enough time to originate in the plasmachemical processes during the expansion-away of the torch plasma.

However, as we mentioned above, we cannot rule out the possibility of the emergence of life in the impact crater, i.e., in a comfortable medium rich in organic compounds. In this case much would depend on the functional and structural properties of organic compounds synthesized in the plasma torch.

The environment where the first living organism could find itself after its origination should have ensured its survival and evolutionary development. To understand how likely and to which extent this could happen in impact processes, let us consider some physical processes and peculiarities of the formation of the impact crater medium. We discussed these processes in more detail in Chapters 2 and 3. The information presented in these chapters suggests that more than 30% of complex organic compounds synthesized after the formation of the crater could end up at a considerable depth under the soil layer. There may be other ways of migration of synthesized organic compounds on the Earth. However, the penetration of the primary organism into a zone rich in organic compounds should be viewed as the most favorable and promising outcome of the process of the origination of life. In such a natural "incubator" rich in organic "food", i.e., simple and complex organic compounds, the primary living being, which initially lacked enzymatic "digestion ", could survive feeding on the products of impact synthesis. On the other hand, the primary organisms that adapted to "austere conditions", where carbon dioxide and hydrogen were the only food, could have become the progenitors of the most ancient terrestrial chemilithotrophic microorganisms – methanogens, which are capable of living under anaerobic conditions deep underground, without solar energy and oxygen.

Methanogens – the best surviving organisms under such conditions – may inhabit other planets, e.g., Mars or Titan, deep under the surfaces of these bodies. Methanogens and other chemilithotrophic organisms may have been the first living beings because their very limited requirements to ambient conditions made their origination more likely. Such a scenario is supported quite well by modern data from the studies of microorganism habitats on the Earth. Some researchers publish scenarios of the origin of life deep down in the rocks of the Earth's crust (Gold, 1992; Jones, 2004). The concept proposed in this book agrees very nicely with these new ideas and can provide mechanisms for the realization of events of this kind.

We managed, while developing this hypothetical scenario of the origin of life, to answer the above questions, which can be formulated as follows:

- Life could have originated on the Earth 30—100 million years after the formation of the planet;
- The emergence of life was facilitated by the plasmachemical processes that develop in the torch produced by a hypervelocity meteorite impact onto the Earth;
- The products of the synthesis of organic compounds and, possibly, the primary forms of living matter that originated in the torch, continued to exist in deep-seated layers of impact craters;
- Life that originated in part as a result of impact-driven plasmachenical processes in the torch plasma adapted to the conditions in deep-seated layers, mostly by evolving into highly resistant chemolithotrophic forms.

The scenario described above may have its weak points, but it is evidently free of many controversies that are inherent to numerous origin-of-life hypotheses. The main inconsistencies in the origin-of-life scenarios could have been overcome by invoking the results of the studies of plasmachemical processes in the torch plasma and the idea that impact may produce an environment that is extremely friendly to vital synthesis processes. These events on the early Earth could occur regularly and thereby secure the efficiency of the proposed mechanisms.

The new scenario can be realized even on planetary satellites with ultralow surface temperatures provided the availability of positive temperatures, water bodies, and moisture in their deep-seated layers. The conditions in the deep-seated layers of these bodies could have differed little from the corresponding conditions of the early Earth and therefore they could resulted in the emergence of biological life and its subsequent subjective evolution.

The hypothetical possibility of the existence of extraterrestrial life considered here makes it necessary to perform a more in-depth analysis of the panspermia theory in view of the new ideas proposed in this book.

7.8. New Opportunities for Panspermia and the Future of Cryptobiology

The sections of some publications dedicated to panspermia are entitled "Panspermia or autogenesis". Such a title should be viewed as not entirely correct because panspermia is now defined as autogenesis of life beyond the Earth and its subsequent delivery to our planet. Clearly from this viewpoint the main problems associated with the origin of life become even more complicated because it is difficult to imagine and even more difficult to find a place in the Universe that would be more comfortable for the emergence and evolution of life than our planet Earth.

The exceptional properties of the plasma torch produced by a hypervelocity impact considered in this book – the properties associated with the synthesis of complex organic compounds – open up new opportunities for the origin of life beyond the Earth an, in particular, on planets, planetary satellites, and small bodies of the Solar System. In this connection the panspermia hypothesis can be viewed as a realistic and explainable process of the origin of life on various objects of the Solar System with the subsequent delivery of arbitrary forms of living matter to the Earth.

In the proposed concept the origin of primary forms of living matter on some bodies of the Solar System can be a result of the injection – as a consequence of in impact processes - of complex organic compounds synthesized in the torch into deep-seated warm and moist rock layers or into subsurface water bodies on the celestial objects considered.

In this case natural processes provide the conditions required for further realization of the chemical stage of evolution. This means that the experimentally confirmed possibility of the synthesis of complex organic compounds in hypervelocity impact processes is a serious argument for the realization or the initial stage of panspermia beyond the Earth. In particular, the allowance for this possibility may change fundamentally the current views by showing that the conditions necessary for the origin of biological life could and did appear in impact processes in the Martian crust, in the depths of Europa's ocean, in the lakes of Titan and Enceladus, and even in cometary cores (Hoover and Rozanov, 2002; Hoover, 2006).

The arguments stated above are based on the results of experimental simulations of impact processes, which show that the synthesis of organic compounds and the formation of a comfortable zone in the impact crater are inevitable if the velocity of the meteorite exceeds a certain critical level. The results of these experiments can be viewed as material evidence for the realizability of the first stage of panspermia. However, the same studies require a revision of some of the earlier arguments that appeared to support the panspermia hypothesis, because these arguments become counter-evidence when interpreted in the light of the new results.

Thus the fact that the chirality "sign" of the symmetry breaking in amino acids found in meteorites coincides with that of bioorganic structures (Chyba, 1995; Anders, 1989) was used as evidence for the extraterrestrial origin of life. However, the explanation of this phenomenon proposed in this book is based on experimental results and consists in the following: the observed coincidence of the chirality "signs" of the symmetry breaking in amino acids found in the bioorganic world on the Earth and in meteorites is due to these organic compounds originating in identical physical, or, more specifically, impact-driven, processes. Or perhaps the plasma medium where these compounds were synthesized contained the initial natural symmetry breaking due to the presence of local unipolar electric and magnetic fields. Such unidirectional, nonequilibrium fields that develop in the plasma torch could facilitate the development of local chiral physical fields and secure the observed symmetry breaking. These unipolar fields were "fixed" in space and should have reversed their direction in the processes of the torch formation. These fields could have reversed their orientation (by 180 degrees) only in the case of the generation of a torch where electron and proton could bear positive and negative charge, respectively.

It follows from the above that taking into account the newly discovered properties of the torch plasma of a hypervelocity impact and of the accompanying processes can change fundamentally the generally agreed and established approach to the problem of the possible emergence of life beyond the Earth. This may make possible the realization in nature of the first stage of panspermia involving the emergence of extraterrestrial life.

As the second panspermia stage it would be appropriate to consider the delivery of extraterrestrial samples of living matter to the Earth. A cosmic body capable of performing this mission may form as a result of impact catastrophes and happen to be within the reach of the gravitational field of the Earth while containing living organisms. In the case of the impact of such a meteorite onto the Earth organic compounds of any degree of complexity would remain preserved provided that the impact onto the surface produces no plasma torch. For this to happen, it suffices for the meteorite to fall into a deep water body located at the Earth's surface.

It is well known (Shuvalov et al., 2008) that the impact produces no plasma torch if $D_{IMP}/H_W < 0.2$, where H_W and D_{IMP} are the water depth and meteorite diameter, respectively. In this case the projectile is destroyed and decelerated without reaching the bottom and remains mostly unmolten. Such an impact ensures "soft" landing of living organisms provided there are any in the meteorite body. Therefore panspermia as a natural phenomenon had all the conditions to be realized in the Solar System whatever the direction of the motion of living beings: either toward the Earth or away from it.

It is important that the efficiency of the delivery of living beings to the Earth in the case of panspermia could have been very low. This may have been due to the following interesting effect: when carried on a meteorite from the Earth to other cosmic objects, organic compounds were destroyed and organisms died in impact processes because of the complete lack of deep water bodies on these objects. Living matter carried by a meteorite to the Earth may have survived. These aliens may have their future fate unfolding two different scenarios.

Terrestrial conditions may have been ideal for the emergence of life allowing aboriginal organisms to build up substantial mass and develop a biosphere. Or, on the contrary, weak extraterrestrial aliens could not compete with terrestrial forms and perished in a short time.

On the other hand, in the light of the new concept the panspermia idea could acquire a different dimension and importance and require an analysis and reevaluation. In the case of diversified and independent origin of organismal life – in particular, on different bodies of the Solar System, - combined with the possibility of transfer of these organisms in cosmic medium without impairing their vitality, the rate of evolutionary processes under favorable terrestrial conditions should have been rather high as ensured by the fast amplification of biological diversity. It is logical to expect that periods of intensified precipitation of cosmic material (meteorites, comets, cosmic dust) onto the planet should be followed, after a certain time interval (the adaptation period) by evolutionary bursts resulting from the perturbation of the equilibrium of homeostatic systems and stressful successive processes in the populations of adapted "aliens". We thus cannot rule out the possibility that the amazing sweeping evolution of life on our planet could have been triggered by panspermia.

One should bear in mind that on the Earth microbial life determines the existence of all other higher forms of life, closely interacts with them, and is the driving force of the biospheric "breathing" of the planet.

It is possible for the effects that arise in meteorite impact processes to permanently reproduce panspermia. They are directly related to another, equally interesting, natural phenomenon referred to as cryptobiology or biology that is hidden from researchers.

The hypothesis about the existence in nature of hidden forms of biological communities is supported by the proposed hypothetical scenario of the emergence of life in the processes accompanying a hypervelocity impact. According to this scenario, the first beings – the progenitors of the modern biosphere – could originate in the plasma torch and continue their life in deep-seated warm and moist soil layers after the meteorite impact. Early in their development these organisms could "feed" on organic compounds – i.e., on synthesis products. This was followed by the development of the great variety of heterotrophic and chemolithotrophic forms. The latter could consume CO, CO_2, and CH_4 as the source of carbon and produce energy using H_2, CO, H_2S, NH_4^+, CH_4, Fe_3^+, and other compounds. We chose methanogens for the scenario proposed above, however, it could equally well operate with sulphate reducing bacteria, acetogenes, denitrifiers, nitrate-reducing bacteria, hydrogenogenes, etc. The particular type of anaerobic microorganism was of minor importance for the hypothetical model considered.

If the proposed scenario of the origin of the first organisms in the depths of an impact crater has any foundation, the development, with time, of full-blown microorganism communities under anaerobic conditions could also be a sure event. Such microorganisms must have resided in deep-seated sedimentary crust layers and, possibly, in ocean-floor sediments. The discovery of microorganisms at large depths on the Earth may count in favor of the proposed scenario of the origination of life in impact processes. Similar hidden extraterrestrial forms of life could emerge in the corresponding processes and exist in deep-seated rock layers or in subsurface water bodies on other planets and planetary satellites in the Solar System.

Various microorganism communities have been discovered with confidence at large depths under the Earth's surface. Microorganisms have been found in sedimentary rocks, ocean-floor sediments (Parkes and Maxwell, 1993), and even in ice sheets (Abyzov, 1993) in permafrost (Gilichinsky et al., 1992; Vorobyova et al., 1997).

The idea of natural underground evolution was proposed by T.Gold, who later also analyzed the possibility of the existence of the subsurface biosphere on Mars (Gold, 1992). At about the same time, M.I.Ivanov and A.Lein (Ivanov and Lein, 1991) studied the possibility of chemolithoautotrophic origin of life on the Earth and the prospects for searching for methanogenes as a biospheric marker on Mars.

Two possible colonization scenarios were considered to explain the high concentration of microorganisms at large depths under the Earth's surface: (1) microorganisms could have gotten into deep-seated layers as a result of tectonic motions of the Earth's crust and subsequent conservation of communities and (2) they may have originated in lithospheric depths and developed a naturally evolving underground ecosystem. The idea of in situ origination of microbes appears more appealing because the conservation of cells and their survival over a long geological operiod should have resulted in the preservation of relict systems characterized by the drop-out of some links. However, no such systems have been found and the observed biosystems proved to be sufficiently full-blown and diversified.

There is also a third hypothesis based on the idea of panspermia. It was proposed by E.A.Vorobyova in 1994. She analyzed the possibility of the capture of "seminated" cosmic bodies in the process of the formation of planets and, in particular, of the Earth, with the incorporation of viable anabiotic forms of the captured body into the planet. The subsequent heating of the planet, on the one hand, drove life to the crust zone and, on the other hand, nonuniform heating of subsurface layers could activate dormant cells thereby triggering a new round of life and creating the deep-seated biosphere of the reviving planet. This hypothesis looks quite plausible, but it needs experimental confirmation of panspermia (transspermia) of microorganisms in space. However, the possibility of the emergence of extraterrestrial life is the main problem that we discuss in this paragraph, and our primary aim is to show that it can be realized in terms of the new concept.

The hypothesis that Mars may possess subsurface chemolithoautotrophic biosphere and the search for methanogenes are now rather popular topics among the researchers. Moreover, laboratory model experiments have been performed whose results suggest that such searches should be very promising. The most encouraging evidence is provided by the studies of extreme natural habitats on the Earth, which are viewed as natural models of extraterrestrial medium. One of such models is the ancient terrestrial permafrost, which has not thawed for several million years. In the Northern Hemisphere it is represented by the ancient permafrost of Eastern Siberia, Alaska, and Canada and in the Southern Hemisphere, by Antarctic frozen rocks. Permafrost studies have been conducted for many years by a joint team of researchers from the Institute of the Physicochemical and Biological Problems of Soil Science of the Russian Academy of Sciences and the Soil Science Faculty of Lomonosov Moscow State University. Back in early 1980-ies this team found and studied viable cold-resistant microorgamisms in ancient frozen rocks of Eastern Siberia (Zvyagitnsev et al., 1985; Gilichinsky, Vorobyova et al., 1992; Vorobyova et al., 1997, 2005; Gilichinsky et al., 2007). They studied Arctic samples retrieved from depths down to 400 m with a permafrost age of up to 3 million years and Antarctic samples from the 20-m depths dated by the permafrost age of up to 2 million years (up to 15 million years according to other estimates). In all these cases many microbial beings (several ten and hundred million bacterial cells and several thousand fungal germs per 1 g of rock) have been found in the permafrost. Arctic communities are quite active (from -7 to -10 C), whereas microorganisms in Antarctic rocks (from -20 to -27 C) are mostly in anabiotic state. Note that both anaerobic and aerobic microorganism forms, chemolithoautrophic and chemolithiheterotrophic bacteria have been found at large depths in perfmafrost.

Microorganisms were found at large depths both in continental and anchor ice in Antarctic (Abyzov, 1993; Vorobyova et al., 2002; Vorobyova et al., 2005) and in Greenland ice (retrieved from a depth of 2 km). The concentration of cells in ice samples ranged from 10^5 to 10^9 cells/cm^3. Currently microorganisms have been found in ice wells at a maximum depth of 3640 m. The relict subglacial Lake Vostok is located below this level. The in situ study of this lake needs to be carefully thought out and made with appropriately prepared instruments. The work in this direction is now under way.

The first evidence for the existence of biosphere at the Pacific floor near hot water discharge sites was obtained in 1979. These natural habitats are opposites of permafrost conditions in temperature terms. Water temperature in underwater volcanoes exceeds 100 C and therefore microorganisms living in these media are called “extremophiles”.

Thirty-year long ocean-floor studies showed that unique natural communities form at the sea bed around underwater volcanoes. These communities too consist of “extremophiles”. “Extremophile” bacteria also use methane and hydrogen sulphide to produce organic compounds. The lack of solar energy at such depths renders photosynthesis processes impossible. Bacteria in such places feed on methane and hydrogen sulphate. All animals living in these communities – worms, crustaceans, and fishes – feed on “extremophile” bacteria and bacteria may live at the surfaces of these animals, inside them, or in sediments. As of now, more than 500 animal species have been described that live in hydrothermal communities, and the estimates of the biomass of these communities leads us to conclude that the Earth must possess extensive underwater biosphere, which owes its life and existence to anaerobic bacteria.

The two opposite extreme temperature domains of subsurface biosphere suggest that extreme temperature conditions do not prevent the existence and metabolism of a number of microbial communities in the absence of solar energy and oxygen. Microorganisms are known to exhibit, in addition to the temperature extremophilia, also high resistance to pressure, radiation, high acidity, salinity, etc.

Therefore there are enough grounds to suspect the existence of extensive underground biosphere, where temperature and other conditions may be much milder than in glaciers or deep-sea sediments. The hypothesis about the possible existence of a powerful underground biosphere becomes increasingly popular because it is supported by increasingly serious experimental evidence. According to this hypothesis, a great variety of hidden, unique ecotopes populated with microbial beings should exist under the Earth’s surface down to great depths, and the total biomass of these ecotopes should be much greater than the total biomass at surface. Perhaps the existence of such underground biosphere might prove that the ancestors of these organisms could have originated in deep-seated layers of the Earth’s crust, subsurface water bodies, and in sea-floor layers. Such processes could occur not only in deep-seated Earth’s layers, but also, hypothetically, on other planets and planetary satellites of the Solar System with moderate subsurface temperatures, which ensure the existence of water bodies or, at least, underground moisture.

Given the temperature distribution in the Earth’s crust (Dvorov, 1976), the very general considerations allow us to identify the possible habitat layer of underground microorganisms.

According to the currently available data, temperature in the Earth's crust increases by about 1 C for every 33 m depth.

In the regions of ongoing volcanic activity temperature increases at a rate that is greater by one order of magnitude, and reaches 100 C at a depth of ~300 m. In regions where the upper part of the crust is frost penetrated down to 1.5 km – e.g., in the permafrost zone, - the temperature at the depth of 1.8 km is equal to ~3.6 C and increases by 1 C for a 500 m depth. We can coarsely infer from this the minimum and maximum depths where microorganisms may live. These depths range from 300 m in volcanic regions to 5 km in permafrost regions. In deeper layers, e.g., 10—15 km under the Earth's surface, water is in gaseous state because of the high pressure at temperatures reaching 700 C, making it impossible to live in for the currently known microbial communities. There is evidence suggesting that microorganisms may exist in sparging waters from oil wells with depths down to 7 km. The concentration of microorganisms at these depths was as high as 10^3 cells/cm^3. However, these data should be treated with caution because the factor of prime importance in microbiological studies is now the correctness of sample recovery, which is primarily determined by the drilling method employed.

In view of the above, the idea that hidden biosphere might exist in deep-seated layers of Mars as proposed by Ivanov and Lein (1991), Gold (1992), Gilichinsky et al. (1992), and Vorobyova et al. (2002) has a sound experimental basis. According to the above authors, Martian interiors may support the existence of microbial communities in the Martian lithosphere, where water was found in substantial quantities. The above authors believe that analogs of terrestrial anaerobes may be the best Martian aboriginal candidates.

The new concept described in this book provides important support for the idea of the possible existence of life under the Martian surface and in deep-seated layers on planetary satellites in the Solar System, in particular, on Europe, Enceladus, and Titan (see, e.g., Gal'chenko, 2003, 2004). The new concept is based on real processes that occur in nature, where complex organic compounds synthesized in the plasma torch may be injected into deep-seated water reservoirs on cosmic bodies in the process of a penetrating meteorite impact. In any case, we have got to face up to the fact that there is hardly a simpler, more convenient, and efficient natural mechanism for the synthesis of organic compounds than a hypervelocity impact when it comes to cosmic bodies with very low surface temperatures. This phenomenon is capable of providing the conditions necessary for the emergence of extraterrestrial life, the synthesis of sufficient amounts of organic compounds, and the formation of a potential survival zone. In this zone the processes of chemical evolution may begin. Chemical evolution on cryogenic planets and their satellites may occur only in deep-seated water reservoirs and rock layers. The processes that accompany the impact can make it possible to circumvent all the contradictions that the attempts to develop a scenario of the origin of life under harsh conditions on very cold planetary satellites have to face. Therefore the hypervelocity impact phenomenon should in this case be viewed as a no-alternatives scenario.

However, the most important consequence of the usefulness of the new concept is that it can be used to logically explain the origin of the powerful underground and underwater biosphere on the Earth via the realization of a hypothetical scenario of its emergence in life-creating processes of penetrating meteorite impacts.

***

Thus the results obtained via observations and laboratory experiments and presented in this chapter lead us to a number of important conclusions directly concerning the origin of life and the extent to which processes that occur in hypervelocity meteorite impacts take part in this phenomenon.

We showed above that the availability of carbon in meteorites combined with the presence of water and hypervelocity impact should be viewed as the key conditions for the emergence of the first organism. Meteorite impacts could facilitate the emergence of life not only on the Earth's surface, but also deep under water and deep in crust rocks.

These two crucial factors – carbon and hypervelocity impact – were always provided by nature during the formation of planets. The third must-have component for the emergence of life - water, which is in abundant supply on the Earth's surface, on planets, and on planetary satellites of the Solar System, – could be found in liquid form only deep in the interiors of these bodies or in the form of ice at their surfaces. The researchers have been always looking for natural mechanisms capable of ensuring efficient synthesis of organic compounds both at the surfaces of cosmic bodies and under kilometers-thick rock or ice layers. Therefore the discovery of a mechanism that makes possible the introduction of organic compounds directly into the subsurface layers of cosmic bodies radically expanded the hypothetical habitat of microbial life on planetary satellites in the Solar System. Moreover, we also show that hypervelocity impact processes create all the necessary conditions for the emergence of life on planets and planetary satellites irrespectively of their surface temperatures provided the availability of a water reservoir or ice in the depths of the corresponding cosmic bodies.

Laboratory studies of the processes of direct introduction into water of organic compounds synthesized in the plasma torch are of great interest and some of these configurations can be modeled. These studies showed that organic compounds are synthesized and accumulated in underwater plasma torches.

The new results obtained by space missions to satellites of giant planets provided evidence for the presence of water and water ice on these cosmic bodies and led the researchers to conclude that extraterrestrial life could originate on these objects. However, no other concrete viable mechanisms except the impact scenario have been proposed for the emergence of life on these cosmic objects. Therefore the demonstration showing that the impact zone can, in principle, be heated for a long time, combined with the experimental confirmation of the synthesis of organic compounds during the generation of a torch under water proved to be very important for justifying the possibility of the emergence of extraterrestrial life. The aforesaid leads us to conclude that microbial life may be a widespread phenomenon in the Solar System and, in particularly, on some satellites of giant planets. This conclusion is based on bona fide natural phenomena and physical mechanisms found and confirmed in laboratory experiments.

If primary forms of life could form in a plasma torch of a meteorite impact then, given the widespread nature of impact phenomena and suitable ambient conditions, one can expect the discovery of extraterrestrial life in space to be a real possibility. On the Earth such a being should have formed during the period of intense meteorite bombardment. In some events of this kind liquid water that appeared in the impact region could have refrozen. Therefore we may possibly find in space, depending on the particular properties of the medium, regions bearing microflora frozen at different stages of its evolutionary development. This is one of the unexpected consequences of the new concept. Time will show the scientific value of the proposed idea.

Chapter 8

SEARCH FOR SIGNS OF EXTRATERRESTRIAL LIFE AND THE "HEAD-ON COLLISION"

8.1. SPECIFICS OF THE APPROACH TO THE PROBLEM

Many prominent researchers believe that the discovery of extraterrestrial life, if it happens, will be the most important event of the third millennium.

The existence of microbial life on other planets or planetary satellites of the Solar System is now viewed as more likely than it was thought until recently. This is due to the recent discovery of water - the key component for the origin of life - on Jovian and Saturnian satellites. Large amounts of water are very likely to be present in the lakes and oceans beneath the extremely cold surfaces of these bodies with the temperatures as low as -200 ° C. However, the presence of water deep beneath the surface of some planetary satellites does not, by itself, provide conclusive evidence for the existence of life in the depths of these cosmic bodies. A number of extra conditions should be met for life to originate deep under the surfaces of cosmic objects. One of these requirements consists in the high concentration of simple and complex organic compounds under the surfaces of these objects and, in particular, in underground water bodies. This may be indicative of natural processes that could make possible abiogenic synthesis "in situ", or of working mechanisms for delivering organic compounds synthesized somewhere else into the deep layers of the object. The concentration of organic compounds must be high enough to ensure their further complexification in the process of chemical evolution.

Today we can imagine no natural phenomenon capable of ensuring efficient synthesis of complex organic compounds under conditions involving extremely low surface temperatures of about –200 C and hard radiation exposure. Mechanisms are even more difficult to find that could deliver organic compounds synthesized at the surface to the deep layers or to water reservoirs located beneath kilometers-thick ice or rock layers. The only exceptions are the processes that accompany meteorite impacts. According to modern concepts, the events mentioned above cannot occur without such processes. It is therefore of particular interest to analyze the natural mechanism of hypervelocity impact, which we believe to be the only possible way to enrich a planet with organic compounds down to a considerable depth.

We showed above that a certain impact configuration, in particular, that of the penetrating impact is capable of ensuring the synthesis of complex organic compounds in the plasma torch and their injection several kilometers deep under the surface, and that synthesis products may include reproductive forms of primary living matter.

We suspect (in the case of a few planetary satellites) or know for certain (for some of them) that the energy of tidal forces increases local temperature to positive values and results in the formation of water in the inner regions of these bodies. When polymer organic compounds synthesized in the course of a meteorite impact are delivered to these regions, the concentration of such compounds may become rather high. In this case, further complexification of polymers synthesized in the torch would continue, and the processes that we do not yet entirely understand could result in the development of primitive forms of life.

The search for and, in the case of success, the study of extraterrestrial life are the most important tasks of modern science. Future detailed and comprehensive studies of the "other" life, which may prove to be very similar to terrestrial life, should consist of several levels of experimental analysis. In particular, from finding signs of life beyond the Earth to the in-depth study of its basic properties and characteristics and conditions of origin.

To search for, detect and perform a detailed study of extraterrestrial life, which is very likely to have the form of microorganisms, from space probes, we will need special new-generation onboard analytical equipment and a highly intelligent onboard robot capable of operating under harsh open-space conditions. This should include highly sophisticated preparation of the sample followed by complex mass-spectrometric studies. The method of measurement is well known and it is commonly employed in medical laboratories on the Earth. One of the most informative methods – the analysis of lipid markers – makes it possible to simultaneously study the basic biological characteristics of up to 200 microorganism strains by analyzing the mass spectra of lipids and hydrocarbons extracted from the cells. Such measurements will be impossible to perform in space in the next few decades, and they will become a reality only after the measurement technique is successfully developed and a laboratory prototype of onboard equipment is ready. Such robotic devices will make it possible to prepare samples and take measurements. Much effort will be needed to ensure high purity of the measurements and increase the sensitivity of mass spectrometric equipment.

We also discuss an alternative approach: to deliver samples to the Earth and analyze them in ground-based laboratories. However, in this case, a number of serious problems should also be addressed. In particular, we have first to analyze the criteria to be applied to each object in order to determine where to take the sample so as to maximize the likelihood of a positive result. Only the future will show which approach wins in this competition of ideas (where to look?) and technology (how to detect?).

Today we must think about relatively simple experiments and techniques that can be implemented in the coming years. These experiments should be less ambitious and their aim should be to address the problem of detecting signs of extraterrestrial life using a minimum sufficient set of methods, which by itself is also an issue of exceptional importance. The successes achieved in such experiments may determine the direction for future works: whether to deliver instrumentation "there" or test samples "here" on Earth.

The methods and instruments proposed in this chapter to be used for detecting signs of life on planetary satellites in the Solar System have unique features of their own. They stem from the fact that we will have to study microorganisms and the products of their metabolism, which remained unchanged and escaped destruction for a long time. To this end, the studies will focus on the search for and detection of microbial biomass or waste products frozen in the ice water. According to Abyzov (1993), in the matrix of ice could protect the sought-for cells and complex organic compounds against fossilization. The surface contamination of microorganisms is known to often prevent accurate mass spectrometric measurements of their elemental composition. By selecting ice as a habitat we may ensure the surface purity of microorganisms, thereby significantly simplifying the preparation of samples onboard the spacecraft. In the presence of an atmosphere and subject to hydrothermal regime microorganisms immobilized at the surface of minerals become fossilized. Fossilization of cells may be quite a fast process, which ultimately leaves virtually no traces of organic compounds initially contained in them. Such forms can be identified only using morphologically direct microscopic methods. However, a conclusive answer will be impossible to obtain in the absence of extra criteria that would allow one to reliably differentiate between cells and mineral particles. Microorganisms that preserve their elemental composition are of greatest interest for astrobiological search, irrespective of whether they are still viable (proliferating cells, anabiotic cells) or dead (mummies) (Vorobyova et al, 2002; Mulyukin et al, 2002).

Microorganisms do not fossilize in the ice matrix. In this book we suggest to use this property, whereby cells preserve their elemental composition for thousands or millions of years (Mulyukin et al, 2002), as a criterion for searching for signs of life. The key measurements are performed using an onboard mass spectrometer, and, in the simplest case, the elemental composition of biological samples or biomass is analyzed.

We now illustrate the key points of the analysis by the example of Europa. Let us assume that Europa’s ocean is inhabited and the bulk of microorganisms are of the same type as those found on the Earth. The existence of the biosphere implies a rather high concentration of cells. In this case, ocean ice should contain not only germs but also products of their metabolism. The studies made within the framework of the Messenger program suggest the possibility of the dredge-up of microorganisms and their metabolic products from the ocean to Europa’s surface. These studies show conclusively that Europa’s ocean communicates with the surface. Water penetrates through cracks, which form from time to time in Europa’s ice shell, flows onto the satellite's surface, and quickly freezes. Water from the ocean can also get to the surface as a result of a penetrating impact of a large meteorite. Thus the configuration of the Tyre crater suggests that water from the ocean flooded the "hole" punched by the impact of a 10-km diameter meteorite and only then froze, producing a kind of a "roof", where the smooth surface of frozen water alternates with surface ice fragments interspersed in it.

Water ejections onto Europa’s surface flood some areas and destroy the earlier produced meteorite craters so that the age of the satellite’s surface is estimated to be no more than several hundred millions years. It follows from the aforesaid that if Europa’s ocean is inhabited then the surface layer of ice can be studied in order to identify eventual traces of ocean dwellers.

Two methodological approaches have been proposed for the future missions aimed at the detection of microorganisms and organic products on Europa’s ice surface.

The first approach consists in remote detection from onboard an orbiting probe of organic molecules that are present in Europa’s microionosphere. These molecules are synthesized in the processes of secondary ion emission when Europa’s ice surface is bombarded by magnetospheric high-energy particles. The discovery of such molecules in the induced Europa’s ionosphere should indicate that the ice surface of the satellite is rich in organic compounds. Because the formation of organic compounds at Europa's surface is unlikely due to the low temperature and intense radiation, it would be natural to associate the detection of these compounds in the microionosphere with their presence of the ice matrix, which, in turn can be explained by assuming that the ocean is inhabited.

In the second approach mass-spectrometric measurements of the elemental composition of the specimen obtained after melting the samples of ice and filtering of water are to be made onboard the lander. In this case the preparation of the sample would involve a number of manipulations that will allow washing small solid particles off the potential cells before filtration. The mass spectra taken using an onboard laser TOF mass analyzer will determine the ratios of the mass peaks of major biogenic elements - C, O, N, and H, – and also between the peaks of biologically important elements P, S, K, and Ca. The ratios of these peaks will allow us to conclude whether terrestrial-type microorganisms are present in the sample and roughly assess their condition. In the following sections we discuss in more detail underlying the physical processes ensuring the viability of the proposed methods, and onboard mass-analytical instruments that can be used to perform the studies mentioned above.

8.2. Emission of Organic Compound Ions from the Surfaces of Cosmic Bodies

Managadze and Sagdeev (1988) were the first to suggest the hypothesis that atmosphereless planetary satellites and small bodies of the Solar System may develop a microionosphere and microatmosphere generated in the processes of secondary ion emission caused by solar-wind ions. Somewhat later Managadze (1994) demonstrated that solar-wind ions may cause not only atomic ions and neutral atoms, but also organic compound molecules, if any, to be emitted from the surface of a cosmic body. The proposed ideas are based on a series of laboratory experiments where the primary flux of hydrogen ions generated the flux of secondary atomic ions and ions of polyatomic organic compounds (Cherepyn, 1992). In these experiments, mostly solid targets were subject to ion exposure.

G.Tantsyrev and his colleagues (Nikolaev and Tantsyrev, 1980; Malyarova and Tantsyrev, 1991) were the first to show the possibility of the emission of organic compounds from the ice matrix in the processes of secondary ion emission caused by ion bombardment. In the proposed method, aqueous solution containing organic compounds was frozen to the boiling point of liquid nitrogen and subjected to ion bombardment. The method proved effective for mass spectrometric analysis of low molecular weight organic compounds. The use of the ice matrix with an admixture of glycerin and mixed with acids resulted in the high yield of high molecular weight organic compounds as well. The results indicated that the bombardment of Europa's ice surface, which is believed to consist of ice with possible acids admixtures, should cause the emission of organic compounds of various types.

This physical effect has been used to develop a scenario for the search for signs of life from onboard an orbiter using a highly sensitive new-generation TOF mass-reflectron.

When developing an onboard device meant for measurements of this kind, we had to address an internally conflicting task – maximize the sensitivity of the instrument while keeping background illumination of its sensitive element - the detector – as low as possible.

The detector assembled on two serially connected microchannel plates (MCP) has high detection efficiency for UV radiation.

To maximize the sensitivity of the device, its geometric factor should be high enough. This could have been achieved by maximizing the two device parameters - the solid angle and the area for particle collection. However, the increase of the geometric factor of the instrument also increases the intensity of solar UV radiation and cosmic background flux onto the detector.

To reduce the illumination of the detector by solar radiation, we had to develop a system of UV absorbing ion mirrors, capable, at the expense of a slight reduction of the ion beam, of decreasing the UV flux by a factor of 10^{12}-10^{13}. Such a system had to be mounted inside the TOF part of the analyzer between the window of the input electrostatic shutter of the instrument and the detector, as shown in the scheme of the instrument in Figure 43.

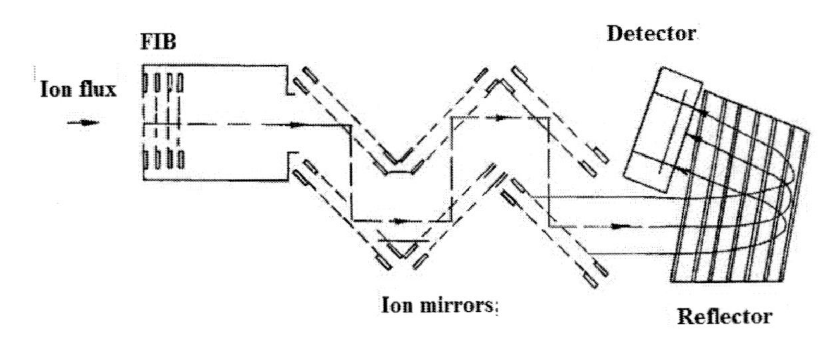

Figure 43. Ion-optical scheme of the MANAGA device for recording secondary ions.

Figure 44. Flight unit of MANAGA device.

We successfully addressed these problems while developing the MANAGA time-of-flight mass-reflectron meant for flux measurements of molecular ions – including organic compound ions - in the "Phobos-Grunt" space mission, and for the study of organic compounds “escaping” from the Earth or from the organic microionosphere of the International Space Station. The device designed for these projects did not have the analytical characteristics needed to search for signs of life from the orbiter, however, it could nevertheless be used as a base tool. The instrument has an input box with an effective ion-collection area of 15-20 cm^2 and a solid angle equal to 60° x 40 °, or 0.74 sr.

The MANAGA onboard unit shown in Figure 44 successfully combined high sensitivity with the possibility of measuring ion fluxes over a wide range of intensities and masses. That is why it was included into the equipment set for future space missions.

The measuring part of the onboard device – a TOF mass-reflectron – consists of the following basic functional units: an electrostatic input window, an ion-drift region with ion mirrors mounted inside it, an ion reflector, and a detector.

The device operates as follows: electrostatic entrance window, which is normally closed, prevents external ions from reaching the analytical part of the device. The window opens with a frequency of 10^3-10^4 Hz for 10 ns, allowing a packet of external ions from the ambient space to reach the analyzer. After extra acceleration the ions enter the TOF region and, after crossing it, reach the reflector. In the decelerating field of the reflector the space-time focusing of ion packets takes place, which ensures that narrow mass peaks form when the packets reach the detector. To reduce the background illumination of the detector, six planar ion mirrors are installed at 90° to each other in the TOF part of the analyzer. This system reduces external illumination by a factor of about 10^{12}. The reduction is achieved by applying special strongly absorbing coatings to the grid electrodes of ion mirrors and other elements thereof, as well as to other structural elements of the instrument, which may be exposed to ambient light. Given the additional reflection of the scattered light in the reflectron, the total flux reduction factor was equal to 10^{13}.

The detector is an assembly of two 90 x 30 mm microchannel plates whose signal is fed to the processing amplifier, and then reaches the histogrammator, which measures the time intervals. The registration system operates in the ion counting mode and provides:

- The distribution of individual ion pulses into 20-ns long time cells;
- The summation of pulses in each cell;
- The storing the data for 10^4 individual time scans per second, i.e., with a frequency of 10^4 Hz.

Thus, the parameters measured by the instrument include: the time of flight of the ion, which allows it's mass to be uniquely determined, and the intensity of ion fluxes used to determine the proportions of different masses. We list the basic analytical properties of the instrument in Table 10.

Further improvements of the instrument should include additional acceleration of ions by 5 to 10 kV before they reach the detector. This fix should allow the instrument to reliably record ions of high-molecular organic compounds.

The installation of an electrostatic deflector in front of the entrance window of the instrument is expected to allow the mass spectra of the ion flux recorded from the upper and lower hemispheres to be compared. This comparison will determine to what extent the ion fluxes from the lower hemisphere are "heavier" than the corresponding fluxes from the upper hemisphere, thereby potentially providing further evidence for the presence of complex organic compounds in the ice matrix of the underlying surface.

Laboratory tests of the prototype MANAGA flight instrument were conducted. During these tests ions of the residual gas generated by an ion accelerator reached the entrance of the instrument. Figure 45 shows typical mass spectra obtained in these tests. It is evident from these spectra that TOF is the best technique to be used for such studies, and that the instruments based on this technique are superior in terms of sensitivity and speed to many modern instruments designed for space studies.

Table 10. Physical characteristics of MANAGA instrument

Parameter	Range of measurements
Mass range	1... 1000 a.m.u.
Mass resolution at the 50% level, not lower than	~ 150
Absolute detection limit for 1 sec for neutral component	$\geq 10^5$ cm^{-3}
Absolute detection limit for 1 sec for ion component	10^5 cm^{-3}
Relative sensitivity	1 ppm
Dynamic range	10^6
Frequency of spectrum output, in 1 s	10^4
Field of view	60° x 40°
Entrance window area	15.43 cm^2
Number of quantization intervals (channels)	2038*
Duration of quantization intervals	20 ns
(Instrumental) duration of the measurement of a single spectrum	40.96 μs*
Maximum signal repetition rate	Up to 256 MHz

* Nominal value. May vary from 512 to 65536.

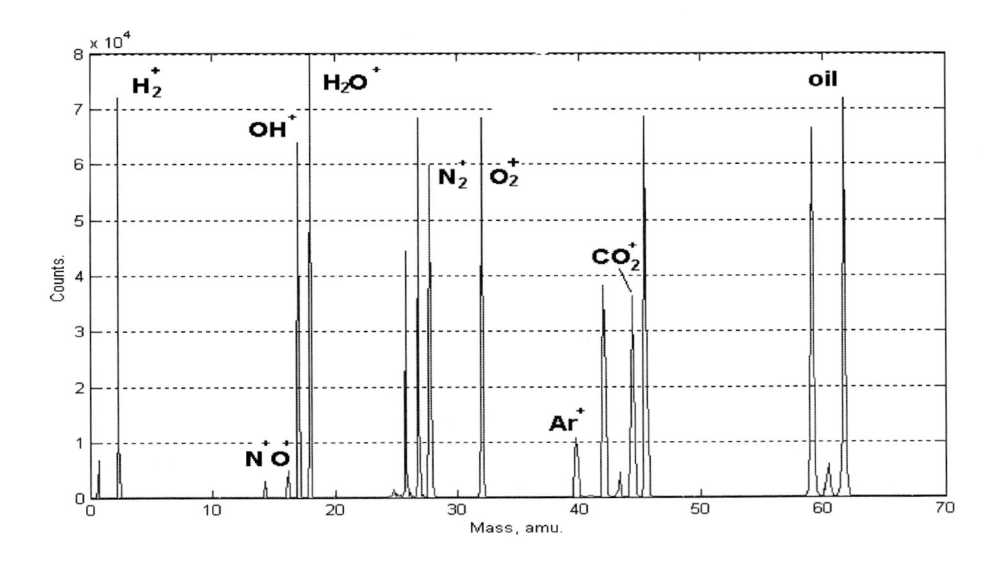

Figure 45. Mass spectrum of residual gas recorded by the MANAGA instrument.

Note in conclusion that the design of the proposed onboard instrument ensures its protection against solar UV radiation in the case of the entrance window area and the effective ion-collection area equal to 20 cm^2.

8.3. Measurement of the Elemental Composition of the Biomass

The technique proposed is based on the fact that living cells not only have a common structure, but also a common elemental composition. This applies both to the basic elements that make up 98% of the cell mass - C, O, N, and H, - and to the trace elements present in the cell. The percentage content of the basic elements that make up the cell is: O - 70%, C - 18%, N - 10%, and N - 3%. However, if water is removed from the cells the proportions become O - 7%, C - 8%, H - 3%, and N-3%. The proportion of carbon (C), nitrogen (N), and organic phosphorus (P) in the biomass is 106: 16: 1 (Zavarzin, 2006a, b). The amplitudes of the mass peaks of these elements, like their ratios, may serve as bona fide markers allowing the substance studied to be classified as biogenic or living matter.

An additional group of elements whose atomic-peak ratios allow cells to be tentatively identified and their condition determined include S, P, K, Ca, Cl, Mg, and Fe. Thus Mulyukin et al. (2002) used the Ca/K and P/S peak ratios to determine the physiological state of Bacillus cereus bacteria cells in a culture. This criterion allowed the above authors to distinguish proliferating vegetative cells from viable resting cells and from their non-viable or mummified analogues. The total content of the above elements in living cells was equal to 1.9%, while the content of individual elements amounted only to several tenths and hundredths of a percent. In a further development of this work the above authors used the same approach for direct detection and analysis of microbial cells in a sample of Antarctic permafrost in situ, and in dust washed from laboratory surfaces (Mulyukin et al., 2002). The method allows cells to be differentiated from mineral particles. In situ samples (obtained without breeding and isolating of microorganisms) are shown to contain mostly resting anabiotic cells.

The content of such elements as Zn, Cu, I, and F, which are present in a living cell in macroscopic amounts, is equal to 0.02% of the weight of a living cell. A detection of these elements in a sample combined with the above-mentioned proportions of the key elements may also be indicative of the biological nature of the sample.

Simultaneous mass spectrometric measurements of all the above elements and their isotopes can be made with LASMA compact onboard TOF laser mass-reflectron developed to analyze the elemental and isotopic composition of regolith onboard the lander for the proposed Phobos-Grunt mission. The design and physical principle of operation of LASMA instrument are very similar to those of a laboratory instrument used to discover the new properties of the plasma torch and described in Chapter 4. However, despite the identical design of these instruments they differ significantly in the mode of their operation, allowing the new device to detect organic compounds synthesized in the plasma torch. This is achieved by reducing the two main initial parameters of the instrument: the energy of plasma ions (decreased by a factor of three) and the power density of laser irradiation (slightly decreased by a ~ 30%). This allows recording low-velocity polyatomic ions synthesized in the plasma torch.

In the nominal mode, when the device operates as a mass spectrometer, polyatomic ions are usually "cut off" and do not reach the detector. In the case of a high power density of laser radiation ions of this type are also subject to thermal destruction. These processes, which are caused by the effect of laser radiation during the impact, are completely excluded. Thus, when the proposed instrument operates in the nominal mode as a mass spectrometer, polyatomic ions are "invisible" for two reasons: first, because of the low efficiency of the process of their formation and because even if synthesized, they are destroyed by laser radiation, and, second, because the remaining low-energy ions are cut off by the analytical unit at the detector input. Hence when it operates in the mass spectrometer mode the device records only chemical elements and their isotopes.

The analytical parameters of LASMA instrument, such as the mass resolution and sensitivity, allow it to simultaneously record both mass peaks of the main components contained in the samples prepared for analysis, and those of trace components. This is evidenced by the main characteristics of the device listed in Table. 11 determined from numerous bona fide calibration tests of the instrument.

We show the onboard prototype in Figure 46. This instrument, like its laboratory analogue, consists of the following functional units: laser, attenuator, focusing lens, and an analyzer with a detector. These elements are easy to spot in the photo of the onboard instrument using the scheme shown in Figure 10 in Chapter 4.

The mass spectra of C3 carbonaceous chondrite shown in Figure 47 demonstrate the potential analytical capabilities of the device.

We developed and operate a special automatic program to process the data of both single and coadded mass spectra. This program records the mass peaks of the elements and their isotopes, estimates the errors of these measurements, and assesses the significance of the results.

Thus the LASMA device is a ready to use onboard instrument, which by its analytical and operational characteristics is capable to perform mass-spectrometric measurements needed to detect signs of life by analyzing the elemental composition of a sample suspected of containing cells of microorganisms and having a clean surface due to their localization in the ice matrix.

Table 11. Main physical characteristics of LASMA instrument

1	Mass range	1-250 a.m.u.
2	Resolution	300-500
3	Relative sensitivity in 1 spectrum	10 ppm
4	Absolute detection limit by mass in 1 analysis	$5 \cdot 10^{-13}$ g
5	Instrument's operation speed at 1 a.m.u.	200 ns
6	Dynamic range	10^4-10^6
7	ADC sampling speed	10 ns
8	Accuracy	10%
9	Mass	1.4 kg
10	Power consumption (average)	5 W
11	Overall size	262 x 110 x 225 mm

The onboard system of sample preparation includes the following functional units and mechanisms: a sample acquisition unit, a container for melting ice and ultrasonic cleaning of cells, a unit for filtration and the preparation of the target, and a unit for placing the samples into the laser beam area.

Consider the sequence of operations needed to prepare the samples after the spacecraft lands on Europa's surface. Upon its acquisition ice must be delivered to the container. Drilling may be the optimum solution for this purpose. According to preliminary estimates, processing about 1 kg of ice would be enough to collect the needed amount of the biomass. This is 10 times less than the amount of the sample to be acquired using the method proposed by K. Chiba (Chiba, 2000), which does not take into account the possibility that complex organic compounds can be synthesized in Europa's ocean or introduced into it the process of a penetrating meteorite impact.

Figure 46. Prototype of LASMA flight instrument.

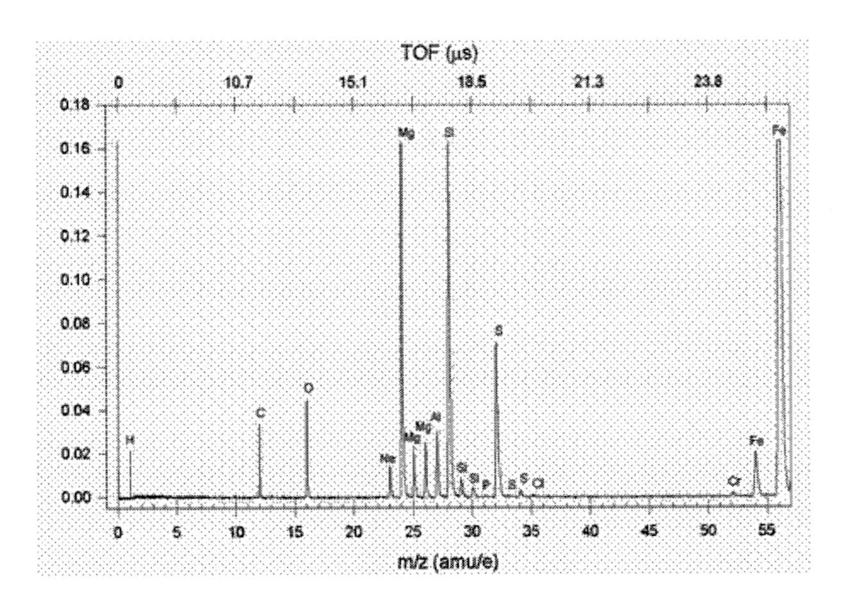

Figure 47. Typical spectrum obtained in the control measurements of a sample of unknown origin and composition. The reduction of the spectra made it possible to draw the correct conclusion that the sample is of meteorite origin and belongs to a C3 carbonaceous chondrite.

Ice in the container will be melted by electric heaters. The aqueous suspension containing microorganisms will be then subject to ultrasound treatment in order to clean the cell surface of external contamination. After deposition and removal of contaminants water is pumped through molecular filters made of high-purity noble metal. By the end of this procedure the bulk of the biomass remains at the surface of the filter. Short heating of the mass obtained to a temperature not exceeding 40-60 ° C will dry the sample and biomass will then be placed into the laser beam zone together with the filter.

The mass spectra recorded by the onboard instrument are to be compared with the spectra from a database to quantify the results and compare the substance analyzed with terrestrial microbial analogues. If the spectra obtained prove to resemble those of terrestrial microbial analogues, the data can be considered as evidence for the presence of terrestrial-type life in Europa's ocean.

The elemental composition of the sample after filtration may differ from the spectra of terrestrial microorganisms. Such an unfavorable outcome would at least yield the elemental ratios in the sample and make it possible to see whether the result can be explained by the presence of a hypothetical model of life.

We can also invent more sophisticated experimental techniques and configurations, as well as costly procedures that would give a definite answer to the question. Such experiments will take much time to prepare.

An important feature of the technique proposed above is that it is relatively simple and currently feasible, allowing it to be classified as a "low-cost mission". The simplified sample preparation procedure employed and the fact that the analytical equipment is ready to fly to Europa "at a moment's notice" should be viewed as an important advantage over other proposals, especially because, to a considerable extent, simplicity implies reliability.

The above considerations are supported by preliminary laboratory calibration tests of the LASMA instrument. The mass spectrum shown in Figure 48 was obtained by subjecting the yeast biomass to a single laser exposure and the registration of the matrix elements of the culture. The curve showing the ratio of the matrix elements in yeast obtained by averaging 30 single spectra was compared to a similar curve representing the elemental ratios in terrestrial microorganisms. As is evident from the plot shown in Figure 49, the two curves agree well with each other. This demonstrates the potential for the widespread use application of the proposed method for the search for signs of life.

An international mission to Europa is now seriously discussed by the space agencies of Russia, Europe, and USA. Moreover, the new methods for the detection of signs of life presented in this book are viewed as the most interesting and competitive.

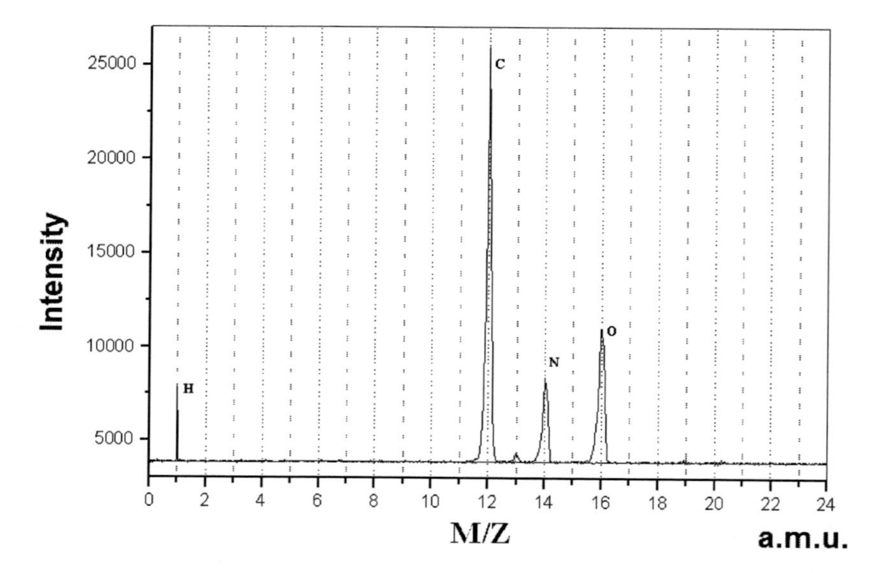

Figure 48. Mass spectrum obtained by subjecting the biomass of food yeast to a single laser exposure.

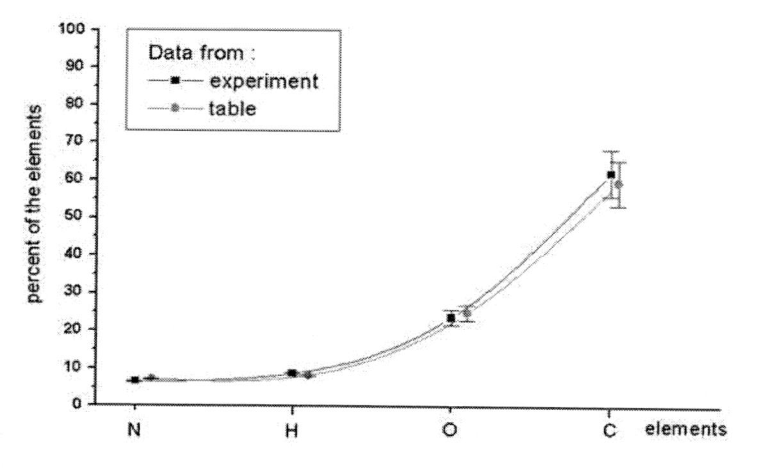

Figure 49. Experimentally determined percentage of matrix elements in food yeast and the average percentages according to tabulated data.

8.4. Hypervelocity Impact in Space without Terrestrial Limitations

The results of laboratory simulations of a hypervelocity impact presented in Chapters 4 and 5 show that the mass of organic compounds synthesized in the plasma torch increases with the size of plasma formation region. Thus an impact of a 100-μm diameter meteorite may cause the formation of a polymeric organic compound with a mass of up to 5000 a.m.u. in the plasma formation region. These results allow us to make an approximation, whereby organic compounds with molecular masses 100 000-500 000 a.m.u., which is comparable to the molecular masses of the primary forms of living matter or DNA and RNA or protein macromolecules, may be synthesized in the case of the characteristic size of the plasma formation region corresponding to that of a ~1 cm diameter meteorite moving at a minimum above-critical impact velocity of 15—20 km/s.

However, such high impact parameters are impossible to achieve in laboratory dust-impact facilities. For example, the heaviest particles that can be accelerated to such velocities in modern dust-particle accelerators have the mass of 10^{-11} g.

Currently, the only possible way to perform an impact experiment of this class is to use space as the medium, meteorites and cometary nuclei moving with high velocities in this medium as targets, and interplanetary probes as projectiles.

The first active experiment of this type – Deep Impact mission – carried out by NASA in 2004, involved the collision of the projectile module of the interplanetary probe with the nucleus of comet Temple (McFadden and A'Hearn, 2004). In this experiment, which was conducted in order to study the properties of the cometary nucleus, optical observations were made during the impact. These observations included imaging of the impact-produced torch and dust outburst by instruments mounted on a detachable module. According to the results obtained, a relatively short and bright flash of the torch plasma was followed by the expansion of the gas-and-dust cloud. The density of the cometary nucleus was found to be low, of about 0.3 g/cm^3, i.e., 30 times less than that of the projectile (projectile). Such a low density prevented the development of a full-blown plasma torch at the surface of the cometary nucleus. Onboard the diagnostic module the radiation from the torch region was recorded only at optical wavelengths and with a rather low temporal resolution. These factors made it impossible to obtain any information about the synthesis of organic compounds in the impact. Yet, the Deep Impact mission should be viewed as the first successful active experiment involving an artificial hypervelocity impact onto a cometary nucleus performed with a man-made projectile moving at a velocity of ~10 km/s.

An analysis of the initial conditions of the Deep Impact experiment and its diagnostic capabilities showed that to study complex organic compounds in the plasma plume of a hypervelocity impact, we must change significantly the initial conditions and expand the range of analytical instruments. Furthermore, to reduce the cost of conducting the experiment in the interplanetary medium, it is worth trying to perform it in the Earth's magnetosphere and in relatively low orbits.

The preliminary analysis of the problem proved to be a success because it yielded acceptable solutions that allowed impact velocities of ~ 15 km/s to be achieved in the Earth's magnetosphere. The realization of such experiments involved the use of modern technical capabilities of artificial satellites and ballistic missiles, which could also meet the requirements imposed. The LIMA-D remote laser TOF mass-reflectron may prove to the best solution for performing the most informative measurements of complex organic compounds synthesized in the plasma torch. This unique device, which was earlier developed to study the elemental and isotopic composition of the regolith of Phobos, was designed for conducting mass spectrometric analysis of samples from a flyby module located 100 meters above the surface (Managadze et al., 1987).

A novel solution has been found to achieve the required impact velocity. The proposed idea involves the use of two artificial earth satellites moving toward each other at close orbits. One of them is to carry the target and a diagnostic module, which should move slowly away from the target during the experiment. Projectiles have the form of 5 to 20-cm diameter carbon spheres filled with weighting material. They will crush into a target at a relative velocity of 15-16 km/s. The plasma torch produced by the impact may rise to a height of 20 to 30 m. Organic compounds synthesized in the plume can be detected at a distance of 5 to 10 km from the target.

We will discuss the above idea of the diagnostic support of the space experiment in more detail in the subsequent sections.

8.5. Control of the Satellites Involved in the Realization of a "Head-On Collision"

The magnetospheric experiment for the realization of a hypervelocity impact is to involve two spacecraft to produce a collision between the projectile and the target in a way that would be most appropriate for the formulated physical task. One of the criteria of appropriateness is the kinetic energy of colliding bodies. It is evident that the maximum energy can be achieved if the orbital velocities of the two spacecraft are collinear and directed toward each other at the time of the encounter.

The following scenario for the realization of a hypervelocity impact can be proposed: the spacecraft are put into orbit in two launches: one of the satellites is to be launched into an orbit with the inclination i, and the other one, into an orbit with the inclination of 180-i. In the simplest case the satellites are launched into circular orbits with the same of semi-major axes and hence with close orbital periods. The gravitational field of the Earth causes the orbit to evolve in such a way (Eliasberg, 1961) that at some time instant their planes coincide. The collision between the projectile and the target is to happen in these orbits, where the orbital planes are close to coinciding. The spacecraft are equipped with tools to control their orbital motion within a limited range of several tens of meters per second. The simplest option is to use gas-jet engines to this end.

These engines, which can be mounted, for example, on the projectile-bearing satellite, are used to correct the orbital parameters in such a way as to ensure that the projectile hits the target. At a certain time before the impact a diagnostic module should detach from the target-bearing spacecraft. The motion of the module with diagnostic equipment is then controlled in such a way as to ensure that the diagnostic module would be located at the same geomagnetic field line as the target at the time of the encounter between the projectile and the target. This will ensure that, given limited separation between the impact point and the diagnostic module, the plasma produced as a result of the impact passes through the impact sensors of onboard measuring instruments.

Such an approach to conducting the experiment imposes certain restrictions on the choice of the orbit and the meeting point between the projectile and the target, as well as the correction pulses for separating the target and the diagnostic module. The projectile can hit the target only if their orbits are determined with sufficient accuracy. A coordinate accuracy of about 1 m seems to be quite realistic for spacecraft equipped with appropriate receivers in the case of the use of a satellite global positioning system (GPS, GLONASS) (Nakamura et al., 2004). Multiple measurements make it possible to determine the orbital parameters accurately enough for the projectile to hit a several meters large target provided that the spacecraft use corrective engines.

Note that all the three space element of the system (the projectile, target, and the diagnostic module) should be equipped with GPS/GLONASS receivers.

It is important that the proposed configuration of the system allows multiple repeated attempts to be made in order to hit the target in the event of failure. The rate of the relative motion of the orbital planes depends on their altitude and inclination. For example, the rate of change of the longitude of the ascending node of a circular orbit with an altitude of 800 km and an inclination of 72 degrees is - 1°/day, and that of its corresponding orbit with an inclination of 98 degrees, +1°/day (the so-called sun-synchronous orbit). This means that the optimum relative position of the orbital planes of the satellites (i.e., when the two planes coincide) repeats with a period of 180 days. In other words, the waiting time for the optimum orbital conditions does not exceed 180 days whatever the initial longitudes of the ascending nodes of the two orbits.

The relative velocity of the two spacecraft in the optimum case described is equal to 14.9 km/s. For orbits with an altitude equal to that of the ISS (400 km) the maximum relative velocity is 15.34 km/s. The rate of change of the longitude of the ascending node of such an orbit (with an inclination of 51.5°) is about 5.1°/day, which means that the plane of this orbit coincides with the plane of the orbit of the oppositely moving spacecraft with an inclination of 128.5° once every 35.3 days.

Note that the collision of the projectile and the target can be achieved in any circuit regardless of the relative position of the spacecraft orbits, however, the relative velocity of the two spacecraft is lower than in the optimum case and is minimal in the case of the minimum angle between the vectors of orbital velocity at the time of the encounter.

In the limiting case (if we maintain the relation between the inclinations of the two orbits as adopted above) this angle is equal to 90° and the relative velocity at the time of the encounter reduces by a factor of 1.4 compared to its maximum value, i.e., becomes 11 km/s, which still meets the experimental requirements.

It should also be borne in mind that the requirement that the projectile and the target should move in orbits with inclinations equal to i and 180 - i is not strict. The two spacecraft can be launched into orbits with arbitrary inclinations, but, as we pointed out above, the result will be a lower relative velocity at the time of the encounter.

We so far considered only circular orbits and assumed that they have equal altitudes. If we abandon this assumption, we must take into account the increase of the characteristic velocity needed to make the two spacecraft meet. Furthermore, we may also consider a more costly and sophisticated version of the experiment with the satellites moving in highly elliptical orbits. In the extreme case a relative velocity of 20.5 km/s can be achieved for the projectile and the target meeting near the perigee at an altitude of 400 km. However, in this extreme case, when the satellite's speed approaches the parabolic velocity, the per kilogram launch cost increases by a factor of about four compared the case of a low circular orbit.

8.6. Onboard Measurements of the Structure and Mass of Organic Compounds

Let us now consider in more detail the set of onboard scientific equipment to be included in the diagnostic module of the magnetospheric experiment.

We already pointed out above the special importance of the onboard mass spectrometer for the measurement of the mass composition of complex organic compounds synthesized in the plasma torch. However, to accomplish the task, we must supplement these measurements with other onboard and ground-based measurements and a number of other ancillary measurements.

High-speed imaging of the development of the plasma torch will provide a wealth of data. Spectrometric measurements of the radiation of the flare plasma in the optical, UV and infrared parts of the spectrum, as well as optical observations with multichannel detectors capable of detecting circularly polarized radiation of the plasma torch will also be of use.

All these ideas concerning the instruments to be incorporated in the diagnostic module are self-evident, however, these basic instruments may be supplemented by other measurement techniques and new devices in the experimental design process.

Optical methods of this type, except for polarization measurements, have been used extensively in the Deep Impact space mission. A description of the onboard and ground-based optical diagnostic of this mission can be found in the paper of McFadden and A'Hearn (2004), where they describe the preparation and realization of this experiment. It is important that in the proposed experiment we can use unchanged some of the methods employed in the Deep Impact mission. Other methods may require certain modifications, e.g., in terms of the speed of operation because of the relatively small characteristic size and short time span of impact experiments in the Earth's magnetosphere.

During the preparation stage for the magnetospheric mission of each of the optical methods should be discussed in detail in order to refine the basic spectral and temporal characteristics of the onboard and ground-based equipment.

The TOF technique and the LIMA-D instrument (Managadze et al., 1987) shown in Figure 50 and developed at the Space Research Institute of the Russian Academy of Sciences for the Phobos mission should be viewed as the best solution for measuring the masses of polymeric organic compounds synthesized in the plasma torch. This instrument was meant for the determination of the elemental and isotopic composition of regolith from the flyby space probe from a distance of ~ 80-100 m. The LIMA-D instrument consisted of the following functional units: an infrared laser operating in the Q-switched pulsed with a tunable lens, laser rangefinder to measure the distance from the surface to the flyby system, a reflector to focus ion packets before they hit the detector, and a detector with a registration system.

The instrument operated as follows. During the flyby of the spacecraft above the surface of Phobos in the height interval from 30 to 80 m the control system, after determining the distance to the surface via rangefinder, focused the lens for this distance to produce a 1.5—2 m diameter laser spot and adjust appropriately the voltage at the grids of the reflector. This procedure was followed by a laser exposure producing the plasma torch where ions were generated and accelerated. Ions sputtered from the surface into the upper hemisphere and entered the reflecting field of the reflector, formed a narrow packet, and reached the detector. The registration system started operating simultaneously with the laser exposure. The first to reach the detector were hydrogen atoms, which were followed by other ions according to their mass, and the last came uranium ions.

It should be borne in mind when replacing the laser TOF mass-reflectron by a TOF impact instrument that in the latter case polyatomic ions consisting of the constituent elements of the projectile and the target should be generated in the impact-produced plasma torch. Therefore, the basic underlying concept of the LIMA-D instrument can also be applied for the development of a remote mass-reflectron for impact shock experiments in space.

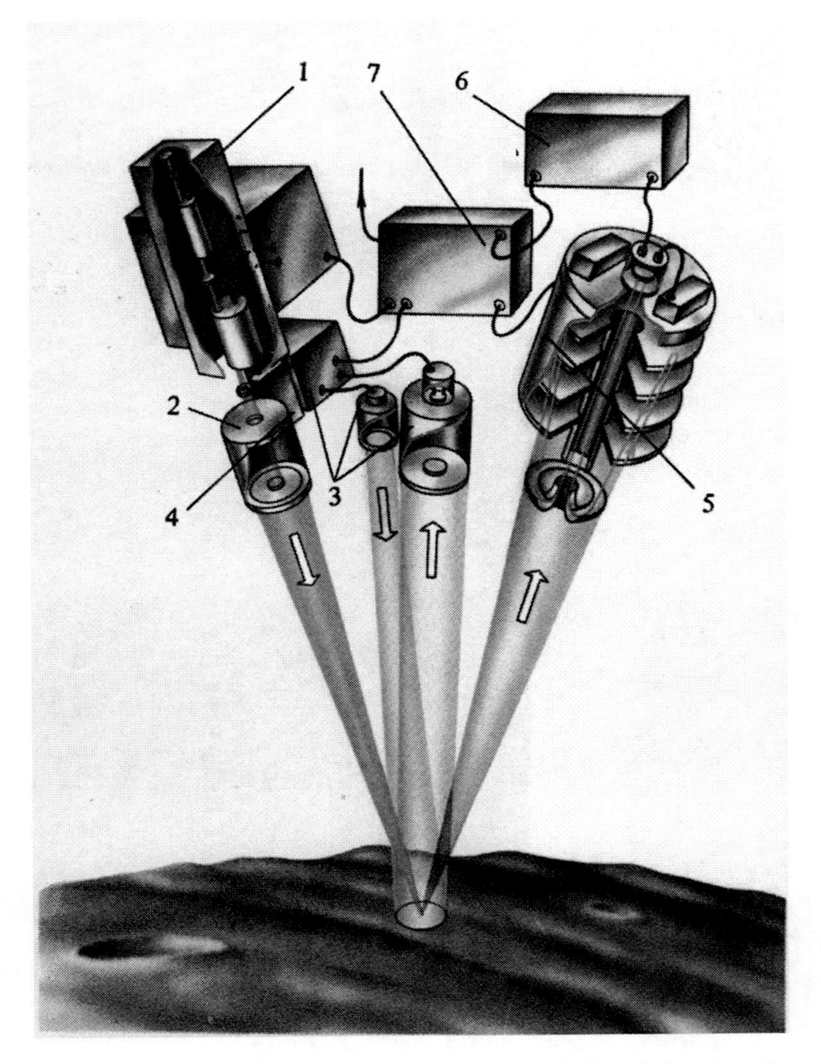

Figure 50. LIMA-D remote laser time-of-flight mass spectrometer. Key: 1. Laser; 2. Focusing lens; 3. Range finder; 4. Aperture lens drive; 5. Reflector; 6. Registration unit; 7. Control unit.

In this case, the distance between the target and the diagnostic module, which is measured using the range finder, should be used only for the control of the reflector grids, whereas the registration system should be started by the light flash generated by the plasma torch. Note that laser and impact TOF mass-reflectrons are identical instruments, which record plasma ions and differ only in the method of the generation of the plasma torch. Note that modification and upgrade of the of the LIMA-D class remote laser instrument to adjust it to the new tasks of the experimental study of hypervelocity impact processes should be a rather difficult task. However, the exceptional exceptional importance and urgency of the problem studied, and the available know-how will allow the task to be performed rather fast if necessary.

Figure 51 shows the possible configuration of the target, the analytical instrumental unit, and trajectory of the projectile. The organic compounds synthesized in the plasma torch will be recorded by the onboard mass analyzer, which will then transmit the data to the Earth.

The new generation of instruments should be developed with the much higher ion flux in the impact experiment compared to the laser experiment in mind. This circumstance, on the one hand, allows the sensitivity, weight, and size of the analyzer to be decreased severalfold, and, on the other hand, increases the operating distance from the target to the diagnostic module to 2 km compared to the maximum range of ~80 m for the LIMA-D instrument.

The realization of an experiment of this class is of great importance. This is because so far no controlled experiments involving hypervelocity impact interaction with a projectile of diameter exceeding a fraction of a micron have been performed under laboratory conditions. An almost millionfold increase in the projectile size in magnetospheric experiments will make it possible to generate a large-scale plasma torch and perform mass-spectrometric measurements of complex organic compounds synthesized in the plasma. This will make it possible to obtain unique and hitherto unavailable scientific results, which, without exaggeration, will trigger a breakthrough in the study of a number of unknown properties of hypervelocity impacts. Space experiments of this class, along with the origin-of-life problem, may be of great interest to many expert researchers and engineers whose work involves hypervelocity impact processes in nature and technology.

Figure 51. Configuration of the controlled "Head-on collision" experiment aimed at the study of the properties of the plasma torch produced by a hypervelocity impact in the near-Earth magnetosphere.

An important feature of the proposed magnetospheric experiments is their relatively low cost. They can be performed with onboard ballistic missiles required to be destroyed. Therefore "head-on collision" type experiments can be carried out within the so-called "Low-cost mission” program. Passive elements of the experiment, which include the target and projectiles, as well as written-off launch vehicles, would not require significant investment. The design and production of the diagnostic equipment can be simplified due to the relatively brief duration of the experiment, like it was done in the series of the first magnetospheric experiments involving artificial effects on the environment carried out within the framework of national (Hess et al., 1971), Echo (Hendrickson et al., 1976) and international projects, ARAX (Sagdeev et al, 1977), "Zarnitsa” (Cambou et al. 1975; Lyakhov, Managadze, 1977), and "Georgia-Spurt " (Managadze et al., 1983).

The urgency and importance of this scientific problem, combined with the low cost of its experimental realization are evident indicators of the realizability of these experiments in the near future.

***

The original on-board instruments for detecting signs of life on Europa discussed in this chapter are based on mass-spectrometric methods. These instruments represent compact new-generation onboard equipment ready to use, if necessary, in the forthcoming space missions.

Mass-spectrometric methods are not a random choice. The decision is determined by high reliability and informative value of these methods whose physical principles ensure the optimum mass resolution combined with extreme sensitivity. The techniques were selected from among the currently available laboratory instruments, after a detailed study of the basic principles of their operation and analytical characteristics. The main selection criteria included, in addition to the analytical properties mentioned above, such characteristics as simplicity of the design, high reliability, and the possibility of reducing the size of the instrument and performing measurements without prior preparation of the sample. The selected instruments were fundamentally redesigned based on their underlying physical principles and adapted to open-space conditions. As a result, an entire series of newly-designed onboard mass spectrometers was made that are meant for addressing a number of space problems of great importance.

The efficiency of the work performed is evidenced by the fact that the new generation of the proposed instruments includes, in addition to the devices mentioned in this chapter, such well-known onboard research facilities as LIMA-D (Managadze et al., 1987) and DION (Managadze and Sagdeev, 1987) of the "Phobos" mission, the MTOF or Managadze-TOF device (Managadze, 1986) used in SOHO, WIND, and ACE missions (see, e.g., Oetliker et al., 1997). These instruments, albeit being rather old, remain popular until now. It is safe to say that they will be used in the future. There are serious reasons for this conclusion. It is safe to say that if the “impact” concept of the emergence of the first living organism proposed here becomes experimentally supported then the “head-on collision” method proposed in this chapter will become the most popular non-alternative way to study this phenomenon because providing the conditions for mass-spectrometric studies of the synthesis products will become a task of prime importance. Furthermore, the novel technique proposed for the generation of the torch plasma torch in impact interactions in the Earth's magnetosphere may also play the crucial part for the investigation of various properties of the impact plasma – a phenomenon, which, as it turns out, remains poorly understood until now.

CONCLUSIONS

In the concluding section of this book we discuss some important features of the new concept and the prospects for its further investigation.

The most important feature of the proposed concept is that its main ideas are based on the results of laboratory experiments and available natural material evidence, and this was mostly due to the fact that the main idea of the concept was suggested by an experiment.

The local nature of the meteorite impact underlying the concept – the fact that the processes that accompany impact actions are totally independent of the conditions that existed on the Earth's surface – allowed us to circumvent or ignore at the very first stage of the work the traditional difficulties and inconsistencies typical of generally agreed early scenarios.

These difficulties were mostly due to the lack of data on the composition and nature of the atmosphere of the early Earth, low concentration of key organic compounds, and lack of clear natural mechanisms for the complexification of these compounds in the presence of radiation, and some other equally important conditions and factors.

The spatially isolated nature of meteorite impact allowed the proposed concept to coexist with reasonable scenarios earlier proposed for the prebiological stage of evolution.

Before the development of the concept began, meteorite bombardment was known to be the most powerful external factor in the geological history of the Earth. This bombardment, which lasted ~600 million years, was the final stage in the formation of the planet. Bombardment was known to have been produced mostly by cometary nuclei and planetesimals, which were the primordial bodies of the Solar System with a chemical composition close to that of carbonaceous chondrites. Furthermore, bona fide information was available about the approximate chemical composition of these bodies. In particular, the concentration of carbon in these bodies was known to have varied from 3 to 15% and to have included all biogenic elements. The maximum characteristic sizes of meteorite bodies that have fallen onto the Earth during the first 100 million years of its existence could have been as large as one thousand kilometers.

It was also known at the time of the early stage of the development of the new concept that hypervelocity impacts of meteorite bodies onto the Earth's surface produced powerful outbursts of high-velocity and dense plasma — plasma torches, which preceded the processes of the formation of impact craters.

Organic compounds were also detected in interstellar gas and dust clouds. Their origin – in the absence of other, more realistic mechanisms – was explained by the synthesis of these compounds at the surfaces of dust particles at a temperature of ~ 4 °K. From the very beginning, this mechanism of the synthesis of molecules at such low temperatures seemed too hard to implement.

It was believed that the long period of catastrophic impacts must have affected the geological history of the Earth. Many researchers believed that impacts from space could have triggered the beginning of the biological history of the planet. However, how could this happen has remained a puzzle.

The authors of a number of earlier published studies presented in Section 1.2 investigated meteorite impact effects. However, they addressed mostly the processes of the interaction between the shocks generated at the time of the meteorite impact onto the Earth's surface and the atmosphere. In some of these experiments involving laboratory modeling of such processes, individual amino acids and similar structures were synthesized.

Other studies of the physical effects caused by relatively low-velocity meteorite impacts assumed with no justification that impacts with velocities from 6 to 10 km/s fail to produce a plasma torch, and that such events may produce only media in a gas-vapor state. It is important that the selection of similarity parameters for laboratory simulation of a low-velocity impact also ruled out completely the possibility of the generation of the torch plasma. That is why the synthesis of simple organic compounds exclusively was observed in these experiments, as well as a slight complexification of these compounds. Section 1 of the Afterword discusses in detail the possibility of the synthesis of new substances under the conditions of the generation of a plasma torch produced by low-velocity impacts. The synthesis is shown to begin at impact velocities of 2 and 6 km/s in the case of iron and carbon projectiles, respectively.

In this connection, it is appropriate to recall that the author of this book was the first to propose the hypothesis that organic compounds potentially conducive to the emergence of living substance could have been synthesized in the plasma torch of a meteorite impact, which may also have triggered the generation of processes resulting in the ordering of matter, its complexification, and breaking of the mirror symmetry of molecular structures.

The first data on the plasma properties conducive to the synthesis of organic compounds became available in early last century. It was then shown in many laboratory experiments that some key monomers could be synthesized in simple configurations of plasma media, e.g., in glow discharges, disruptive discharges, lightnings, or cold ionospheric plasma.

Plasma torch has a unique position in the hierarchy of plasma media creatable on the Earth's surface. This is due to the following important advantages of the torch:

- High occurrence rate of this phenomenon on the early Earth;
- Enormous input of energy into plasma due to high initial velocities of meteorites and the processes of their further acceleration in the gravitational field of the planet.;
- Great variety of the initial conditions in the torch due to the inhomogeneity of meteorite bodies and large scatter of their chemical composition;
- The large number of various plasma instabilities associated with the three-dimensional expansion of nonequilibrium, dense, and hot plasma.

All these properties are of crucial importance for the synthesis, assembly, and ordering of prebiotic macromolecular structures needed for the emergence of living matter.

No such combination of properties can be found in other natural or artificial formations except laser plasma.

Before the development of the new concept we did not know that impact-produced plasma torch is, as a remote analog of plasmachemical reactor, an ideal medium for the efficient synthesis of organic substances. These properties of the torch medium were found experimentally in the process of the development of the new concept. It was also found experimentally during these studies that ordering of matter occurs in the torch plasma in the process of the synthesis of macromolecular structures.

Note that the authors of earlier published experimental studies of the torch plasma found evidence indicative of the possible breaking of mirror symmetry in local physical fields and also in spontaneous processes occurring in torches. This information of greatest importance has remained unnoticed for many decades. It proved to be of great use after the discovery of new properties of the plasma torch as a medium for efficient synthesis of organic compounds.

No one has thought before that hypervelocity collisions of dust particles in the interstellar medium, which result in their destruction, must have been accompanied by the efficient synthesis of a considerable amount of simple organic molecules. Laboratory simulations of hypervelocity particle impacts showed that 70% of organic molecules observed in the interstellar medium can be reproduced in laboratory experiments. These results were indicative of the correct choice of similarity parameters in the experiments simulating the processes of plasma synthesis in nature and in laboratory experiments.

Some of the aforementioned and hitherto unknown unique properties of the plasma torch have received reliable experimental confirmation in the process of the development of the new concept, whereas some of these properties proved to be less reliable because they only indicated the presence of the signs of certain processes. This was achieved not only by using the results of original, dedicated, and specially performed experiments, but also by analyzing a wealth of data, which, at first sight, had nothing to do with the problem considered, but proved to be naturally related to it. These results – e.g., those concerning the properties of electromagnetic fields in the torch or linearly polarized radiation emitted by the torch, – which were earlier published by other authors, included the data obtained in many plasma laboratories in the Soviet Union, Russia, USA, Germany, France, and Japan from 1960-ies and until now. The results of these studies often played the crucial role in the independent experimental confirmation of individual constructs of the proposed concept.

Thus the in-depth studies of the unique properties of the plasma torch concerning the assembly and ordering of matter in the case of purely carbon plasma showed that such a plasma medium maintains its capability for the organization of matter if a substantial amount of biogenic elements are introduced into it. This result allowed us to pass from the synthesis of purely carbon structures, e.g., carbines, fullerenes, and hyperfullerenes, to the synthesis of organic macromolecules. The experimentally discovered new properties are of special interest because they indicate that the processes of ordering occur in the case of the synthesis of organic molecules whatever the initial ratio of biogenic elements.

Thus multivariate analysis of mostly experimental data led us to the important conclusion that currently we can find no natural phenomenon which, by the combination of the major properties needed for creating the conditions for the emergence of living matter, would be comparable to meteorite impact. That is why meteorite impact has no alternatives as a phenomenon providing processes ranging from the delivery of the necessary biogenic elements into the formation region of the plasma torch to the introduction of the macromolecular structures produced in the torch into deep-seated layers of cosmic bodies, and creating of a «survival zone». This conclusion is corroborated by the properties of both the meteorite impact and torch plasma. The concentration of energy in a meteorite impact is at least several hundred billion times higher than the concentration of solar-radiation energy. The mass density in the plasma is several million times higher than the atmospheric density. Because of plasma-driven catalysis, the rate of plasmachemical reactions in this medium exceeds the average rate of chemical reactions under natural conditions by a factor of several hundred million. The torch plasma has unique ordering capabilities, which result in the assembly of complex macromolecular structures of organic compounds in irreversible plasmachemical reactions that occur in the process of plasma expansion-away.

The above properties of the meteorite impact and plasma torch allow us to view the natural phenomenon of impact as an event that is capable, without fundamental difficulties or inconsistencies, of creating – in extremely rare cases and after countless trials – the conditions for the synthesis of macromolecular structures possessing the properties of living matter.

The identical nature of processes occurring in the impact- and laser-produced plasma torches was an important factor that ensured the fast progress in the study of the problem. This similarity allowed many results obtained in the studies of laser plasma to be extended to the impact-produced plasma.

Before the work began on the development of the new concept, experimental and theoretical studies of the torch plasma were limited to viewing the torch as a source of ions for mass spectrometry or as a medium for the realization of laser fusion reactions.

Note that before we began the development of the new concept no one viewed the plasma torch as a medium related to the origin-of-life problem. Neither could we find any publications either on abiogenic synthesis of organic compounds in the torch plasma or on the problem of the breaking of mirror symmetry in such a medium. This lack of activity in the field could be due to the fact that no one viewed (and many researchers do not view even now) the high-temperature plasma of the torch as a medium where "fragile" and ordered macromolecules of organic compounds could be synthesized. Hence this book covers all the research that has been done concerning the plasma concept. The material presented in this book, which covers the initial stage of the work, contains only a small fraction of the planned theoretical and experimental studies of plasmachemical processes occurring in the torch.

We chose Altstein's protoviroid as the primary form of living matter because the original Altstein's hypothesis (the hypothesis of progenes) provided a concrete and clear description of the mechanism of the simultaneous origin of the first gene, its coded protein, and the simplest genetic code. According to the proposed model, the protoviroid consisted of two macromolecules — a polynucleotide-gene (~ 300 monomers) and a polypeptyde — a processive polymerase (~ 100 amino acids). Protoviroid could proliferate via replication, transcription, and translation. These processes were fundamentally similar to the corresponding present-day processes, and they involved nucleotide triplets associated with nonrandom amino acids. Protoviroid evolved in accordance with Darwinian principle: «inheritance — variability — natural selection». It was proposed as the first living being on the Earth — the progenitor of the biosphere. The molecular mass of the protoviroid could be of about $\sim 10^5$ a.m.u. The extrapolation of the results of laboratory simulations of the impacts of 10- and 100-μm diameter micrometeorites showed that the impact of a 1000 μm projectile in the absence of the atmosphere would make the resulting plasma formation region sufficient for the synthesis of macromolecules with the mass and structure comparable to those of the corresponding «protoviroid» macromolecules.

This may mean that a meteorite impact has unlimited capabilities for the synthesis of protoviroid-like macromolecular structures. Thus, according to coarse estimates, about 10^{45}—10^{46} carbon atoms have been delivered to the Earth during the first 200 million years of its existence. Efficient synthesis of macromolecular structures in the plasma torch within a small plasma-formation volume could ensure the production of at least 10^{40} molecules with masses close to that of the protoviroid via the realization of the above-mentioned «countless number of trials».

It is important that the concept was meant to overcome the contradictions that have plagued many early and modern scenarios of the origin of living matter. It can overcome many difficulties associated with the conditions on the early Earth and inconsistent with the underlying ideas of the «RNA world» hypothesis. It also makes sense «trying» to use the concept to eliminate the difficulties of other modern scenarios where life appears underground or in the depths of an ocean.

The existence of life not only on the Earth's surface, but also in kilometers-deep crust layers and in benthic ocean has been leading an increasing number of researchers to adopt the idea that life could have originated in these very places. In this connection, the proposed «impact» hypothesis may easily and clearly explain the role that the "penetrating properties" of the meteorite and "synthesizing properties" of plasma could have played in this process. However, the main consequence of the existence of subsurface life on the Earth is that the approach in question expands considerably the potential habitat of extraterrestrial or alien life. The main idea consists in the following. Many Solar-System bodies with extreme surface temperatures may be heated to moderate temperatures at large depths by the energy of tidal forces, volcanic activity, or meteorite impacts, and these processes which may provide the «underground» layers of the bodies not only with heat, but also with water. Such regions on other planers or planetary satellites do not differ from our Earth's «underground». Hence they may host «deep-seated life» like that found on the Earth, which, like the life on the Earth, could have been "conceived" by meteorites. The possibility of the realization of such a scenario under natural conditions is consistent with the basic laws of nature and is the most important of consequence of the proposed concept.

Note that the proposed concept includes a scheme of successive realization of the natural processes that accompany the meteorite impact and that are currently known with different degree of confidence. The important stages of this sequence include:

- Bona-fide known processes of the delivery of raw materials by meteorites;
- Experimentally confirmed processes of the synthesis of macromolecules with elements of ordering and organization of matter;
- Experimentally observed signs of the development of chiral physical fields and spontaneous processes, which may result in the symmetry breaking;
- The undisputed possibility of the formation of a «survival zone» in the impact crater.

The impact processes meeting the most important requirements concerning the synthesis of homochiralic macromolecular structures could have created the conditions for the emergence and preservation of protoviroid-like living substance and ensured its evolutionary development during the initial period of time.

The new concept inspires hope that the plasma processes discussed above could have created the conditions for the «animation» of inorganic substance and formation of protoviroid-like macromolecular structures. However, currently we have only hypothetical outlines of such a transition from «dead» to «live» matter with no bona fide information as to how this could have happened in natural impact processes.

The existence of life indicates that there must be such a way. In this connection, it would be reasonable, in order to find the «lifeline road», to consider, first and foremost, the processes that may occur in the plasma torch.

Many researchers suggest that «stochastic chemistry» should be used to cross the «chasm» between «nonliving» and «living» matter. This very chemistry could ensure the formation of primary forms of living matter. In the opinion of V.Goldanskii and A.Avetisov, the emergence of life should have been followed by the emergence of the new «algorithmic chemistry» typical for biological processes – a precise and «preprogrammed» chemistry. The «operation» of «algorithmic» chemistry may have been preceded by that of an analog of «stochastic» chemistry — the «fluctuation chemistry» proposed by I.Prigogine. In this case, the above hypothetical properties of the processes capable of producing the required macromolecules could have determined the main properties of «fluctuation chemistry». Thus the processes that characterize «such» chemistry should have been highly nonequilibrium and unstable and be far from the thermodynamic branch of equilibrium. This could ensure the formation of «dissipative structures» with intense fluctuations, which result in ordering, spontaneous symmetry breaking, and bifurcation processes. All this could have contributed to the development of an entire class of nonlinear phenomena, where fluctuations could have played the crucial part. Such processes usually occur in chemical explosions involving free radicals and a large number of molecules. These phenomena could, in the opinion of I.Prigogine, result in the development of a new field of science — the «fluctuation chemistry».

A comparison of the properties of the processes of «fluctuation chemistry» with those that occur during the explosive expansion-away of the torch plasma suggests that the two types of processes can be viewed as similar with one exception. In particular, the processes that occur in the torch are of plasmachemical nature and should be described in terms of plasma physics and therefore the corresponding field of science should be classified as the «fluctuation, or stochastic plasma chemistry».

The experimentally discovered possibility of applying plasmachemical laws to describe the processes that create the conditions for the emergence of primary forms of living matter may change substantially the main direction of future experimental and theoretical studies. Hence we will therefore possibly have to invoke the «fluctuation chemistry» to understand the processes of the synthesis, assembly, and ordering of organic molecules and the breaking of mirror symmetry during their emergence.

Studies in this direction should harmonioously combine dedicated experimental and theoretical research. In the nearest future, experiments aimed at the study of hypervelocity impacts should not be limited to direct impact experiments or to the simulation of this process under laboratory conditions. Experiments should be «carried» into space in order to increase the characteristic projectile size by several orders of magnitude. Therefore the novel technique of direct, controlled, impact experiments in the Earth's magnetosphere should provide unique results not only in the field of the synthesis and ordering of macromolecular structures in the plasma torch, but also in the study of the dynamics of impact processes for relatively large meteorites. The results of these experiments will contribute to setting new tasks for theoretical research in the plasma chemistry of the torch, and ensure fast progress in this promising direction.

Given the viability of the proposed concept for the conditions needed for the emergence of primary forms of living matter, it would suffice for a system of planets to form around the newly-born star. If there are planets, everything else, e.g., meteorite bodies moving at very high velocities, water, biogenic elements, a plasma torch as a region of the synthesis of macromolecules and as a «zone» of the emergence of living matter, would be created automatically in impact processes. The properties of these processes can be considered to be inevitable and unavoidable. Does the concentration of the crucial conditions necessary for the emergence of primary forms of living matter in a single natural phenomenon mean that life is bound to emerge? Or, perhaps, that extraterrestrial forms of life – close analogs of terrestrial forms – should be widespread on many cosmic objects of the Solar System and beyond?

Life was very likely to have emerged in random processes, but in strict agreement with the laws of nature. The insurmountable difficulty of solving the puzzle of the Eternal Problem stems from the fact that the most important among these laws remain unknown because they have not yet been discovered.

If the proposed concept is correct, then microbial life may be found in the future in the depths of many Solar system bodies. This primarily concerns Mars, Europa, Enceladus, and Titan, and, moreover, various forms of living matter may also be found in the depths of Mercury and Venus provided that these planets have subsurface water reservoirs.

Thus, given moderate temperature and the availability of water in the depths of any cosmic object, everything else that is needed for the emergence of living matter can be provided by two factors. One of them ensures the synthesis of complex macromolecular structures of organic compounds in the plasma torch. The second factor ensures the delivery of the products of synthesis to the depths of such objects via an impact. We consider it very important that the proposed concept faces no insurmountable or fundamental problems as far as the implementation of the proposed scenario under natural conditions is concerned, both on the Earth and beyond. We also expect the proposed concept to be experimentally confirmed in the next few years.

The most important results obtained in the process of the development of the new concept can be briefly formulated as follows:

- A new concept of the prehistory of life at the stage of prebiotic evolution, and a novel scenario of its realization in natural processes. According to this concept, the conditions needed for the emergence of primary forms of living matter and the media making possible the survival and evolutionary development of living substance could form in natural processes accompanying the hypervelocity impact of meteorite bodies onto the Earth's surface at early stages of the formation of the planet.
- The fact that the experimentally discovered and hitherto unknown properties of the torch plasma generated by a meteorite impact could, in the process of the expansion-away of this plasma, make possible the synthesis of various substances ranging from intermediate low-molecular reactive compounds to complex organic macromolecular structures. The masses and structural properties of the synthesis products were thoroughly and synchronously studied during model experiments performed using a time-of-flight mass-reflectron. To deliver the synthesis products to more complex measuring instruments, these products were collected on supports. In the process of

the analysis of the synthesis products we identified such compounds as carbines, fullerenes and their "onion-like" modifications, and up to 12 biological amino acids. A number of compounds were, by the distribution and periodicity of the mass peaks in the spectrum, interpreted as nucleoitides and their oligomers, highly branched acetylenic hydrocarbons, and fourth-generation dendrimers.

- Experimental evidence indicating that the processes that occur in the impact-produced plasma torch can be very reliably reproduced, as far as the synthesis of organic compounds is concerned, in the plasma torch generated by the radiation of a laser operating in Q-switched mode. This means that impact- and laser-produced torches are highly similar, and that laser-produced torch, as the best model of the impact-produced torch, can be used to study impact-produced plasma.
- The experimentally discovered mechanism of the synthesis of organic compounds may operate not only on the Earth, but also in hypervelocity collisions of dust particles in interstellar gas-and-dust clouds. We justify the assumption that the plasma mechanism is the only hypothetical mechanism of the synthesis of organic compounds in interstellar clouds where a close to absolute zero ambient temperature cannot prevent the high efficiency of the synthesis of organic compounds.
- The experimentally observed ordering effect in the synthesis of molecular structures ranging from carbines to hyperfullerenes in a purely carbon plasma also shows up in the synthesis products obtained when a substantial amount of biogenic elements are added to carbon. This result makes it possible to ensure the ordering of both the carbon structures and organic compounds, which is the crucial factor for the assembly of macromolecular structures. This is evidenced by the experimentally observed possibility of the synthesis of highly branched molecular structures of organic compounds.
- Experimental studies of the main properties of the torch plasma showed that its high catalytic activity can be explained by the special state of the highly overheated plasma medium. The strongly nonequilibrium plasma medium that forms in the process of a hypervelocity impact and that is far from the thermodynamic branch of equilibrium and has the properties of "dissipative structures" may result in high degree of ordering of the the medium and synthesis products and contribute to the breaking of mirror symmetry in the process of the formation of enantiomers.
- The conclusion based on the results of earlier published experimental studies of electromagnetic fields and plane-polarized radiation of the torch that nonequilibrium electric and magnetic fields of the torch plasma have constant orientation in space and therefore are «intrinsically» asymmetric. They meet the main requirements to local chiral physical fields and may cause a minor breaking of the mirror symmetry (but with a constant sign). The well-defined predominance of a certain direction in the distribution of linearly polarized radiation may, in turn, be indicative of the high degree of ordering of the medium and provide indirect evidence for the tendency for spontaneous symmetry breaking in the plasma torch. Of the two hypothetical factors discussed above the former may have determined the "sign" of the asymmetry of the bioorganic world, whereas the latter could have ensured the homochiralic nature of the macromolecular structures synthesized in the torch.

- A joint study and a comparison of the experimentally confirmed properties of the torch plasma that ensure the ordering of macromolecular structures in the process of their synthesis with the hypothetical possibilities of the breaking of symmetry in the case of simultaneous realization of these processes. Such conditions are believed to be able to result in the apperance of factors necessary for the emergence of protoviroid-like primitive forms of living matter. Such a structure consisting of two oligonucleotide and peptide molecules connected by primitive genetic code may have been able to replicate in the case of the molecular mass of $\sim 10^5$ a.m.u. According to the estimates performed, such a molecular structure could have formed in the plasma torch during its expansion-away and later in the impact crater before the end of its formation.
- The unique properties of the plasma torch concerning the synthesis of macromolecular structures, combined with the exceptional properties of the meteorite impact capable of ensuring the generation of the torch plasma and the synthesis of organic molecules in the depths of cosmic objects, can be viewed as the most likely scenario of the emergence of living matter in the depths of the Earth's crust. Such processes could have also occurred on other cosmic objects with extreme surface temperatures provided that there was a source of heating sufficient for the emergence of water. The processes of a meteorite impact, which are similar to the corresponding processes on the Earth, could have ensured the emergence of extraterrestrial life by substantially expanding its habitat.

The understanding of the physical bases of the new concept allows the deep-rooted perceptions to be changed fundamentally: the impact of a major meteorite involving the formation of a giant plasma torch should not be viewed only as the most powerful and destructive planetary-scale influence with catastrophic consequences, which is capable of destroying the entire civilization in a flash. This knowledge allows us also to see the constructive effect of such a grand phenomenon, which may have been responsible for the very existence mankind and the unique form of matter, which the people tried to enjoy in the past and, we hope, will continue to enjoy in the future for many millennia to come.

AFTERWORD: PLASMA PROCESSES IN THE DEEP IMPACT MISSION, MECHANISMS OF THE GENERATION OF CIRCULAR POLARIZED EMISSION AND A HOMOCHIRALIC MEDIUM

In this section we present the results of experimental and theoretical studies from various fields, which indicate that the properties of the plasma medium created in the impact process are by no means limited to the processes already discussed in this book. In particular, the properties found in recent years may facilitate the discovery of so far unknown mechanisms, which expand the potentialities of the plasma medium in providing the conditions for the emergence of living matter. These processes and mechanisms should be viewed as further material evidence for the realizability of the proposed concept.

1. PLASMA PROCESSES IN THE DEEP IMPACT MISSION

1.1. Introduction

The first active space experiment where a cometary nucleus was subject to a high-velocity impact was carried out on July 4, 2005 within the framework of the Deep Impact mission (A'Hearn, 2005). During this mission a 370-kg impactor with a density of 3 g/cm^3 released by the main module crushed into the nucleus of comet Tempel 1 at a velocity of 10.2 km/s.

The scientific program of the mission focused on the study of the structure and subsurface composition of the cometary nucleus, the processes that accompany the impact, and the formation of the impact crater. The mission also addressed ballistic problems related to preventing asteroid (cometary) danger.

Of special interest for further development of the concept proposed in this paper are the results of laboratory experiments involving the simulation of impact processes at the particle accelerator in Ames. The results of the natural experiment in space should have become the culmination of all these studies.

The special interest in the results of the Deep Impact mission was due to the plasma processes that develop in relatively low-velocity impacts. Thus the results presented in

Section 4.4 of this book show that in laser simulations of impact processes the maximum mass of synthesized molecular structures – 6500 a.m.u. – was achieved for the power density of $W_L \sim 5\ 10^8$ W/cm^2. This W_L value allowed only the surface ionization of the target to be achieved, which for impact processes corresponded to shock compression produced by ~ 2g/cm^3 projectile moving at a velocity of 5-7 km/s. Such impacts are known to have produced only surface ionization of the target.

Note that long before the Deep Impact mission was carried out, independent experiments performed in Ames and Heidelberg to reproduce impacts using the particle accelerators had shown conclusively that such processes are accompanied by the formation of a plasma torch. Plasma formed if the projectile moved at velocity higher than or equal to 2 to 5 km/s and had density of 8 to 2 g/cm^3, respectively. In Heidelberg experiments the formation of the plasma torch was accompanied by the (simultaneous) synthesis of new compounds.

In their series of works, Crawford and Schultz (1988, 1991, 1993, 1999) observed plasma effects to occur simultaneously with the generation of a plasma torch in the cases where 2-3 g/cm^3 and denser projectiles moved at the velocities of 4-5 km/s. The above authors observed plasma glow to measure the temperature of the medium and used these data to determine the plasma ionization degree. Plasma processes in the torch ensured the generation of a magnetic field whose magnitude and distribution were recorded by dedicated probes. Of special value for the concept proposed in this book are the results of the measurements of the electric charge at the surface of the dust component of crater ejecta.

The results of these experiments were of great scientific value and therefore deserve very close attention. During these experiments all the important processes have been recorded, which accompany the formation and expansion-away of the plasma torch except the detection of new compounds synthesized in the plasma medium.

Göller and Grün (1989) performed experiments on the particle accelerator in Heidelberg and obtained bona fide data about the generation of plasma outbursts over a wide range of impact velocities. These data agree with the results obtained by Crawford and Schultz (1988, 1991, 1993, 1999). However, the specificity of the work in Heidelberg allowed the synthesis of new compounds to be observed since the very onset of the plasma torch. These results have since then been repeatedly corroborated by the studies of Stubig et al., (2001, 2002); Stubig (2002), and Goldsworthy et al., (2002) carried out in Heidelberg and Canterbury.

Such convincing results concerning the possibility of the generation of a plasma torch in a low-velocity impact should have stimulated the search for and the discovery of plasma formations in the natural experiment performed during the Deep Impact mission. However, no such search has been carried out. This was supposedly due to the fact that the study of plasma processes was not considered to be a task of primary importance in the mission program. Such neglect can be explained by the small energy fraction contributed to the plasma processes – it is usually ~1% of the total impact energy in the case of above-critical velocity impacts (Zel'dovich and Raizer, 2002).

Given the exceptional importance of the results of the Deep Impact mission for the development of some aspects of the plasma concept, I decided to show that not only the formation of a plasma torch, but also that of some plasma structures linked organically to this natural phenomenon were observed during the natural experiment of the mission. The characteristic dimensions of the space and laboratory experiments allowed a number of the effects to be "examined closely" for the first time, and therefore the results should be viewed as unique.

Note that the main results of these studies are based on the experimental data obtained in the process of the Deep Impact space mission. These studies also used the data obtained in various high-precision laboratory simulations of impacts. Therefore the conclusions based on experimental data should be highly reliable.

Despite the small energy contribution to the processes of plasma formation or the possibly low ionization degree in the case of low-velocity impacts plasma processes should nevertheless play the crucial part in the dynamics of the expansion-away of matter as a result of the impact, in the formation of the torch, in the synthesis of new chemical compounds and, therefore, in the determination of the chemical composition of final products.

A number of published research papers focused on specific features of the dynamics of the formation and determination of the main parameters of the plasma torch produced by an impact with a velocity of $V_{IMP} \sim 50$ km/s (Kovalev et al., 2008). No dedicated studies of the plasma processes for V_{IMP} ranging from 1 to 10 km/s have been made with exception of those performed by Crawford and Schultz (1988, 1991, 1993, 1999). Moreover, many specialists who studied impact processes are unaware of the possibility of the generation of a plasma torch in the case of such low velocities.

Important results on this problem have been obtained in laboratories during the preparation for the Deep Impact mission. In these "direct" impact experiments low-velocity impact processes were reproduced using a particle accelerator. Thus the processes of the formation of an impact crater and mass ejection produced by a pyrex projectile (A'Hearn, et al., 2005, 2006) were studied using NASA's accelerator in Ames (Schultz, 1996; Schultz et al., 2005, 2006, Schultz and Anderson, 2005; Ernst and Schultz, 2003, 2005; Eberhardy and Schultz, 2004). The accelerator could drive a several-mm large projectile to a maximum velocity of $V_{IMP} \sim 7.5$ km/s and was equipped with state-of-the-art optical spectroscopic instruments, a high-speed camera, and other analytical instruments.

The experiments carried out within the framework of the Deep Impact mission using the accelerator in Ames involved a 0.635-cm diameter pyrex projectile moving at a velocity of $V_{IMP} \sim 5.5$ km/s. This impact velocity V_{IMP} combined with the projectile density of ~ 3 g/cm^3 corresponded to $W_{IMP} \sim 2 \cdot 10^{10}$ W/cm^2 and provided a shock compression of $\sim 10^6$ bar. According to optical measurements, the ambient temperature amounted to 5500 K, which, in the opinion of P.Schultz, could ensure an ionization degree of up to 6% for alcali ions of surface contaminations. The experiments also involved the measurements of the magnetic field generated in the process of the expansion-away of crater ejecta, and such field was indeed detected.

The above results and published data served as a starting point for my study of the plasma processes that occur in laboratory experiments and in the space experiment of the Deep Impact mission.

Below the main results of these studies are reported.

1.2. Peculiarities of the Expansion-Away of the Torch Plasma in Laboratory Experiments

Let us first consider the important information about the properties of the ejection from the crater obtained using a high-speed camera. These results included observational data and were therefore easier to interpret. That is why we analyze them first. We show these results in Figs. 7 and 8 (McFadden and A'Hearn, 2004).

The series of images shown in Figure 7 records radiating ejections from the crater corresponding to frames 2-6. They demonstrate how the structure of these ejections varied with time. In particular, the glow of the ejection shown in frame 2 had a blurred or diffuse boundary. The next images corresponding to frames 3 through 6 show a well-defined change in the structure of the ejection, because numerous fine filamentary jetlike features appeared.

Also of interest were the frames shown in Figure 8. The radiating ejection in the first frame was highly similar to a laser-produced plasma torch both in configuration and the nature of the glow. The velocity of the radiating ejection as determined from these frames was equal to ~3-5 10^5 cm/s, which is typical of plasma formations. The radiating ejection was running ahead of the expansion-away of the nonradiating cloud. As a result, the ejection broke into two fragments with different velocities. The intensity of the bright glow of the first ejection depended on the physical properties of the target and this glow lasted several hundred microseconds. It is evident from the third frame in Figure 7 that in the process of its expansion-away the high-velocity, radiating part of the ejection broke into individual vertical arms. A similar break-up, albeit on a smaller scale, also occurred in the nonradiating part of the ejection.

The results of a visual analysis of the frames obtained by a high-speed camera taking into account the presence of a magnetic field indicated that the impact-produced and experimentally observed radiating ejecta had many properties of a plasma medium and could be identified with a plasma torch. This was also evidenced by the structural peculiarities of these ejecta. In particular, the fact that they broke up into vertical arms and flew away as separate threadlike jets. The typical features of a plasma torch were its glow, dynamics of the expansion-away, and velocity of motion. The typical features of pulsed plasma formations also included their break-up into two fragments.

To more thoroughly and consistently analyze the above properties of the emerging medium, we determine the ionization degree of the ejection plasma in the cases where the measured temperature of the glowing medium was, according to the data of R.Schultz, as high as 5500 K.

In plasma physics the ionization degree of a plasma formation is usually determined using the Saha equation, which allows this quantity to be computed if the temperature of the medium and ionization potential of ejecta are known. Thus, according to estimates based on this equation (Raizer, 2009), ionization potential does not exceed 10% when a medium consisting of alcali metals with the ionization potentials of about ~ 5 eV heats up to 5500K. P. Schultz estimates the ionization degree at 6%, which can be viewed as a good agreement. Such a medium can be viewed as plasma with no restrictions.

The ionization degree for matter with a higher ionization potential – e.g., hydrogen, nitrogen, or oxygen – can be substantially lower and amount to 10^{-2} – 10^{-3}. However, even this formation should possess and manifest all the properties of a plasma medium.

We show below that for the 5500-K plasma produced as a result of an V_{IMP} ~ 5.5 km/s impact in a laboratory experiment performed in Ames the Saha equation yields an ionization degree that agrees with the experimental results obtained in similar experiments performed using a dust-particle accelerator in Heidelberg. However, we begin by commenting some of the structural features of above plasma formations as observed in experiments.

1.3. Filaments and Strata

The threadlike structures recorded in frames 3-6 in Figure 7 during the expansion-away of the radiating ejection can be explained by the emergence of the plasma jets of special structures, which were first observed during the expansion-away of a laser-produced plasma torch, and which are referred to as filaments in the research literature.

Filamentation (Bol'shov et al. 1987, Afonin et al., 2001; Borisenko et. al. 2008) is a phenomenon that is characteristic of adiabatically expanding plasma torches. It was discovered in a laser-produced torch and can be seen in Figure 52. The development of filaments or jets breaks the initially uniform plasma into individual threadlike structures. Filaments develop because of the nonuniform energy distribution on the laser focal spot. However, the same effect may be a result of the surface inhomogeneity of the target. Because of the identical nature of the impact- and laser-produced torch, filamentation may also develop in an impact-produced plasma torch if the projectile or the target are not enough homogeneous. The frames of a high-speed filming obtained in Ames can be viewed as the first-ever evidence for the formation of filaments in an impact-produced plasma torch. It is important that plasma filaments or jets occur in nature only where there is plasma because they are associated with plasma instabilities.

This also largely applies to the processes of the formation of "arms" shown in Figure 8. However, in this case they form via other plasma mechanisms. In particular, like in the previous case, such formations develop only in a plasma medium and are referred to as strata. Plasma stratification is known to develop as a result of the ionization-recombination instability and is a phenomenon well studied and easily recognized by specialists in the physics of weakly ionized plasma (Limpoukh and Rozanov, 1984; Afonin, 2001).

Upon a closer view, the configuration of the mass ejection from the crater shown in Figure 8 can be seen to exhibit signs of breaking into individual arms and nonluminous strata-like component. This process may occur if dust particles accumulate a substantial charge and the so-called dust plasma forms. This usually happens when the plasma medium is combined with the dust medium. Below we describe some of the properties of this unusual medium.

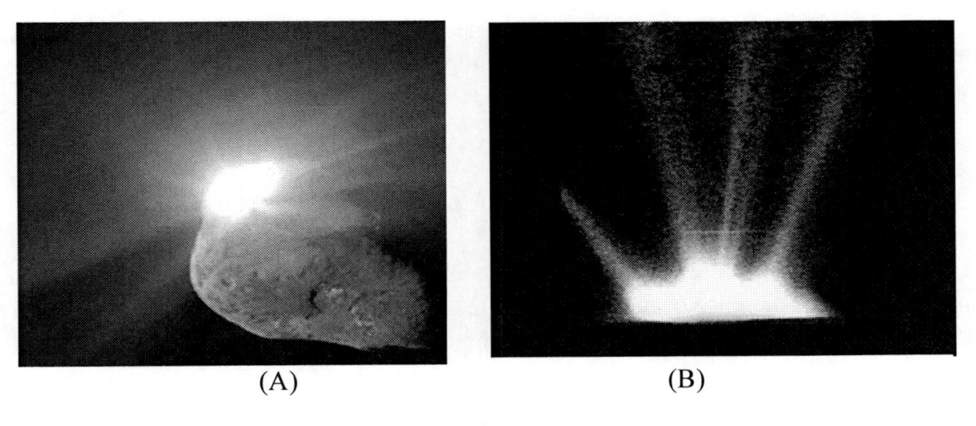

(A) (B)

Figure 52. Image of jets ejected at the late stage of the expansion-away of the matter from the crater in the Deep Impact mission (Schultz et al., 2006) (panel (A)). For comparison we show in panel (B) the image of the filaments that formed in a laboratory experiment (Bol'shov et al. 1987) involving a target subject to laser radiation with a nonuniform energy distribution in the focal spot.

The process of the breaking of plasma into two blobs recorded in experiments performed in Ames and shown in Figure 7 is also a characteristic feature of a pulsed and high-velocity plasma medium if it is created within a very short time interval. This process was observed not only in experiments performed with the accelerator in Ames, but also in the space experiment.

The breakup of the plasma ejection into two blobs can also be observed in the laser-produced torch. This effect was discovered in the early years of the torch plasma studies and the first results were published by Kim and Namba (1967) and Namba et al. (1966, 1967a, 1967b). Note that the second blob has usually escaped the attention of researchers. This was due to the second, cool blob containing a large amount of impurities in the form of ions of various elements and dust particles charged to a substantial potential. It is important that laser plasma always had a dust component to develop in it (Gurevich, 1940; Raizer, 1959; Anisimov and Luk'yanchuk, 2002), although its density was usually much lower than that of the dust component in the impact-produced plasma. The researchers were also unsatisfied with the very low temperature of the plasma of the second blob, which amounted only to ~ 1000K. All these factors prevented the use of the plasma of the second blob in addressing the important tasks of the plasma physics.

However, as far as the problems considered in this book are concerned, it is the cool part of either the laser- or impact-produced torch that may be most favorable for the synthesis of new compounds at low temperatures. The presence of the dust component, in turn, could have created the conditions for the generation of circularly polarized radiation, which is an important factor to cause the breaking of the mirror symmetry in the process of the synthesis of enantiomers.

An analysis of the experimental results mentioned above suggests that plasma media could form in the "hot spot" in the impact process. These plasma media can be characterized as follows. The first medium is classical ideal plasma characterized by a relatively high expansion-away velocity and radiating capacity. The second, nonradiating medium is low-temperature plasma of relatively high density. The third medium can be classified as nonideal dust plasma. The former fraction is easy to identify during the expansion-away process, whereas the latter two fractions cannot be spatially resolved because of their close expansion-away velocities. The observed variety of the media should create the necessary conditions facilitating not only the synthesis of organic compounds, but also the breaking of symmetry in this process.

1.4. Mechanisms of Ionization

Consider now the most important natural mechanisms capable of ensuring the formation of a plasma torch in the case of an impact of a 2—3 g/cm^3 projectile moving at a velocity of 5-6 km/s. In these cases the density-related and thermal effects of the ionization of the medium are usually studied (Fortov et al., 2003). Thus laser exposure and impacts are usually associated with thermal and density-related effects, respectively.

To understand the density-related effect of heating, we must first consider the results reported by Zel'dovich and Raizer (2002). They showed that the impact process is accompanied by intense heating of the medium and that the plasma formations observed in the experiments conducted in Ames could have developed as a result of the shock compression of the solid body.

Thus a substantial increase of the temperature of interacting bodies is hardly to be expected if the shock compression of the solid body produces a pressure of 10^5 bar. The processes that occur at such pressures are referred to as elastic or "cold" and are determined by repulsive forces. However, in such cases individual, irregular cases of plasma formation are observed. In the case of a higher impact velocity of 5 - 6 km/s the shock compression produced by a 3 g/cm^3 projectile raises the pressure to ~ 10^6 bar and the process is accompanied by the heating of matter due to the thermal motion of the lattice atoms. Such a pressure is referred to as "thermal" to distinguish it from elastic or "cold" pressure. Hence the increase of the shock amplitude results in the increased amplitude of shock compression and in such cases we must consider "thermal" pressure and expect the medium to heat up to 20 000 – 30 000 K and become ionized.

Experimental evidence for the ionization of the medium as a result of shock compression was obtained in a series of studies carried out using the dust-particle accelerator in Heidelberg (Stubig et al., 2001, 2002; Stubig 2002) and Canterbury (Goldsworthy et al., 2002). The results of these works include the experimentally obtained evidence of the development of the conditions for the generation of a plasma torch in low-velocity impacts. The results are based on numerous direct impact experiments involving the measurement of the concentration of plasma ions and plasma ionization degree as a function of impact velocity.

These studies were performed using a dust-particle accelerator where the potential difference of $2 \cdot 10^6$ V was created by an electrostatic Van de Graaf belt generator. This potential difference was needed to accelerate pre-charged microparticles to 60 km/s.

In these studies reliable results have been obtained concerning the formation of a plasma torch for impact velocities V_{IMP} ranging from 1 to 60 km/s. These results agree with those obtained in independent experiments carried out using a similar accelerator in Canterbury (Goldsworthy et al., 2002).

The measurements in the above experiments were made using highly sensitive TOF mass-spectrometric instruments meant to study of the properties of plasma ions that form in impact processes in space. These instruments are designed to determine the mass and isotopic composition of micrometeoritic streams near Jupiter using the onboard TOF dust-impact mass-spectrometric instruments (Göller and Grün, 1989).

It was shown, based on the results of these studies, that the flow of plasma ions in the mass spectra appears starting from an impact velocity of V_{IMP} ~ 2-3 km/s in the case of an iron projectile. When plasma appears, the synthesis of new compounds is observed. When the impact velocity reaches 5-6 km/s the ion flux increases. A further increase of the impact velocity translates into increased ion flux, which becomes a factor of 10 higher when V_{IMP} reaches 15-20 km/s. These measurements also showed that in the case of a carbon projectile the plasma torch begins to form at V_{IMP} ~ 5-6 km/s, mostly in the processes of surface ionization. The plasma contains mostly the elements of the projectile and target matter and also the contaminant elements present at their surfaces.

It is important that in the case of low and intermediate impact velocities the mass spectra exhibited not only the peaks of atomic ions, but also peaks of the ions corresponding to the constituent compounds of the projectile and the target and peaks corresponding to chemical compounds that form as a result of a combination of these ions in the synthesis processes.

Thus an important natural phenomenon has been experimentally demonstrated where the synthesis of new compounds in impact processes was observed after the development of plasma formations. This occurs when the compression shock pressure reaches 0.3 to 0.5 10^6 bar and the effect has been confirmed conclusively for iron, aluminum, carbon, and pyrex projectiles. It follows from this that in impact processes the generation of the plasma and the synthesis of new compounds are inseparably associated with each other. Note that the formation of plasma precedes the synthesis and the two processes constitute a necessary and characteristic combination that accompanies an impact.

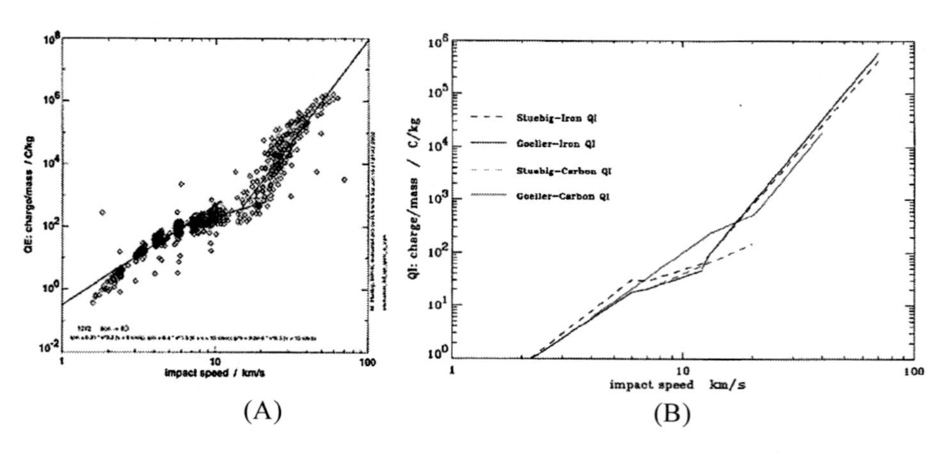

Figure 53. Dependence of the yield of plasma ions on impact velocity in the case of an iron projectile (A). Panel (B) compares the dependences of ion yields on impact velocity for iron and carbon projectiles as observed by Stubig (2002) (curves 2 and 4) and Goller and Grun (1987) (curves 1 and 3).

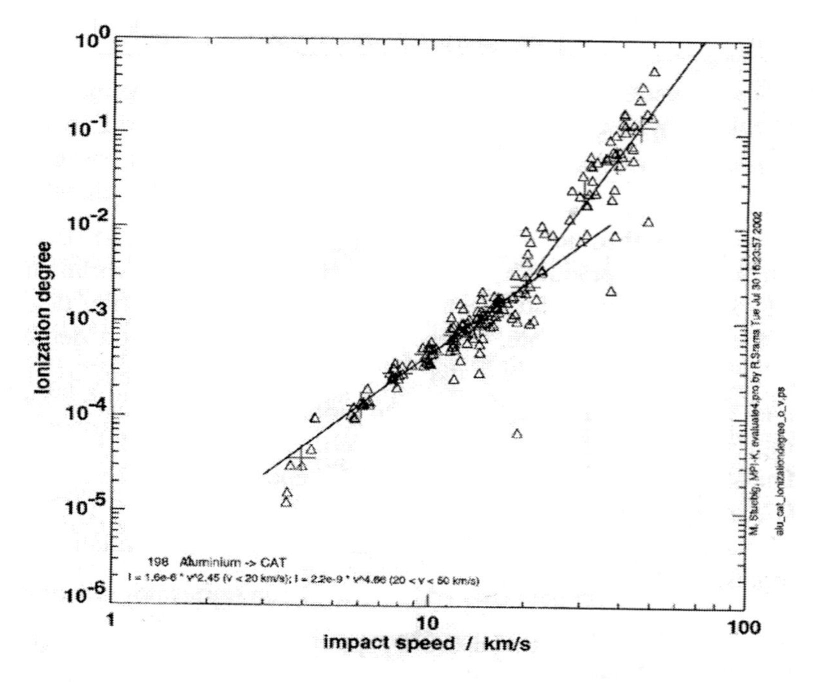

Figure 54. Dependence of the degree of ionization on impact velocity for an iron projectile.

Volume ionization of the projectile and the target was observed at impact velocities of $V_{IMP} \sim 15$-20 km/s and higher because at these velocities the fraction of energy spent for the processes of plasma formation became sufficient for the complete atomization and ionization of the matter of the projectile and of a small part of the matter of the target.

We used the results of these measurements of ion density, ion composition, and the ionization degree of the torch plasma to make the plots shown in Figs. 53 and 54. These plots demonstrate that the factor-of-ten difference in the ion density at $V_{IMP} \sim 5$-6 km/s and V_{IMP} ~15-20 km/s can be explained by a change in the ionization degree in the plasma. The resulting mass spectra of the ion composition of the plasma torch indicated that at low impact velocities the ion component included Na and K ions of surface contaminations whose mass peaks dominated in the spectra. However, at higher V_{IMP} the mass spectra began to be dominated by the peaks corresponding to the materials of the projectile and the target.

Figure 28b shows the mass spectrum taken at an impact velocity of V_{IMP} ~16 km/s in the process of interaction between a carbon projectile with a rhodium target for W_{IMP} corresponding to ~ $4 \cdot 10^{11}$ W/cm^2. This mass spectrum is indicative of the involvement of both the projectile and target material in the processes of plasma formation. The spectrum exhibits the mass peaks of both carbon and rhodium. It also shows a well-defined mass peak of CRh (115 a.m.u.), which arises as a result of the combination of Rh (103) and C (12). Recall that the mass spectrum shown in Figure 28a was obtained for the torch plasma generated by subjecting a mechanical mix of C and Rh to laser radiation. In this case too CRh was synthesized.

The above processes of the formation of a plasma torch agree with the results of the simulation of impact processes via laser exposure carried out at the Space Research Institute of the Russian Academy of Sciences and described in this book. This agreement was achieved after the quantity W_L, had been empirically determined that is equal to the power density of laser radiation corresponding to the fraction of power spent in the impact for the formation of the torch plasma. This quantity was found to depend on impact velocity. According to the results obtained for $V_{IMP} \sim 15$ km/s or W_{IMP} ~3 10^{11} W/cm^2, the processes of plasma formation consume ~ 1-2% of the impact energy and hence the above W_{IMP} value corresponds to $W_L \sim 3 \cdot 10^9$ W/cm^2. For $V_{IMP} \sim 5$-6 km/s W_{IMP} is equal to 10^{10} W/cm^2. In this case the processes of plasma formation should consume ~ 5% of the impact energy and the corresponding W_L value should be equal to ~ $5 \cdot 10^8$ W/cm^2. In laser experiments of impact simulation such a power density ensures with high degree of confidence the surface ionization of the target material. The observed effect can be associated with a substantial reduction of the crater volume in the case of low impact velocities, which is due to the possible redistribution of the impact energy not spent for the processes of cratering and plasma formation.

If combined, the above results are indicative of the generation of a plasma torch at $V_{IMP} \sim$ 5-6 km/s for a carbon projectile. These conclusions are based on the results of direct impact experiments carried out using accelerators of different type in Ames and Heidelberg, which employ different measurement techniques. They are also based on our own results obtained in a simulation of an impact via laser exposure. Moreover, these results agree with the results of theoretical and experimental studies of very high-amplitude shocks in solid bodies. Therefore the fact that the results of numerous and independent experimental and theoretical studies agree with each other can be viewed as conclusive evidence for the generation of a plasma torch under the conditions mentioned above.

Prior to discussing the results of the Deep Impact natural experiment it would be appropriate to briefly mention other low-threshold mechanisms of the ionization of the medium. Such processes produce a low-temperature, weakly ionized plasma where the ion component may include molecular ions. Such a formation has all the properties of a plasma medium.

Other mechanisms capable of ensuring the ionization of the medium include the well-known mechanisms of associative and Penning ionization (Biberman et al., 1982; Smirnov, 1982).

The former mechanism is based on the formation of a molecular ion and an electron as a result of the combination of two neutral molecules. The mechanism of associative ionization is known to have a low energy threshold of realization. However, in a number of cases associative ionization may have the properties of a nonthreshold mechanism and ensure the formation of a plasma medium almost with no energy consumption.

Penning ionization occurs in a low-temperature plasma containing atoms of different elements with an important role played by collisional ionization of an excited atom whose excitation energy exceeds the binding energy of the electron of the ground-state electron of the other atom. This nonthreshold reaction is very efficient because its products are stabilized automatically via the transformation of excess energy into kinetic energy. The rate of this reaction is rather high for resonantly excited atoms, however, the competition on the part of the optical transition reduces the importance of the reaction. However, this may become the dominating process if long-lived metastable states form in the gas.

The well-known low-threshold ionization mechanism resulting in the formation of molecular plasma is also of certain interest. This mechanism is used to ionize the sample in the MALDI technique of mass-spectrometric analysis (Hillenkamp et al., 1975).The physical mechanism of the formation of ions is associated with the processes of explosive desorption of the matrix when exposed to laser radiation with a duration of 10^{-8} s and low power density $W_{L,}$ no greater than 10^5-10^6 W/cm^2. The mechanism of ionization caused by laser radiation with such a low W_L has not yet been entirely understood (Zenobi and Khochenmuss, 1999, Karas et al., 2000, Khochenmuss et al., 2000). Moreover, the potentialities of this mechanism, which is capable of ionizing the medium in impact processes, have been completely unexplored. It is important that this mechanism works when special, exceptional materials are subject to laser radiation, and that is why it is by no means easy to reproduce in a laboratory. However, the effect of the formation of weakly ionized plasma in the experiments of Crawford and Schultz (1988) and Srama et al. (2009) with impact velocities below 4-5 km/s can be explained by the realization of a MALDI-type mechanism.

A comparison of the results of laboratory experiments with the results of the space experiment showed (Ernst et al., 2006) that, as it is evident from the plot shown in Figure 53, the plasma ion density could have increased by a factor of 4 for V_{IMP} ~ 10.2 km/s. The ionization degree of the resulting plasma should have increased by a bout the same factor (Stubig, 2002). Given the direction of the impact, which should have occurred at an angle of 25 to 35^0 with respect to the horizon, W_{IMP} for this velocity should correspond to $3 \cdot 10^{11}$ W/cm^2. Given that the density of the projectile – a space probe – was equal to 3 g/cm^{3}, , the shock-compression pressure could be as high as 3 10^6 bar. In this case the specific W_{IMP} spent for the processes of plasma formation could have been no less than ~ 1.5-$2 \cdot 10^{10}$ W/cm^2. According to the results of laboratory experiments made in Ames, this value should have been sufficient to produce impact plasma and generate a plasma torch.

It is now appropriate to show that the results obtained in laboratories for micron-sized projectiles can be applied to impacts of several hundred meters large bodies.

1.5. Limits of Applicability of the Results of Laboratory Experiments

The results of laboratory determinations of the degree of ionization of an impact-produced plasma torch carried out by Stubig et al. (2001, 2002) and Stubig (2002) for projectiles with sizes greater than 200 nm indicate that these data can also be used, with no limitations, to determine the degree of ionization of a plasma torch produced by an impact of a several hundred meters large projectile. This is because in both cases the duration of the impact is sufficient for the resulting plasma to reach local thermodynamic equilibrium. The same authors suggest that plasma may fail to reach equilibrium if the characteristic size of the projectile is smaller than 10 nm.

There is also other evidence that, as far as the determination of the degree of ionization in the plasma torch is concerned, shows that the results obtained for projectiles with characteristic sizes greater than 100 nm can be applied to projectiles with the sizes of several tens or several hundred meters.

Thus Drapatz and Michel (1974) studied computationally the interaction of a 50-nm diameter iron projectile with a tungsten target. The above authors showed that at $V_{IMP} \sim 50$ km/s the degree of ionization in the resulting plasma can be as high as 60%. Koval' et al. (2004) performed magnetohydrodynamic computations using as initial data the results of the gas-dynamic computations of the process of the projectile's entry into the atmosphere. The above authors studied ionospheric and magnetospheric effects of the impacts of comets and asteroids. They showed that the maximum degree of ionization in the plasma torch produced by an impact onto the Earth of a 200-m diameter meteorite moving at a velocity of 50 km/s can be as high as 80%.

Let us now compare these results with those obtained by Stubig (2002) assuming that the yield of ions made of the matter of the aluminum projectile was, according to Kissel and Krüger (1987a), about four times greater than the yield of ions made of the matter of the target. In this case the degree of ionization in the torch plasma for $V_{IMP} \sim 50$ km/s should vary from 40 to 75%.

The results of the theoretical and experimental studies mentioned above indicate that the results of dust-impact experiments can be applied – if the diameter of the projectile exceeds 50 nm – to determine at least the degree of ionization of a torch produced by an impact of relatively large projectiles.

1.6. Results of the Space Experiment

Let us now return to the space experiment carried out within the framework of the Deep Impact mission and briefly discuss the results of observations made onboard the measuring module using various optical instruments. These instruments recorded rather reliably the sequence and configurations of mass ejections from the crater. Thus the first dome-shaped ejection, which, as it is evident from Figure 55, appeared 186 ms after the impact, has been observed for 0.7 s and was described in detail by Melosh (2006). It was recorded by the MRI onboard telescope. The velocity of the visible part of the ejection was shown to be ~ 10^6 cm/s and its spatial size corresponded to 3 km in diameter and 10 km in length. Photometric observations recorded optical emission of the outburst in the process of the expansion-away starting from 200 ms after the impact with a total duration of ~ 200 ms (Ernst et. al, 2006). However, the intensity of the emission proved to be five orders of magnitude lower than the computed intensity. This discrepancy, like the observed delay of emission, could be due to the penetration of the projectile into the loose regolith of the cometary nucleus.

The highly porous ice target had a density that did not exceed ~ 0.3 g/cm^3. According to the computations of Klumov et al. (2005), the characteristic initial temperature of the ejection could be as high as ~15000 K for a ~0.5-porisity target moving at a velocity of ~10 km/s. Such conditions could ensure the generation of a plasma torch with a large margin. However, the porosity of the target used in the experiment was about twice higher and therefore the ambient temperature could be substantially higher and amount to 25000 K.

Starting from ~420 ms, the observed expanding feature began to shine with reflected light. As it is evident from Figure52 A, when the diameter of the feature reached 1 km, threadlike features began to appear whose expansion-away velocity exceeded substantially be expansion-away velocity of the disk. Shultz et. al (2006) called these features «curved ejecta ray systems» and did not directly associate them with plasma processes. Lisse (2006) interpreted the spectral properties of the late outburst observed in the IR as compounds ejected from deep down in the cometary nucleus and unobserved in the spectra of the matter at the surface of the comet recorded before the impact.

In this connection, it would be appropriate, first and foremost, to consider the possibility of the synthesis of new chemical compounds in the plasma torch produced by a low-velocity impact.

It was shown in a number of published papers (Managadze 2001, 2002a, 2010a, 2010b; Managadze and Eismont, 2009; Stubig et al., 2001, 2002; Stubig, 2002) that highly efficient synthesis of new compounds is observed when V_{IMP} or W_L reaches a certain critical limit. However, no bona fide evidence has so far been available that would indicate the possibility of the synthesis of new compounds in the case of very low and intermediate V_{IMP} , e.g., amounting to 2-3, 5-6, and 9-10 km/s, respectively, i.e., for impact velocities below the critical thresholds.

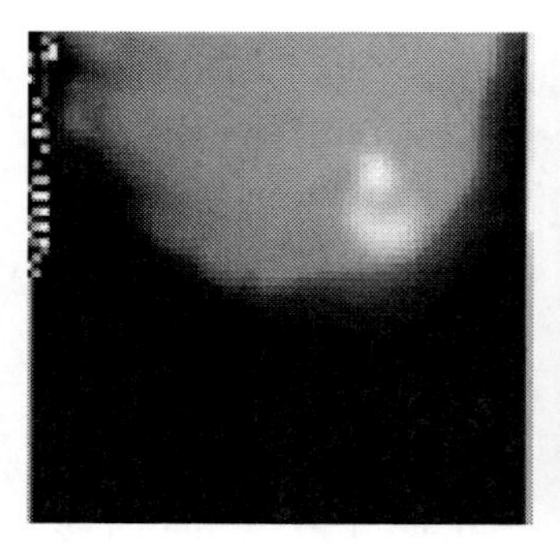
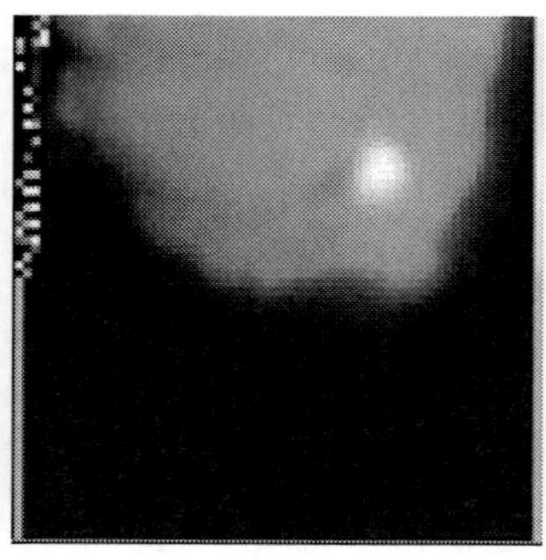
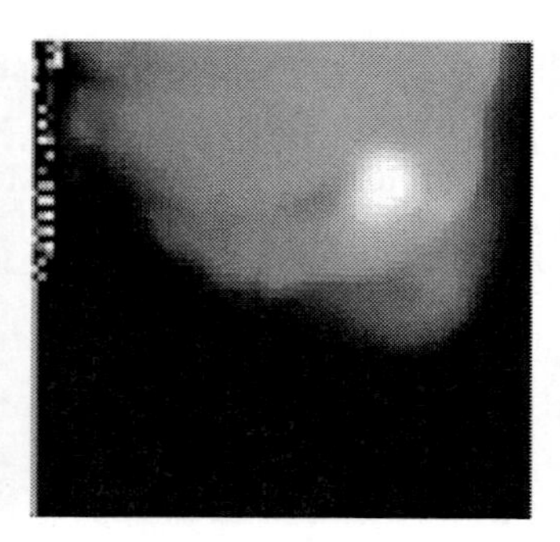

Figure 55. Dynamics of the expansion-away of a high-velocity luminous ejection from the crater. The frames were recorded 310, 372, and 434 ms after the impact. This feature is very likely to be the first, hot blob of the plasma torch.

This problem is of special interest also because we try to demonstrate in this paragraph that a plasma torch may be produced by a low-velocity impact. We can expect that the generation of plasma could also be accompanied by the detection of compounds not initially present in the projectile and the target. Such detection may be interpreted to confirm, to a substantial degree, the formation of a plasma torch.

1.7. Discussion of Results

The experimental results shown in the plot in Figure 56 A demonstrate conclusively that in the case of the impact of an iron projectile onto a rhodium target the synthesis of $RhFe^+$ begins starting from a velocity of V_{IMP}~2 km/s. The plot in Figure 56 B shows that an impact of a carbon projectile moving at a velocity of 6 km/s onto a rhodium target results in the synthesis of RhC^+. This compound is especially difficult to synthesize in the absence of plasmachemical reactions. It can be easily seen from the plots shown that the yield of $RhFe^+$ varies from 0.9 to 0.001, and that of RhC^+, from 0.1 to 0.002.

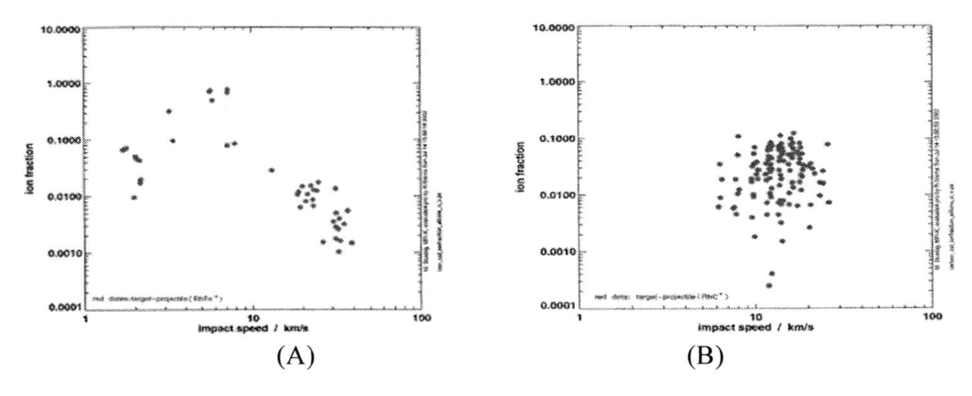

Figure 56. Dependence of the efficiency of the synthesis on impact velocity in the vase of a rhodium target hit by an iron (A) and carbon (B) projectiles, which ensure the synthesis of RhFe and RhC, respectively.

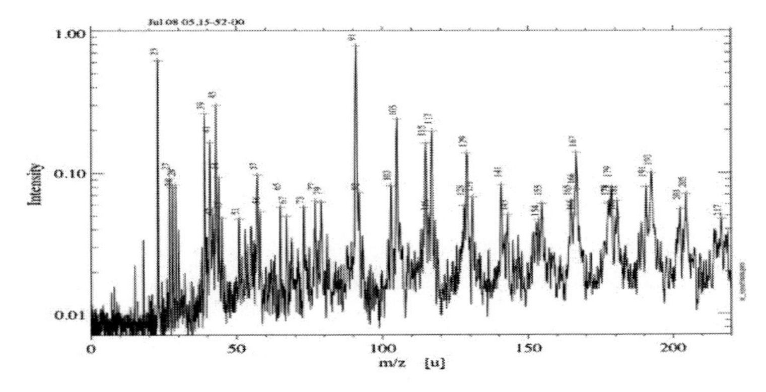

Figure 57. Spectrum of the mass peaks of surface contaminants and molecular ions of organic compounds recorded in the case of an impact of a 1.81-μm latex projectile moving at a velocity of 5 km/s (Srama et al. 2009).

Of special interest is the mass spectrum of atomic ions and organic compounds shown in Figure 57 and recorded in an impact of a latex projectile moving at a velocity of 5 km/s (Srama et al. 2009). This spectrum exhibits, along with the mass peaks corresponding to fragments of organic compounds that enter the composition of the projectile, also the mass peaks of new compounds.

Note that the spectra shown in Figure 57 are the first direct evidence for the possibility of the synthesis of organic compounds in impact processes. It is important that these results were obtained in direct experiments that fully reproduce the natural phenomenon in question in a laboratory and not just in impact simulation experiments.

The experimental detection of chemical compounds synthesized in the plasma torch produced by a low-velocity impact should be viewed as exceptional and new properties of the emerging medium and a fact of extreme importance.

These results can explain the mechanism of the origin of unusual compounds of various metals on the Moon, which are usually attributed to «low-temperature synthesis in the gaseous phase». Such a denomination can be viewed as partially correct because the synthesis in the case considered occurs in an impact produced low-temperature plasma.

These results may also indicate that spectral lines of the compounds hitherto undetected in the surface layer of the comet and found by the onboard IR telescope in the matter ejected from the crater (Lisse, 2006) may belong, in part, to new compounds synthesized in the plasma torch.

The above results of laboratory experiments carried out on the particle accelerator in Ames and the results of earlier experiments carried out to study the processes of the generation and expansion-away of the laser-produced plasma torch (Kim and Namba, 1967; Namba et al., 1966, 1967a, 1967b) show that two blobs form as a result of the expansion-away of the impact-produced plasma. The first blob consists if relatively hot and high-velocity glowing plasma. Cool or low-temperature plasma does not emit radiation and flies away at a low velocity. One must also bear in mind that such plasma structures make possible the accumulation of a substantial potential at the surface of dust particles and the formation of dust plasma, which, under certain circumstances, may produce regular spatial structures (Fortov et al. 2004). Because of the similarity of the processes that occur in the plasma media considered, these results can be used to interpret the data obtained in space experiments including the experiment carried out within the framework of the Deep Impact mission.

The proposed scenario of the impact consequences, which takes into account the plasma processes, may reasonably explain a number of phenomena.

According to the preliminary «plasma scenario», which can be corrected based on the results of further studies, the processes of the impact and mass ejection in the natural experiment could have occurred in the following sequence. The low density of the target allowed the projectile to penetrate to a substantial depth. At this depth the projectile could transfer its energy to the «compressed target» with the process accompanied by the formation of a plasma torch. The substantial penetration of the projectile into the target could have resulted in a delay of the ejection of the first plasma blob with the intensity of emission decreasing by a factor of 10^5 compared to the intensity computed based on the results of a laboratory experiment. The ejection of the first blob was followed by the ejection of a second, low-temperature plasma blob coincident with the the dust outburst of the matter of the cometary nucleus. The presence of plasma in the second blob was detected only because of the formation of threadlike structures, or filaments, of the plasma flow, which may have been detected by the onboard telescope. Figure 52B compares laser plasma filaments and similar structures recorded in the Deep Impact space experiment. In the proposed scenario the observed spectra of chemical compounds not found prior to the impact could contain substances synthesized in the process of the expansion-away of the plasma torch produced by the impact.

Note that a more in-depth analysis of the results of the experiment should involve the study of raw scientific data and especially of the results of optical and magnetic measurements.

It would be appropriate to point out that the only hypothetical component in the proposed scenario is the sequence of the processes that occur during the realization of the natural experiment. The possibility of the generation of the torch plasma in low-velocity impacts is now a well established fact that is beyond doubt.

Note that in this book we treat the expansion-away of the torch as a single continuous process, which is capable – owing to the emerging local chiral physical fields – of causing the initial breaking of mirror symmetry. We discussed the possible symmetry breaking mechanisms in Chapter 6. However, the development of this idea led us to suggest that the process in question could consist of two stages. This idea was later corroborated by the results of the Deep Impact space mission. We therefore believe that the correct version of the process has the following form. The first blob – the hot part of the plasma torch – makes possible the formation of unidirectional asymmetric electric and magnetic fields, which interact with the cool, and, possibly, also with the dust plasma to produce local physical chiral fields. New chemical compounds can in this case be synthesized in the low-temperature part of the torch via the known plasmachemical synthesis mechanisms discussed in Chapters 4 and 5. This mechanism is, to a substantial degree, consistent with the synthesis processes that occur in the laser plasma when complex compounds form in the low-velocity part of the plasma torch. Such a medium may result in the breaking of the mirror symmetry of synthesized enantiomers in local physical chiral fields and in the formation of homochiralic molecular structures of organic compounds in spontaneous processes, which discussed in detail in Chapter 6.

The experimental results mentioned above indicate that the formation of plasma torches and, consequently, the synthesis of new compounds, are observed on a regular basis in impacts with velocities exceeding 2-6 km/s and involving projectiles with densities 8-2 g/cm^3. The most important and interesting results include the simultaneous formation of the plasma and new compounds synthesized from the constituent elements of the projectile and the target. In impact processes these element are initially separated in space and therefore the formation of such a compound can be explained only by their synthesis in the impact-produced plasma.

The results obtained were not surprising. In their book published back in 1966 in Russian and (see Zel'dovich and Raizer (2002) for the recent English edition) Zel'dovich and Raizer (2002) described theoretically the possible ionization of the medium in processes of shock compression in low-velocity impacts. The possibility of such a process was later confirmed experimentally by Crawford and Schultz (1988) and Göller and Grün (1989).

A comparison of the results of different experiments and a comparison of experimental and theoretical results leaves no doubt of their credibility. Thus in the experiments of Stubig (2002), where a Rh target was subject to impacts of Fe, Al, and C projectiles, the processes of plasma formation and synthesis start at about the same level of shock compression, when pressure reaches ~ 0.3 – 0.5 10^6 bar. At such pressures the medium is ionized via the compression mechanisms discussed above (Fortov et al. 2003).

The results of the experimental studies described in Chapters 3, 4, and 5 of this book show that impact is not the only way to combine these two processes of great importance - ionization and synthesis – and reproduce them in a laboratory.

Thus we showed in Chapter 4 of this book that an impact-produced plasma torch can be reproduced in a laboratory using a laser. In this case the ionization of the medium that makes possible the onset of the synthesis can be achieved via thermal ionization mechanism by exposing the target to a laser beam. To this end, the laser must operate in the *Q-switched mode* and provide a light pulse with a duration of ~5 10^{-9} s, and this is a necessary condition.

Under such conditions, starting with W_L~ 5 10^8 W/cm^2 surface ionization of the target is ensured along with the generation of the plasma torch, which facilitates the synthesis of new compounds. Note that the resulting plasma torch should be very similar to the torch produced by a 5 - 10 km/s impact. The results presented by Managadze and Eismont (2009) clearly demonstrate that in the laser-produced plasma torch the main plasma processes are reproduced that accompany an impact. Starting from the intensity density of W_L~10^9 W/cm^2 and above the volume ionization of the target occurs and ensures highly efficient synthesis of new compounds. However, the efficiency of the synthesis is usually higher at W_L~ 5 10^8 W/cm^2.

The above pattern demonstrates the possibility of the synthesis of new compounds as a result of an impact, as well as a technique for recreating this process in a laboratory by exposing the target to the radiation of a laser operating in the Q-switched mode provided the correct choice of the similarity parameters (Managadze, 2009, 2010a). It demonstrates with extreme clarity the possibility of a correct set-up of laboratory experiments simulating impact action as well as their realizability. Ultrafast input of energy of external influence to ensure fast heating of the medium, which, in turn, results in the formation of plasma and generation of a plasma torch, should be viewed as the crucial similarity parameter.

Needless to say that it is totally impossible, without following the above recommendations and without correctly choosing the similarity parameters, to adequately describe the physical processes that accompany a meteorite impact, or to correctly recreate such processes – including low-velocity impacts – in a laboratory.

Yet, such studies have been carried out.

Thus the investigations of Gerasimov (1984) and Gerasimov et al. (1999, 2003, 2007) , which have been carried out since 1979, were, according to their authors, dedicated to the study of the processes of degassing of matter during the early stage of the formation of the Earth and to abiogenous synthesis of organic compounds as a result of a meteorite impact for impact velocities ranging from 6 to 10 km/s.

A laser facility has been developed to simulate impact processes in a laboratory in the velocity interval from 6 to 10 km/s (Gerasimov (1984) and Gerasimov et al. (1999)). The laser used in this facility operated in the Q-switched mode and provided an intensity density ranging from W_L = $3\cdot10^4$ W/cm^2 to $7\cdot10^7$ W/cm^2 with a pulse duration of ~ 10^{-3}s. The substance synthesized as a result of such a laser exposure was then analyzed.

The main inadequacy of the above studies is that their authors have never considered plasma processes in their publications covering 30 years of research (Gerasimov et al., 2003, 2007), whereas the basic parameters of their laboratory facility ruled out the possibility of plasma generation and the users of the facility have been well aware of the fact.

Therefore plasma processes, which are the most important factors of the impact and determine the composition of the final product, were not addressed in the above studies and certainly were not reproduced in the laboratory facility employed.

At the same time, the experimental results mentioned above and obtained in direct impact experiments in Ames, Heidelberg, and Canterbury as early as in late 1980-ies, indicate conclusively that regular generation of a plasma torch is observed in impacts with velocities ranging from 5 to 9 km/s.

In such a case, what did Gerasimov (1984) and Gerasimov (1999, 2003, 2007) study? Evaporation?

As is well known, the expansion-away of the plasma torch has nothing in common with evaporation because plasma and vapor-and-gas media are different aggregate states. Therefore the expansion-away of the torch has very little in common with thermal evaporation of matter. Hence the final products synthesized in these processes should also differ fundamentally.

Hence the discrepancy consists in the following.

It is clear that only the presence of a plasma torch would make possible the processes that result in true shock degassing and abiogenous synthesis of organic compounds.

However, neglecting this process and the fact that the conditions of laboratory experiments whose authors claim to reproduce the processes of shock degassing and abiogenous synthesis do not ensure the generation of a plasma torch, which is the universally recognized factor conducive to the emergence of new compounds, prevent adequate analysis and recreation of the processes considered.

A comparison of the data shows that both the set-up of the experiment and the results obtained using a laser facility with a low rate of energy input (Gerasimov 1984, Gerasimov et. al. 1999) fail to meet the most important requirements that must be fulfilled to correctly reproduce the processes that accompany a meteorite impact. Therefore these results (Gerasimov 1984, Gerasimov et al. 1999, 2003, 2007) cannot be viewed having anything to do with the processes of shock degassing and abiogenous synthesis of new compounds produced as a result of the impact.

1.8. Conclusions

The results of the study Deep Impact mission data may have interesting ramifications because currently not all unusual results of the mission are available. However, this work has already resulted in an interesting continuation, albeit in a different and equally interesting direction, which we discuss in the next section and which concerns the well-known properties of dust structures that form in space and in laboratory.

Thus the main results obtained by studying the plasma effects arising during the preparation and implementation of the Deep Impact space mission can now be formulated as follows.

It was shown that if a carbon projectile moves at a velocity of V_{IMP}, ~5-6 km/s then the shock compression of the medium results in its high-temperature overheating. This, in turn, causes the surface ionization of the interacting bodies and regular development of a plasma torch. The plasma medium produced in these processes meets, according to its properties (degree of ionization, density, Debye radius), the main requirements to ideal plasma medium by a large margin in terms of density and temperature (Galeev and Sudan, 1983).

Under the conditions where surface ionization of the matter is achieved in colliding bodies, the plasma density is about ~10-20 times lower than the density of the plasma produced as a result of volume ionization of the projectile and the target, which develops if the impact velocity V_{IMP} is equal to or higher than the critical value of ~15 - 20 km/s.

During the expansion-away of the torch the emerging plasma structures should break into individual blobs with different expansion-away velocities. The hot part of the plasma torch should move ahead of the dust ejection, whereas the low-temperature part of the torch should emerge together with the dust. This effect was first observed during the preparation and implementation of the Deep Impact mission owing to the large characteristic size of the impact experiment.

The plasma outburst emerging in the case of impact velocities equal to 2-3 km/s (for Fe), 5-6 km/s (for C and Al), and 10-11 km/s (for ice) is, according to the results of laboratory modeling of impact processes, capable of providing the conditions necessary for the origin of living matter on many cosmic bodies with masses smaller than or equal to the Earth's mass.

The experimentally found relatively low plasma-formation threshold allows plasma diagnostic and mass spectroscopy to be used to study the synthesis of organic compounds in direct impact interactions carried out under laboratory conditions with 0.3 to 0.6-mm diameter projectiles. Such studies can be performed using the NASA particle accelerator in Ames and other, lower-velocity and therefore more easily available accelerators, which allow particles with substantial projectile masses to be accelerated to 5 – 6 km/s.

Earlier studies of the processes of shock degassing and abiogenous synthesis of organic compounds in a vapor-and-gas medium excluded the possibility of the formation of a plasma medium. The consequences of an impact and the products of shock degassing and abiogenous synthesis are totally different in the absence of a highly catalytic plasma medium. Therefore the results obtained in such experiments may be totally unrelated to the mechanisms and final products, and to natural processes of degassing and synthesis in impact interactions.

2. Mechanisms of the Generation Circular Polarized Emission in The Space

During the analysis of the results of the Deep Impact a hypothesis was suggested, which made it possible to apply the well-known natural mechanism that ensures the generation of linearly and circularly polarized radiation in the interstellar medium to the physical processes that develop in the torch of the impact-produced plasma. This mechanism was of great interest because it could ensure substantial symmetry breaking during the synthesis of organic compounds not only as a result of pure plasma processes, but only in the processes of the interaction between radiation and the dust component. One of the important features of this mechanism was that the possibility of its realization was experimentally confirmed in the case of the interstellar medium. Given its special importance, let us analyze this process in more detail for the conditions where $V_{IMP}>V_{CR}$.

It became clear from the studies of the symmetry breaking that further development of the concept proposed in this book requires a bona fide mechanism capable of ensuring the generation of a «unipolar», or, more precisely, levorotatory circularly polarized radiation. This mechanism should be, by analogy with the mechanism realizable in the interstellar medium, operate via the interaction of the UV radiation of the plasma with the dust structures produced as a result of the impact.

In particular, this mechanism could resemble the scattering of stellar radiation with the optical properties identical to those of the radiation of the torch plasma on interstellar dust grains (Chrysostomou et al., 1997, 2000; Bailey et al., 1998, 2000) and stimulate the generation of circularly polarized radiation, which is the crucial physical factor leading to the symmetry breaking in the process of the synthesis of enantiomers. Such a mechanism could also operate in the impact-produced torch during the simultaneous formation of atomic and dust plasma in the same region. It is important that no one has so far proposed such a mechanism for the plasma torch despite the fact that the results obtained for the interstellar medium might be indicative of the possibility of the development of such a mechanism in the impact-produced plasma torch.

The problem of the generation of circularly polarized radiation in interstellar gas clouds has been thoroughly studied in recent decade (Gledhill and McCall, 2000; Lucas et al., 2005). Two mechanisms have been proposed for the realization of this process. In both processes the crucial part was played by dust particles that are present in the interstellar gas-and-dust cloud and in the stellar neighborhoods.

In the first of these mechanisms unpolarized radiation of the star could scatter on the particles of the dust cloud and thereby acquire linear polarization. Circularly polarized radiation could then arise as a result of the superposition of two linearly polarized waves.

The second mechanism is based on the interaction of unpolarized radiation of the star with elongated dust particles aligned in the magnetic field. In this case circularly polarized radiation could be generated immediately as a result of the scattering of unpolarized radiation on dust particles.

Evidence for these mechanisms was found in IR studies of young stars. Thus the polarization measurements in the neighborhood of the young star IRc2 (Chrysostomou et al., 2000) showed that the fraction of circularly polarized emission amounted to 20% in two regions of the reflection nebula. A reasonable explanation was found for the observed, anomalously high fraction of circularly polarized radiation. The above authors believe that this effect is due to the fact that in the case considered unpolarized radiation interacted not with spherical, but rather with slightly elongated dust grains aligned in the magnetic field in the vicinity of the star (Gledhill and McCall, 2000).

Further development of such studies showed that circular polarization may develop not only in the IR, but also in the UV domain of stellar radiation and, depending on the sign of polarization, it may destroy, e.g., one of the two possible isomers of amino-acid molecules. This may result in the preservation of only L or D amino-acid molecules.

In this connection of special interest was the paper by Meierhenrich et al. (2005), who experimentally demonstrated the symmetry breaking when various racemic mixtures of amino acids were exposed to circularly polarized radiation. In the experiments reported by the above authors the exposure of a thin amino-acid film to dextrarotatory circularly polarized radiation with a wavelength of ~ 180 nm resulted in a 2.8 % symmetry breaking with the decreased number of L-isomer molecules. Thus the above authors showed that circularly polarized radiation is very likely to cause symmetry breaking.

The results of these studies served as the basis for the new idea (Lucas et al. 2005) whereby the circularly polarized stellar radiation must have been the factor that played the crucial part in the breaking of the mirror symmetry of bioorganic world and determined the «sign» of its chirality. Thus, the numerical simulation of these processes performed using the Monte Carlo technique (Lucas et al. 2005) showed that the conditions that develop during the formation of Young Stellar Objects may ensure the generation of circularly polarized radiation of substantial intensity at wavelengths ranging from 0.22 to 230 nm, which are needed to destroy amino acids. The above authors also showed that the scattering of the UV radiation emitted by a star on nonspherical dust particles residing in the magnetic field of the emitting star may cause the circular polarization of the initially unpolarized radiation. The main five parameters that determine this process were considered to be: the scattering angle, maximum size of a dust grain, the longitudinal-to-transversal particle size ratio, the optical depth, and the orientation of the magnetic field. However, the above authors did not state explicitly which parameters determined such properties of circularly polarized radiation as the direction of the rotation of the electric vector.

The degree of the asymmetry of the medium considered could have been determined by the balance between the left and right orientation of circularly polarized radiation, and greater asymmetry degree makes it easier to ensure the required symmetry breaking. In the studies reported by the above authors this asymmetry degree was in many cases close to zero and therefore it did not meet the conditions for a chiral medium. This could be due to the complete symmetry in space of the stellar objects considered. Thus circular polarization of about the same intensity in different parts or in different hemispheres of these objects had different directions (left and right) of the electric field of radiation and this circumstance prevented any substantial symmetry breaking.

Plasma torch is known to be inherently asymmetric both in terms of the electromagnetic fields generated and the expansion-away geometry. Therefore there were reasons to expect that the circular polarization arising in such a system could be unipolar. In this case the results obtained for the interstellar medium could also be applied to an impact-produced plasma torch.

Let us now demonstrate this. A plasma torch is known to develop in the process of the impact and to ensure the synthesis of simple and complex organic compounds and their ordering. However, the torch plasma is also a source of vacuum ultraviolet radiation. Unidirectional magnetic fields are generated during the expansion-away of the plasma torch, and the impact produces dense dust clouds. The combination of these conditions should be sufficient for the generation of circularly polarized radiation, which could arise during the scattering of plasma radiation on dust particles and facilitate the breaking of the mirror symmetry when such radiation acts on synthesized enantiomers. The local physical chiral field so generated may have served as the «controlling field» for spontaneous processes and it may have determined the «sign» of the chirality of the emerging medium as a whole.

It is important that according to Crawford and Schultz (1999) dust bears significant charge. Therefore in the case of impact, this medium should be, unlike the interstellar medium, better referred to as dusty plasma. The point is that the medium that develops as a result of an impact is rather dense and individual charged dust particles interact with each other over the short time interval in the process of the expansion-away of the torch. Hence this circumstance should be taken into account and the appropriate terminology should be used when we study the processes that occur in the impact-produced plasma torch.

In the case considered dusty plasma should have the form of micron-sized charged particles residing in relatively hot plasma. Particles are charged by electron and ion flows. Because of the high mobility of electrons particles may possess negative electric charges amounting to 10^5 electron charges and interact electrically with each other.

When exposed to UV radiation, dust particles may, as a result of photoelectric effect (e.g., in space), acquire a positive charge.

Under certain conditions, stationary, regular, and, possibly, asymmetric structures may develop in impact-produced plasma. Under natural conditions – e.g., in microgravity or weightlessness – they may be maintained over a long time, and when in the terrestrial gravitational field, such structures may be maintained as strata by electric field.

In this connection, the main idea associated with the possibility of the symmetry breaking during the generation of dusty plasma in hypervelocity impact processes consists in the following. The expansion-away of a plasma torch is accompanied by the generation of unidirectional magnetic fields having a toroidal or, possibly, axial configuration. These fields may ensure the development of certain structures made up of elongated dust particles. According to Fortov et al. (2004), the dust component may rotate about the longitudinal axis. The scattering of unpolarized plasma radiation on such structures may result in a considerable yield of circularly polarized radiation. The yield of circularly polarized radiation can in this case be expected to be more efficient compared to the yield provided by other, e.g., purely plasma mechanisms, which also produce polarized radiation of this kind (see Section 6).

The discovery of the possibility of the development of the medium and mechanisms in an impact-produced plasma that ensure the generation of circularly polarized radiation is a fact of exceptional importance. This phenomenon requires the most thorough experimental investigation because its possible realizability, if confirmed, may simplify substantially further studies of the problem of the nature of the symmetry breaking. In particular, this mechanism may cause a slight breaking of mirror symmetry and the formation of the controlling field. The dissipative structures that develop under such conditions, i.e., in the presence of a strongly non-equilibrium torch plasma, in spontaneous processes may bring such a weak symmetry breaking to the formation of a homochiralic medium needed for the appearance of chirally "pure" molecular structures.

In this case it is safe to assume that the «sign» of chirality may, like for the processes of symmetry breaking discussed in Chapter 6, be determined by the direction of magnetic and electric fields of the plasma torch, which «maintain the direction », i.e., do not change it from one external action to another nowhere within the media currently known to exist in the Universe.

Currently, preliminary work is under way aimed at addressing the problem and involving an analysis of the papers already published on the subject. This may prove to be heavy and very promising work. However its outcome is now impossible to predict. The point is that at present we do not know which of the two possible mechanisms of the generation of circularly polarized radiation – the «dust» or «plasma» -driven – is more efficient for both laser-radiation and impact-produced torch plasma. Note that, as is evident from the plot shown in Figure, the fraction of linearly polarized radiation observed from laser-produced torch plasma is quite substantial (Kieffer et al., 1992, 1993). This may be interpreted as evidence suggesting that purely plasma processes in plasma medium may also produce a high fraction of circularly polarized radiation because in this case the density of dust component may be substantially lower compared to the case of an impact-produced torch. This radiation is rather difficult to detect because of high expansion-way velocities of the torch and the lack of appropriate instruments for plasma diagnostics. However, indirect methods are known that allow the «polarity» of circular polarization to be determined, and they are easily available.

According to the available data, in the case of the interstellar medium the "dust"-driven mechanism outperforms the "plasma"-driven mechanism by at least a factor of ten. Cyclotron radiation from the regions of supernova remnants after the formation of the pulsar is viewed as a possible plasma source of circularly polarized radiation for the interstellar medium (Rubenstein et al., 1983). Optical observations of the radiation of the Crab nebula pulsar showed that the fraction of detected circularly polarized radiation from this region does not exceed 1 %, which is indicative of the relatively low efficiency of this process if driven by plasma interactions exclusively (Roberts, 1984).

Experimental results have been obtained, which make it possible to estimate the energy needed to produce symmetry breaking and determine the necessary physical characteristics of the action that produces such effect. Thus it is well known that a substantial enantiomer excess can be produced by the exposure of a racemic mixture of amino acids to circularly polarized UV radiation with an intensity of $\sim 2\ 10^{17}$ photons/cm^2 at a wavelength of several hundred nm (Greenberg et al., 1994). However, in the case considered already formed amino acids are exposed to radiation. Circularly polarized radiation is known to cause the photolysis or destruction of L or D amino acids depending on the direction of the rotation of the electric vector of circularly polarized radiation. In particular, laevorotatory circular polarization ensures the photolysis of D amino acids (Meierhenrich et al., 2005).

It is hoped that the production of an enantiomer excess in the case if such exposure occurs in the process of their synthesis, i.e., simultaneously and in parallel with the formation of enantiomers, would require a substantially smaller number of photons compared to the number of photons spent in the experiment of Greenberg et al. (1994). This is due to the fact that the energy required to create the conditions «preventing» the formation of certain enantiomers should be substantially smaller than the energy ensuring the photolysis of the «final product» in the form of an amino-acid molecule.

An analysis of the physical factors arising in the process of the expansion-away of an impact-produced plasma torch shows that:

- The plasma of the torch is the source of the generation of radiation and, in particular, of the radiation in the wavelength interval from 100 to 200 nm. This radiation is directed toward the upper hemisphere and propagates at angles ranging from 0 to 90^0 with respect to the vertical direction;
- Plasma processes in the torch ensure the generation of unidirectional electric and magnetic fields of significant magnitude, and these fields have toroidal and, possibly, axial configuration;
- In the plasma medium that forms in such a way the most important and well-known magnetooptical configurations are realized, where the vector of radiation is orthogonal or parallel to the vector of magnetic field. Such configurations were analyzed in detail by Barron (2008) as the most important mechanism causing appreciable symmetry breaking of a circularly polarized wave – the most efficient external factor to ensure the photolysis of amino acids;
- According to the available results of experimental and theoretical studies, the scattering of UV radiation on dust particles residing in magnetic field may provide an

important, and even possibly the crucial contribution to the process of symmetry breaking, and it may prove to be the most efficient source of circularly polarized radiation.

An objective comparison of the above facts suggests that impact-produced plasma torch is an exceptional environment making possible the achievement of asymmetry in this medium via circularly polarized radiation. Exposure of the synthesis region to such radiation may provide the excess of one of the forming enantiomers and cause the initial, weak symmetry breaking.

However, this medium with exceptional characteristics is also a difficult object for theoretical and experimental study or detection of symmetry breaking or circularly polarized radiation. The problem is that this environment is too complex for both theoretical investigations (because of a large number of independent processes that occur in it) and for experimental studies (because of its very short lifetime).

Therefore the study of the chirality of the final products and, in particular, amino acids, in direct laser simulations of impact processes, is possibly the easiest way of detecting the weak symmetry breaking sufficient for triggering spontaneous processes.

However, to explain the mechanisms of symmetry breaking, theoretical investigations will be needed involving numerical simulations of the processes that occur in the impact-produced torch, as well as experimental studies addressing not only the properties of purely plasma processes, but also those of dusty plasmas when modeling the phenomenon of symmetry breaking under laboratory conditions.

There is serious evidence suggesting that natural mechanisms leading to circular polarization of stellar radiation during the scattering of light on dust particles may also operate in the plasma torch produced by a hypervelocity impact. Impact processes provide all the necessary conditions for this mechanism to develop. In particular: the presence of dust and hot plasma, which facilitates the generation of magnetic field, and of unpolarized, and, possibly, also of linearly polarized plasma radiation with optimum wavelength.

The results obtained are of special value for the concept proposed in this book. They show that at the initial stage of the torch formation plasma media with high and low temperatures form simultaneously along with with dusty plasma. These media form not only simultaneously, but also in the same region. Hence the processes that occur during the expansion-away of the torch can be subdivided into three interacting media with different physical properties, allowing both different and common problems to be addressed. It is important that such an exceptional combination of media with different physical properties, but capable of performing crucial functions, may have allowed the simultaneous occurrence of many different processes. These processes include: the synthesis of new compounds, assembly and ordering of matter, generation of radiation conducive to symmetry breaking in the processes of the formation of enantiomers. These processes are also capable of producing dissipative structures, which may result in spontaneous symmetry breaking, which, in turn, is a necessary condition for the formation of homochiralic molecular structures, and many other things that are so far beyond human fantasy.

3. Properties of Torch Plasma and Homochirality

In this, last section of the book I will venture to undertake the first attempt to confront the main requirements that should be met for chirally pure medium to emerge in nature and for homochiralic molecular structures described by (Avetisov, Goldanskii 1996) to form on the one hand, and the physical capabilities of the plasma medium produced as a result of the expansion-away of a hypervelocity meteorite impact torch on the other hand. The main properties of the plasma medium considered in the book include both those experimentally confirmed and the hypothetical properties that may show up in one way or another, but that have been found experimentally and are consistent with the fundamental laws of nature.

Our adopted hypothetical model of the most primitive form of life is Altstein's protoviroid, which could have assembled from chirally pure medium.

As a result of the proposed comparison some inaccuracies and inconsistensies may transpire, which, hopefully, will be removed after further experimental studies of impact as a natural phenomenon, and after the appropriate discussion of the results of these experiments with expert researchers in various fields.

In the previous paragraphs of this section we discussed relatively new and important properties of the plasma medium, which develop in the process of a hypervelocity impact. These properties, which are of special interest, were associated with the possibility of the formation of a plasma torch and the synthesis of new compounds under the conditions of a low-velocity impact. The results indicating that an intensive circularly polarized radiation can result from the interaction of short-wavelength plasma radiation with dust particles located in pulsar magnetic fields, proved to be equally interesting.

These hitherto poorly known properties of impact processes have expanded substantially the potential use of the effects accompanying a hypervelocity impact for creating the conditions necessary for the emergence of living matter. However, such properties may prove not to be crucial for the emergence of life.

In this connection, the development of the capability for the replication of genetic code in the most primitive forms of living matter has been viewed as the crucial problem for many decades.

In the first chapter of the book we attempted to briefly analyze the main difficulties faced by the researchers addressing the problem of the emergence of life. In the 1980-ies the problem was cleared up to a certain extent, but, at the same time, it became evident that the conditions necessary for the animation of matter are extremely difficult to meet. Thus homochirality of the medium, which must have developed at the prebiotic stage of evolution, is generally believed to be the most important condition for the replication of genetic code to appear in nature. This remains the mostly dominant view until now.

The view in question was supported by serious experimental evidence. Thus the experiments of Joyce (1984), Visser (1986), and Visser et. al. (1984), who studied the effect of the chiral composition of the medium on the process of matrix oligomerization, showed that the process of assembling of a complimentary chain on a chirally pure matrix is strongly suppressed if the monomer solution is racemic. These and similar experiments made it possible to develop a qualitative model, which associates the process of replication with the chiral purity of the medium (Avetisov et al. 1985, Goldanskii et. al. 1987.)

The qualitative estimates obtained by Goldanskii and Kuz'min (1989) showed that the primitive self-replicating systems could have originated only in a medium with a strong mirror symmetry breaking, or, more precisely, only in a chirally pure medium. Indeed, certain processes could have occurred within a restricted space region during the prebiotic stages of evolution and they could have caused a sufficiently strong breaking of mirror symmetry in this medium.

The results of these studies led the researchers to conclude that the biogenic scenario of the origin of the chiral purity of the biosphere could not be realized in the course of evolution even in principle. The explanation was that without the chiral purity the very apparatus of self-replication could not appear, which is the basis of the process of self-reproduction of any organisms. The above authors believed that life could not emerge in a racemic medium.

It is important that the above authors also deemed unacceptable the scenarios that associated the chiral purity of the prebiosphere with the stage of the capture of monomers during the formation and selection of the precursors of the most important biopolymers. Such processes also could not occur in a racemic medium. And, finally, the above authors believe that the abiogenic scenario must be the only possible scenario of the emergence of chirally pure medium.

A fact stands out particularly: all possible processes that may result in the formation of precursors of biostructures were generally bound to a medium where only chemical reactions could occur.

For the concept proposed in this book we adopted the protoviroid (Altstein, 1987) to be the most primitive form of living matter. It is therefore important to recall how, according to the above author, such a structure could have developed abiogenically.

According to A. Altstein, the assembly and animation of the protovioroid at the chemical stage of evolution should be viewed as a rare event, which is a result of a combination of progenes. We described this process in detail in Section 1.4 of this book, where we showed that every new addition of progenes to the macromolecule could have been a random process. However, according to A.Altstein, the results of such additions of individual progenes could have been controlled by the system of progenes already incorporated in the structure of the macromolecule.

Altstein estimated the minimum probability of the formation of a protoviroid to be 10^{-30}-10^{-35} and believed that this process could have been limited by the number of progenes that had formed on the Earth over several hundred million years. The chirality «sign» in this model could emerge accidentally and the homochirality of molecular structures was ensured in the process of the assembly of the molecule. Monomers were selected from a racemic medium.

According to Altstein's estimates, the duration of the assembly of a protoviroid under favorable conditions on the Earth should not have been longer than 10 minutes. Hence 10^{30}-10^{35} trials should have been sufficient for a protoviroid to form in the processes described above.

It is especially apparent that in the scenario of A.Altstein during the formation of a protoviroid monomers could have been selected from a racemic medium in the process of the assembly of the macromolecules. According to the estimates mentioned above (Goldanskii and Kuz'min, 1989), such a scenario should be viewed as unacceptable. This is inconsistent with the statement of the above authors that such a process could be realized only in a chirally pure medium.

Such a «sentence» handed to protoviroid is regrettable because the hypothetical protoviroid model had from the very beginning been viewed as a well thought-out macromolecular system. It has no parallel so far. This primarily concerns the very constructive and easy-to-grasp idea of its formation in the process of the assembly of the primitive genetic code. Moreover, if assembled, the "ready-made" protoviroid would be, by its structure and functional capabilities, a smallest-size macromolecular system capable of self-replication. It is important that no complaints have been made concerning the possibility of the functioning of the system.

That is why it is difficult to «part» with the protoviroid. This was the more difficult because, as we pointed out above, during the assembly of a protoviroid each new addition of a progene to the molecule occurred, according to A.Altstein, in a random way, but its results were controlled by the system of progenes. The emergence of such a «control parameter» in the process of assembly could be of special interest.

In this situation an interesting idea was suggested, which, albeit not thoroughly «computed», was evidently very promising if implemented. The idea was to try to analyze the hypothetical emergence of a protoviroid in the chirally pure medium of the impact-produced plasma torch, while maintaining the main ideas of the formation of the genetic code and, possibly, also the effects of the «control parameter» proposed by A.Altstein.

To this end, we will first analyze the process of the assembly of a protoviroid-like macromolecular structure in a hypothetical homochiralic medium that formed in the plasma torch of a meteorite impact assuming that the genetic code in the emerging system forms in accordance with the schema proposed by A.Altstein. Note that the main "unacceptability" that follows from the results reported by Goldanskii and Kuz'min can be overcome if the protoviroid is assembled in a homochiralic medium. It is also important that the availability of a chirally pure medium should not have hindered the assembly of the protoviroid and, on the contrary, it should have sped up the process. Note also that because of the high catalytic activity of the plasma medium the rate of the assembly of a protoviroid-like system in plasma-chemical reactions could have been much higher.

However, for the above processes to occur, we must answer the questions how a chirally pure medium could form in the process of the expansion-away of a plasma torch and what signs may confirm the possibility of the occurrence of such a process? Direct answers to these extremely complex questions are difficult to obtain. However, simple methods of multivariate analysis can be used to successfully persuade the reader that mechanisms that make the medium chirally pure are very likely to be realized in the processes of a plasma torch.

To analyze this problem in more detail, let us study the original results of experimental investigations presented in this book and also the results of theoretical studies of other authors. Also of use will be the results of earlier estimates.

Let us first consider the results of Pechernikova and Vityazev (2008), who estimated the total amount of carbon in the primitive matter of planetesimals falling, onto the Earth as high as $\sim 1.2\ 10^{23}$ g.

If the entire meteorite carbon was spent for the formation of protoviroid-like molecular systems then their number would have amounted to 10^{42}. This quantity is derived with the allowance for the fact that the percentage of carbon in organic molecules is $\sim$ 60% and the average mass of such formations is close to the mass of a protoviroid and amounts to $\sim$ 100 000 a.m.u.

However, the results of laboratory experiments presented in this book show that the yield of organic compounds in the torch plasma is equal to ~ 0.1%. The formation of structures of limiting size for the given projectile size occurs once in 10^3-10^4 interactions. Therefore with these corrections taken into account, the number of macromolecules with the mass of 100 000 a.m.u. reduces to 10^{35}.

This is the number of macromolecules that could have been synthesized in impact processes during the first 200 million years of the Earth's existence in the case of a single use of each carbon atom of cosmic origin. It is important that the number of trials mentioned above should, according to Altstein (1986), have been sufficient for the formation of a protoviroid.

However, it is only a part of the necessary information about the conditions needed for the emergence of such macromolecules. Of special interest may then be the information about the time needed for the formation of a macromolecular system with the mass and properties resembling those of a protoviroid. It is also important for these characteristics to be possibly determined by the impact velocity and the size of the plasma-formation region, which is proportional to the diameter of the projectile.

To this end, we use the results of model experiments presented in Figure 26 in Chapter 4. The plot shows the dependence of the maximum mass of organic molecules synthesized during the impact on the effective diameter of a 50%-carbon content projectile in the interval from 0.1 to 1 mm. This plot can be used to perform an approximation in the case where the diameter of the projectile amounts to ~1 mm. These results show that a 1-mm diameter projectile with above-critical impact velocity may ensure the synthesis of organic molecules with the masses comparable to that of the protoviroid.

This estimate was further verified by comparing the rates of chemical processes of the synthesis of dendrimers under laboratory conditions and in plasmachemical reactions in the process of laboratory modeling of an impact.

It was shown that because of the high catalytic activity of plasma the rates of plasmachemical reactions may be 10^7-10^9 times higher compared to classical chemical reactions. In this case, when in a plasma medium, it would take $6 \cdot 10^{-5}$ to $6 \cdot 10^{-7}$ s for classical 10-minutes long classical chemical reactions to complete.

These time estimates agree well with the time scales of the main plasma processes in a torch produced by the impact of a ~1 mm diameter projectile. Thus, according to experimental results, a macromolecule with a mass of ~10^5 a.m.u may form in $6 \cdot 10^{-6}$ s under such conditions.

It is important that many of the above estimates are based on experimental data. Their comparison with the results of experiments modeling impact interaction suggests that the main similarity parameters are preserved when such processes are reproduced in a laboratory.

Thus the estimates mentioned above, the data of model experiments, and the results of their approximation suggest that a macromolecular system with a close-to-protoviroid mass, i.e., ~10^5 a.m.u., may form in the process of the expansion-away of the torch plasma generated by the impact of a 2-3 g/cm^3, 1-mm diameter projectile moving at a velocity of 10-15 km/s.

Can such a system be assembled in a chirally pure medium? Most likely, yes. Would genetic code be generated under such conditions? Possibly yes, because nothing seems to prevent this at the first sight. Would the "control parameter" act during the process of assembly? Possibly yes, but the answer to this question remains unknown and experimental proofs will be needed to confirm it. However, it is clear that the number of trials needed for the creation of such a formation could have been provided by meteorite bombardment already during the first 200 million years. And this is a fact of considerable importance.

To confirm the possibility of the realization of the scenario mentioned above, it is important to show that the above possibility of strong symmetry breaking in the processes of the expansion-away of the plasma torch has been confirmed experimentally. To this end, let us analyze the results of Avetisov and Goldanskii (1996b), whose paper is dedicated to revealing the relation between two unique properties biological macromolecules - homochirality and capability for replication. In this paper we pay special attention to the constraints associated with the error catastrophe (Eigen and Shuster, 1979).

According to Avetisov and Goldanskii (1996b), the key idea of these constraints is that homochiralic molecules, which, by their properties, meet the criteria for molecules of biological level of complexity, would, when assembling more than 150 monomeric fragments, contain an exorbitant number of defects. Such an excess of the limit of defects in such macromolecular structures should with time result in the complete loss of genetic information. A molecule of such a level of complexity is a biochemical molecule, which cannot form in the process of a large number of random trials. At the same time, there are grounds to believe that the most primitive form of live on the Earth appeared as a result of abiogenous process, but we so far have failed to find out how the "right" algorithm of such assembly could have developed.

Avetisov and Goldanskii (1996b) analyzed the conditions that make it possible to avoid the error catastrophe, and within the framework of the scenario of asymmetric origination this can be achieved only in the case of chiral purity of the organic material of which macromolecular carriers are made. The conditions for the formation of a macromolecule of biochemical level of complexity also include strict constraints on the length of the time interval over which the homochirality of the medium should be ensured. These constraints are difficult to meet in chemical processes on the Earth and they consist in the following:

«First, the stage of polymer capture should be preceded by the formation of not just asymmetric, but chirally pure medium, and hence mechanisms should be available that could produce a strong breaking of mirror symmetry in geochemical and cosmic habitats».

«Second, the chiral purity of the medium should be maintained not only during the stage of polymer capture and formation of homochiralic macromolecules, not only during the stage when a certain class of structures that can serve as information and functional carriers (e.g., RNA-like structures) set apart among all the homochiralic molecules, but also during the further evolution of these carriers until the development of enantiospecific functions making possible the replication of homochiralic structures. Only after the development of such functions the need for the chiral purity of the medium disappears».

The requirements to the level of spontaneous symmetry breaking are equally difficult to meet. Let us refer to them as «tertiary». They consist in the following.

« For the chiral purity to be achieved that is needed for the evolution of homochiralic structures of biological level of complexity (N>150), the catalytic processes that determine spontaneous symmetry breaking should have a high level of enantioselectivity comparable to that of biochemical functions».

The above requirements practically totally prevent the possibility of realization of the conditions needed for the synthesis of a macromolecular structure of biochemical level of complexity in classical chemical processes on the Earth. However, it is possible that for plasmachemical processes, and, especially, for those that develop during a hypervelocity impact, this conclusion would apply only after dedicated studies of such processes. Note that so far no such studies have been performed.

It is also important because certain results of some experimental studies indicate that signs of a strong symmetry breaking have been found. This can be viewed as a natural result, and it is due to the fact that plasma torch, which, as a medium, is far from the thermodynamic branch of equilibrium, is ideal domain for the occurrence of spontaneous processes. Moreover, the natural asymmetry of physical fields in the torch favors the emergence of truly chiral factors, which are capable of producing a slight symmetry breaking and make possible asymmetric formation of enantiomers. This may happen owing to local chiral physical fields, which may be generated in the torch via magnetooptical processes. Local chiral field may serve as the control field for spontaneous processes and determine the «sign» of the asymmetry of the system as a whole.

Spontaneous symmetry breaking may, in turn, produce a strong symmetry breaking in exceptional cases and result in the emergence of a homochiralic medium needed for polymer formation. Such a sequence of processes occurs during the formation of the plasma torch and constitutes the basis of the new concept.

Let us now analyze the duration of the interdependent processes of symmetry breaking and the synthesis of macromolecular structures that develop in the impact plasma and the consequences of such an action. To this end, let us analyze the effects that occur during the impact of a 10-km diameter meteorite with a 5% carbon content. A total of ~ 10^7 meteorites of that size have fallen onto the Earth's surface during the first 200 million years since the formation of our planet (Pechernikova and Vityazev, 2008). If we assume that the density of these meteorites was 2 g/cm^3 then each of them should have carried 10^{14} g of carbon, which would be sufficient to ensure the synthesis of 10^{33} molecules with the mass of 10^5 a.m.u. if the entire carbon was used to produce such molecules. However, as we pointed out above, only 0.1% of the entire carbon mass is used for the formation of organic compounds in natural impact processes, and therefore, with this correction taken into account, the number of macromolecules should be reduced to 10^{30}.

The time needed for the assembly of such molecules in a plasma torch is estimated at ~ 10^{-5}s for a 1-mm diameter projectile. It is safe to assume that this time should be the same for the case of a plasma-formation region corresponding to a 10-km diameter meteorite.

For a meteorite of such a size the domain where new compounds are synthesized may have a diameter of ~50 km, and, given an average expansion velocity of 10^6 cm/s, its lifetime may amount to ~ 5s.

Consider now the dynamics of plasma processes assuming that the symmetry breaking caused by local factors triggered spontaneous symmetry breaking. A homochiralic medium, if it forms in such processes, remains homochiralic over a long time and ensures chirally pure polymer selection for a substantial number of protoviroids. This process meets the conditions mentioned above and listed as «first» by Avetisov and Goldanskii (1996b).

The medium may remain homochiralic until the formation of compounds capable of supporting the replication of homochiralic structures. The time over which these conditions are provided exceeds the time needed for the assembly of a protoviroid by almost of a factor of one million. We do not know whether this process meets the conditions listed as «second». The point is that we do not know what, in the case of the origination of a protoviroid, may happen during the time interval from time of its formation and the existence of the homochiralic medium. However, this time is long enough for multiple replication of the protoviroid. Moreover, if necessary, this time may increase with increasing meteorite diameter. Therefore the above factors and their duration do not rule out the possibility that several generations of protoviroids may form during the expansion-away of the torch with the properties comparable to biochemical complexity level.

It would be especially interesting to determine whether the requirement from the «third» list – stating that there must be a highly chirally pure homochiralic medium – can be met.

No direct evidence is so far available concerning this issue. However, the available experimental results (Kieffer et. al. 1992, 1993) of the registration of linearly polarized radiation from torch plasma presented in Fig. 36 may contain important information. In particular, the linearly polarized radiation from plasma torch observed in these experiments is highly reproducible and is characterized by a high degree of spatial concentration of the irradiation within the polarization plane. This effect can be viewed as an indication of strong anisotropy and extremely high degree of ordering of the medium. It may point not only to the onset of the processes of spontaneous symmetry breaking, but also to the high degree of such a breaking. If this is true then spontaneous processes causing a strong symmetry breaking may be not too rare in the impact-produced plasma torch and amount to at least 0.1% in the case of the average efficiency of such a process. Therefore in extremely rare cases media of high chiral purity can be expected to form. This may result in a substantial number of homochiral macromolecular systems with the masses of 100 000 a.m.u. where polymer forms in a homochiral medium. Such a process does not need a chirally pure medium to form in a substantial volume of the plasma torch. Such a process may occur in the case if a homochiral medium forms inside a sphere of a diameter several centimeters in diameter.

A multivariate analysis addressing the properties of laser- or impact-produced plasma torches combined with the information mentioned above shows that the processes that occur during the expansion-away of the torch may, to a first approximation, meet the conditions formulated in the «three clauses» of Avetisov and Goldanskii (1996b). These results suggest that plasma torch is the optimum medium capable of ensuring the development of homochiralic macromolecular structures, and, possibly, has no alternatives in such a capacity. This conclusion is based on the experimentally discovered unique properties of torch plasma, which make it possible to move from just discussing the problem to thorough, in-depth experimental studies. Such studies should help us answer the principal question of whether homochiralic medium can form in the plasma torch of a meteorite impact – at least, in a nonextended plasma-formation region. Such investigations will also make it possible to explore the relation between the anisotropy degree of the impact-produced plasma torch and the chiral purity of the emergent medium.

We believe that the studies of the processes of the weak breaking of mirror symmetry possibly followed by strong spontaneous breaking of the symmetry of the medium should be conducted along two directions. First, as we showed in Chapters 4 and 5, these processes should be studied in laboratory simulations of impact processes via laser exposure. To reveal the degree of symmetry breaking, synthesis products are to be accumulated in quantities allowing the separation of L and D enantiomers using available instruments.

Such experiments will also be conducted in studies of the second kind reproducing natural impact processes in a laboratory. In this case, new compounds will be synthesized in the head-on collision configuration involving a projectile accelerated to velocities that ensure the formation of a plasma torch.

An analysis of the results of these studies including the comparison of the data obtained in laser and impact experiments and the results of purely plasma experiments addressing the main physical properties of the torch may provide the necessary information about the key processes that occur in the impact-produced torch plasma and result in the symmetry breaking in the medium.

Thus the results presented in Chapter 6 and in the Afterword of this boom are indicative of the possibility of the breaking of mirror symmetry in the synthesis products that form in a plasma torch in the processes that accompany a hypervelocity meteorite impact. This conclusion is based on the results obtained in many various laboratory plasma experiments. A comparison and joint analysis of these results shows that in the case of both laser- and impact-produced plasma torch the breaking of mirror symmetry in the process of the synthesis of enantiomers may be caused by magnetooptical effects. Such a relatively weak symmetry breaking in the plasma torch may be caused by the experimentally found combination of linearly polarized radiation and magnetic field. Such a combination is known to be recognized as the most efficient "true" asymmetric factor. This symmetry breaking could have determined the «sign» of the chirality of the bioorganic world. The point is that they could serve as the controlling field for spontaneous processes that develop in the hot and nonequilibrium plasma medium of the torch that are far from the thermodynamic branch of equilibrium and could therefore result in a strong symmetry breaking. Hence the processes that develop during the expansion-away of the torch plasma produced by a meteorite impact can be viewed as the most likely natural phenomenon conducive to the formation of a chirally pure medium and, possibly, in the formation of homochiralic macromolecular structures needed for the emergence of primary forms of living matter.

Hopefully, in the nearest future these crucial assumptions will be backed by new, more reliable experimental evidence.

GLOSSARY

Asteroids	small bodies of the Solar System with sizes ranging from hundreds of meters to hundreds of kilometers and moving mainly between the orbits of Mars and Jupiter. Asteroids are one of the sources of meteorites falling onto the Earth.
Catalyst	a substance that accelerates a chemical reaction without being consumed or altered in the process.
Chirality	the property of the geometric structure of a molecule to have a non-superimposable mirror image. Such molecules are said to be chiral. Chiral molecules with the structure of two non-superimposable mirror images are referred to as enantiomers.
Comets	small bodies of the Solar System (along with asteroids and meteoroids) moving in highly elongated orbits. Comets consist mostly of snow and water ice, H_2O, CO, CO_2 and rocky material. When a comet approaches the Sun it develops, in the process of the evaporation of its matter, a gaseous (plasma) and a dust tail.
Destruction	destruction, for example, of organic compounds when heated.
Fullerene	one of the allotropic forms of carbon. Its molecule consists of 60 atoms located at the surface of a sphere. A fullerene molecule forms of five or six carbon rings with a single carbon atom located at the junction of four rings. The structure of such a fullerene is identical to that of a football used in late last century. The number of carbon atoms may differ in different fullerenes and varies from 50 to 70, 76, 82, 84, etc. Their molecular masses may be as high as 5000-6000 a.m.u. or even more. Such compounds are called onion-like or hyperfullerenes.
Homochirality	the property of a complex molecule to consist of monomers with the same "sign" of chirality (see *chirality*).
Ion	An atom or a group of atoms that has acquired a net electric charge by gaining or losing one or more electrons.
Ionization	the process in which one or more electrons are torn off a neutral atom or molecule, and, as a result, the latter acquires a charge. Ionization is achieved by thermal heating or exposure to radiation.

Isotope	a species of a chemical element that differs from other species of the same element only by the mass of its atomic nucleus. Isotopes of the same element have similar chemical properties, but some of them have unstable nuclei and are prone to radioactive decay.
Meteorites	objects that have fallen onto the Earth's surface or on the surfaces of other planets and their satellites from the interplanetary space. During the early stages of the formation of planets meteorites consist mostly of planetesimals, as well as asteroids and planetary cores.
Monomer	a molecule (e.g., nucleotide or amino acid), which usually serves as a building block for linear chains of polymers (in particular, nucleic acids and proteins).
Nucleotide	a monomer subunit of nucleic acids with the common structure: nitrogen base - sugar - phosphoric acid.
Oligonucleotide	a polymer that incorporates many nucleotides. Oligonucleotides are constituent parts of RNA or DNA.
Planetesimals	objects that formed at the last stage of the formation of planets from protoplanetary (protoplanetary) gas-dust matter. Planetesimals are actually planet embryos, or protoplanets.
Plasma	gaseous substance consisting of an equal number of positive and negative charges, which ensures that the substance as whole is electrically neutral. Plasma can be fully ionized, which means that it contains no neutral atoms, like, e.g., the solar corona. Plasma can be partially ionized with one ion for up to 1000 and more neutral particles, e.g., the plasma in some interstellar regions. The degree of ionization depends on electron temperature, total ionization is achieved at 10^5-10^6 K, or 10 to 100 eV.

About the Author

Professor George G. Managadze, Doctor of Sciences, Head of the Laboratory of Active Diagnostics of the Space Research Institute of the Russian Academy of Sciences, Member of the International Academy of Astronautics, Honored Scientist of the Russian Federation.

Professor G. Managadze is known as a specialist in experimental physics whose research interests include the study of planets and small bodies of the Solar System, space and laboratory plasmas. He has mastered to perfection the methods of modeling the space processes in a laboratory and methods of space research involving active influence on the medium. G. Managadze has been a major driving force in the establishment and development of this direction in space research. He had actively participated in the Soviet-French "Arax-Zarnitsa" project by developing new instruments for plasma diagnostics used in the experiments carried out within the project, which made it possible to generate, record and study plasma-beam discharge under space conditions for the first time.

G. Managadze was the principal investigator of the "Gruziya-Spurt" experiment, which involved injection of electron beams from onboard "Vertical" rocket at an altitude of 1500 km. A new physical phenomenon - "superhigh potential" of the space probe – was discovered and studied in this experiment.

G. Managadze is the co inventor and the principal investigator of the widely known LIMA-D and DION onboard remote mass spectrometers used to determine the chemical composition of the regolith of small bodies from onboard a ~100 meters distant flyby probe. These instruments were meant for the study of Phobos and were developed with the active participation of researchers from Bulgaria, Austria, Germany, France, and Finland.

G.Managadze is the inventor of the foil mass-reflectron, which is known as MANAGADZE-TOF(MTOF) in the research literature. This instrument has been patented in the USA, Austria, Germany, and France. While operated as a part of the onboard research instrument in the SOHO, WIND, and ACE interplanetary space missions, it allowed the isotopic composition of heavy ions in the Solar wind to be measured for the first time.

G. Managadze is the author of 220 research papers and 35 inventions of which 25 are used by research organizations in Russia and abroad. G. Managadze has been closely cooperating in joint space research projects with many research organizations and universities in France, Austria, Bulgaria, USA, Switzerland, Germany, Poland, and Finland. From 1993 through 2001 he was a project manager for APTI (USA) and a consultant for the Applied Physics Laboratory of Johns Hopkins University. These works are continued through affiliation with Fenix Technology International (Fenixtec, USA) and through collaboration with the NASA Goddard Space Flight Center.

G. Managadze currently heads the development of MANAGA и LASMA instruments (in cooperation with specialists from the University of Bern) and onboard mass spectrometers for the «Phobos-Grunt» mission. The flight units have been already completed.

REFERENCES

A' Hearn M.F. and the Deep Impact Team (2006). Deep Impact: excavating Comet Tempel 1. *Lunar and Planetary Inst. Technical Report*, 37, no. 1978.

A'Hearn M.F. et al. (2005). Deep Impact: excavating Comet Tempel 1. *Science*, 310, 258-264.

Abyzov S.S. (1993). Microorganisms in Antarctic ice. In *Antarctic Microbiology*. Friedmann EI (ed.), New York: Wiley-Liss, pp. 265-297.

Afonin V.I. (2001). Stratification and filamentation mechanism for a multicharged plasma of a Z-Pinch. *Plasma Phys. Rep.*, 27 (7), 576-581.

Afonin V.I., Potapov A.V. and Ugodenko A.A. (2001). Study of the effect of plasma filamentation on the shock formation in a flat target. *Izvestiya Chelyabinskogo Nauchnogo Tsentra*, 3 (12) (in Russian).

Altstein A.D. (1987). Origin of the genetic system: the progene hypothesis. *Mol. Biol.*, 21 (2), 309-322 (in Russian).

Alvarez W.T. (1997). *Rex and the Crater of Doom*. Princeton, NJ: Princeton Univ. Press.

Anders E. (1989). Prebiotic organic matter from comets and asteroids. *Nature*, 342 (6247), 255-257.

Andrews S.R., Harris F.M. and Parry D.E. (1992). A combined experimental and theoretical investigation of $C_2H^{2+}{}_2$ electronic-state energies. *Chem. Phys.*, 166, 69-76.

Anhalt J.P. and Fenselau C. (1975). Identification of bacteria using mass spectrometry. *Anal. Chem.*, 47(2), 219-225.

Anisimov S.I., Imas Ya.A., Romanov G.S. and Khodyko Yu.V. (1970). *Effect of high-energy radiation on metals*. Moscow: *Nauka*, p. 272 (in Russian).

Anisimov S.I. and Luk'anchuk B.S. (2002). Selected problems of laser ablation theory. *Phys.-Usp.*, 45, 293–324.

Artsimovich L.A. (1969). *Elementary plasma physics* (3-rd ed.), Moscow: Atomizdat (in Russian).

Avetisov V.A. (1985). *Izv. Akad. Nauk Arm. SSR, Fiz.*, 20, 174.

Avetisov V.A., Anikin S.A., Goldanskii V.I. and Kuz'min V.V. (1985). *Dokl. Biophys.,* 282, 115.

Avetisov V. and Goldanskii V. (1996a). Mirror symmetry breaking at the molecular level. *Proc. National. Acad. Sci. USA,* 93, 11435-11442.

Avetisov V.A. and Goldanskii V.I. (1996b). Physical aspects of mirror symmetry breaking of the bioorganic world. *Phys.-Usp.*, 39(8), 819-835.

Avorin E.N., Anuchina N.N., Gadzhieva V.V. et al. (1985). *Preprint of the Institute of Applied Marthematics, USSR Academy of Sciences,* No. 117.

Avorin E.N., Anuchina N.N., Gadzhieva V.V. et al. (1996). *Fiz. goreniya vzryva*, 32(2), 117-123 (in Russian).

Babina V.M., Boustie M., Guseva M.B., Zhuk A.Z., Migault A. and Milyavskii V.V. (1999). Dynamic synthesis of crystalline carbyne from graphite and amorphous carbon. *High Temp.*, 37, 543-551.

Bada J.L. and Lazcano A. (2003). Prebiotic soup—revisiting the Miller experiment. *Science*, 300, 745–746.

Bailey J. (2000). Circular polarization and the origin of biomolecular homochirality. In: *Bioastronomy 99: A New Era in the Search for Life in the Universe, Proceedings of a Conference Held at the Kohala Coast, Hawaii, 2-6 August, 1999. Lemarchand G. and Meech K. (eds.). ASP Conference Series*, 213, p. 213-243.

Bailey J., Chrysostomou A., Hough J., Gledhill T., McCall A., Clark S., Menard F. and Tamura M. (1998). Circular Polarization in Star-Formation Regions: Implications for Biomolecular Homochirality. *Science*, 281(5377), 672-674.

Bar-Nun A., Bar-Nun N., Bauer S. H. and Sagan C. (1970). Shock Synthesis of Amino Acids in Simulated Primitive Environments. *Science,* 168, 470-473.

Barak L. and Bar-Nun A. (1975) The Mechanism of Amino Acid Synthesis by High Temperature Shock-waves. *Origins Life,* 6, 483-503.

Baronova E.O. (2007). Z-pinch Plasmas. In: *Plasma Polarization Spectroscopy*. Fujimoto T. and Iwamae A. (eds.). Berlin: Springer, p. 154.

Baronova E.O., Sholin G.V. and Jakubowski L. (1999). Study of polarized argon lines in plasma-focus device. *Pis'ma Zh. Eksper. Teor. Fiz.*, 69(11-12), 870-873.

Baronova E.O. Sholin G.V. and Jakubowski L. (2003). Application of X-ray Polarization Measurements to Study Plasma Anisotropy in Plasma Focus Machines. *Plasma Phys. Controlled Fusion*, 45(7), 1071-1077.

Baronova E.O. and Stepanenko M.M. (2007). Novel Polarimeter-Spectrometer for X-Rays. In: *Plasma Polarization Spectroscopy*. Fujimoto T. and Iwamae A. (eds.). Berlin: Springer, p. 334.

Baronova E.O., Stepanenko M.M. and Stepanenko A.M. (2008). X-Ray Spectropolarimeter. *Rev. Sci. Instrum.*, 79(4), 83-105.

Baronova E.O and Yakubowski L. (2007). X-Ray polarization Measurements. In: *Plasma Polarization Spectroscopy*. Fujimoto T. and Iwamae A. (eds.). Berlin: Springer, p. 327.

Barron L.D. (1986). True and false chirality and parity violation. *Chem. Phys. Lett.*, 123, 423-427.

Barron L.D. (1994). Can a Magnetic Field Induce Absolute Asymmetric Synthesis? *Science*, 266, 1491.

Barron L.D. (2008). Hirality and life. In: *Strategies of Life Detection. Space sciences series of ISSI.* Botta O., Bada J.L., Gomez-Elvira J., Javaux E., Selsis F. and Summons R. (eds.), p. 380.

Basov N.G., Boiko V.A., Dement'ev V.A., Krokhin O.N. and Sklizkov G.V. (1967). Heating and Decay of Plasma Produced by a Giant Laser Pulse Focused on a Solid Target. *Sov. Phys. JETP*, 24, 659.

Basov N.G., Boiko V.A., Drozhbin Yu.A., Zakharov S.M., Krokhin O.N., Sklizkov, G.V. and Yakovlev, V.A. (1970). Study of the Initial Stage of Gas-Dynamic Expansion of a Laser Flare Plasma. *Sov. Phys. Doklady*, 15, 576.

Bernal J.D. (1969) *Vozniknovenie zhizni* (The origin of life). Moscow: Mir. Russian translation edited by A.Oparin with an Appendix including papers by A.I.Oparin (1924) and J.Haldane (1929). Original English edition: Bernal J.D. (1967). *The Origin of Life*. Cleveland: World.

Biberman L.M., Vorob'ev V.S. and Yakubov I.T. (1982). *Kinetics of Non-equilibrium Low-Temperature Plasma*. Moscow: Nauka (in Russian).

Blank J.G., Miller G.H., Ahreus M.J. and Winans R.E. (2001). Experimental Shock Chemistry of Aqueous Amino Acid Solutions and the Cometary Delivery of Prebiotic Compounds. *Origins Life. Evol. Biospheres*, 31, 15-51.

Bochkarev N.G. (1992). *Basics of the physics of the interstellar medium*. Moscow: Izd. Mosk. Gos. Univ. (in Russian).

Bol'shov L.A., Burdonskii I.N., Velikovich A.L., Gavrilov V.V., Gol'tsov A.Yu., Zhuzhukalo E. V., Zavyalets S.V., Kiselev V.P., Koval'ski N.G., Liberman M.A., Mkhitar'yan L.S., Pergament M.I., Yudin A.I. and Yaroslavsky A.I.. (1987). Acceleration of foils by a pulsed laser beam. *Sov. Phys. JETP*, 65(6), 1160.

Bonner W. (1984). Experimental Evidence for β-decay as a Source of Chirality by Enantiomer Analysis. *Origins Life*, 14, 383-390.

Bonner W.A. (1991). The Origin and Amplification of Biomolecular Chirality. *Origins Life Evol. Biospheres*, 21, 59-111.

Borisenko N.G., Bugrov A.E., Burdonskiy I.N., Fasakhov I.K., Gavrilov V.V., Goltsov A.Yu., Gromov A.I., Khalenkov A.M., Kovalskii N.G., Merkuliev Yu.A., Petryakov V.M., Putilin M.V., Yankovskii G.M. and Zhuzhukalo E.V. (2008). Physical Processes in Laser Interaction with Porous Low-Density Materials. *Laser Part. Beams*, 26, 537-543.

Borucki J.G., Khare B. and Cruikshank D.P. (2002). A New Energy Source for Organic Synthesis in Europa's Surface Ice. *J. Geophys. Res.*, 107, 5114.

Briand J., Adrian V., Tamer M. El., Gomes A., Quemener Y., Dinguirard J. P. and Kieffer J. C. (1985). Axial Magnetic Fields in Laser-Produced Plasmas. *Phys. Rev. Lett.*, 54, 38-41.

Brinckerhoff W.B., Bugrov S.G., Kelner L., Managadze G.G., Managadze N.G., Saralidze G.Z., Srama R., Stubig M. and Chumikov A.E. (2005). Universal Mechanism of Abiogenous Synthesis of Organic Substances in the Processes of SHVI. *Geophys. Res. Abs.*, 7, 11171.

Brinckerhoff W.B., Managadze G.G., McEntier R.W., Cheng A.F. and Creen W.J. (2000). Laser time-of-flight mass spectrometry for space. *Rev. Sci. Instr.*, 71, 536-545.

Bronshten V.A. (1987). *Meteors, meteorites, meteoroids*. Moscow: Nauka, 1987 (in Russian).

Busarev V.V. and Surdin V.G. (2008). *Small bodies of the Solar System. The Solar System*. Surdin V.G. (ed.). Moscow: FIZMATLIT, 400 pp (in Russian).

Bychenkov V.Yu., Kas'yanov Yu.S., Sarkisov G.S. and Tikhonchuk V.T. Mechanisms for magnetic-field generation in laser plasmas (1993). *JETP Lett.*, 58, 184-190.

Bykovskii O.A. and Nevolin I.N. (1985). *Laser Mass Spectroscopy*. Moscow: Energoizdat, (in Russian).

Bykovskii, Iu.A., Vasilev, N.M., Laptev, I.D. and Nevolin,V.N. (1975). Neutral particles from the interaction of laser radiation with a solid target. *Sov. Phys. Tech. Phys.*, 19, 1635.

Bykovskii Yu.A., Zhuravlev G.I. and Belousov V.I. (1979). Lazernyi mass-spektrometricheskii metod analiza veshchestv osoboi chistoty (Laser mass-spectrometric method of the analytsis of high-purity substances). In: *Obtaining and analysis of high-purity substances*. Zorin A.D. (ed.). Moscow: Nauka, p. 276 (in Russian).

Cambou F., Dokoukin V.S., Ivchenko V.N., Managadze G.G., Migulin V.V., Nazarenko O.K., Nesmyanovich A.T., Pyatsi A.Kh., Sagdeev R.Z. and Zhulin I.A. (1975). The Zarnitza rocket experiment on electron injection. *Space Research* 15.4918.

Carr M.H. (2004). The proof is in: Ancient water on Mars.*The Planetary Report*, May/June. 2004, 24(3), 6-11.

Cech T.R. and Baas B.L. (1986). Biological catalysis by RNA. *Ann. Rev. Biochem*, 55, 599-629.

Chernavskaya N. M. and Chernavskii D.S. (1975). Some Theoretical Aspects of the Problem of life Origin. *J.Theor. Biol.*, 53, 13.

Chernavskii D.S. (2000). The origin of life and thinking from the viewpoint of modern physics. *Phys.-Usp.*, 43(2), 151-176.

Chetverina H.V., Demidenko A.A., Ugarov V.I. and Chetverin A.B. (1999). Spontaneous rearrangements in RNA sequences. *FEBS Lett.*, 450, 89-94.

Chrysostomou A., Gledhill T., Ménard F., Hough J., Tamura M. and Bailey J. (2000). Polarimetry of young stellar objects - III. Circular polarimetry of OMC-1. *Mon. Not. R. Astron. Soc.*, 312(1), 103-115.

Chrysostomou A., Ménard F., Gledhill T., Clark S., Hough J., McCall A. and Tamura M. (1997). Polarimetry of young stellar objects - II. Circular polarization of GSS 30. *Mon. Not. R. Astron. Soc.*, 285(4), 750-758.

Chyba C.F. (2000). Energy for microbial life on Europa. *Nature*, 403(6768), 381-382.

Chyba C. F. and McDonald G.D. (1995). The origin of life in the Solar system: current issues. Ann. *Rev. Earth. Planet. Sci.*, 23, 215-249.

Chyba C., Thomas P., Brookshaw L. and Sagan C. (1990). Cometary delivery of organic molecules to the early Earth. *Science*, 249, 366-373.

Crawford D.A. and Schultz P.H. (1988). Laboratory observations of impact-generated magnetic fields. *Nature*, 336, 50-52.

Crawford D.A. and Schultz P.H. (1991). Laboratory investigations of impact-generated plasma. *J. Geophys. Res.*, 96(E3), 18807-18817.

Crawford D. A. and Schultz P. H. (1993). The production and evolution of impact-generated magnetic fields. *Int.J. Impact Eng.*, 14, 205-216.

Crawford D. A. and Schultz P. H. (1999). Electromagnetic properties of impact-generated plasma, vapor and debris. *Int.J. Impact Eng.*, 23 169-180.

Crick F.H.C. (1968). The origin of the genetic code. *J. Mol. Biol.*, 38, 367-379.

Dallman B. K., Grün E., Kissel J. and Dietzel H. (1977). The ion-composition of the plasma produced by impacts of fast dust particles. *Planet. Space Sci.*, 25, 135-147.

Davankov V. (2006). Chirality as an inherent general property of matter. *Chirality*, 18, 459-461.

Davankov V.A. (2009). Inherent homochirality of primary particles and meteorite impacts as possible source of prebiotic molecular chirality, *Russ. J. Phys. Chem. A.*, 83(8), 1247-1256.

De Duve C. (1995). *Vital Dust*. New York : Basic Books.

Delone N.B. (1989). *Interaction of laser radiation with matter.* // Moscow: Nauka, p. 278 (in Russian).

Dickerson R.E. (1978). Chemical evolution and the origin of life. *Sci. Am.*, 70.

Drapatz S. and Michel K.W. (1974). Theory of shock-wave ionization upon high-velocity impact of micrometeorites. *Z. Naturforsch.*, 29a, 870.

Drobyshevski E.M., Zhukov B.G., Sakharov V.A., Studenkov A.M. and Kurakin R.O. (1995). Head-on collision opens 15–20 km/s opportunities. *Int.J. Impact Eng.*, 17, 285-290.

Dulieu F., Amiaud L., Fillion J.-H., Matar E., Momeni A., Pirronello V. and Lemaire J.L. (2007). Experimental evidence of water formation on interstellar dust grains. In: *Molecules in Space and Laboratory, meeting held in Paris, France, May 14-18, 2007.* Lemaire J.L. and Combes F. (eds.). Publisher: S. Diana., p.79.

Dvorov I.M. (1976). *Geothermal energetic*). Moscow: Nauka, 192pp (in Russian).

Eberhardy C.A. and Schultz P.H. (2004). Probing impact-generated vapor plumes. *Lunar Planet. Sci.*, 35, no. 1855.

Egholm M., Buchardt O., Christensen L., Behrens C., Freier S.M., Driver D.A., Berg R.H., Kim S.K., Norden B. and Nielsen P.E. (1993). PNA Hybridizes to complementary oligonucleotides obeying the Watson-Crick hydrogen-bonding rules. *Nature*, 365, 566-568.

Egholm M., Buchardt O., Nielsen P.E. and Berg R.H. (1992). Peptide nucleic acids (PNA). Oligonucleotide analogues with an achiral peptide backbone. *J.Amer. Chem. Soc.*, 114, 1895-1897.

Eigen M. (1973). Molecular self-organization and the early stages of evolution. *Sov. Phys.-Usp.*, 16, 545-589.

Eigen M. and Schuster P. (1979). *The Hypercycle. A Principle of Natural Self-Organization.* Berlin: Springer-Verlag.

Eigen M. and Winkler R. (1975). *Das Spiel. Naturgesetze steuern den Zufall.* Munchen: Piper und Co. Verlag (in German).

Elyasberg P.E. (1961). *Introduction into the theory of motion of artificial Earth satellites.* Moscow: Nauka, p. 491 (in Russian).

Eletskii A.V. and Smirnov B.M. (1995). Fullerenes and the structures of carbon. *Phys. Usp.*, 38, 935–964.

Engel M.H., Macko S.A. and Silfer J.A. (1990). Carbon isotope composition of individual amino acids in the Murchison meteorite. *Nature*, 348, 47-49.

Ernst C.M. and Schultz P. H. (2003). Effect of initial conditions on impact flash decay. *Lunar Planet. Sci.*, 34, no. 2020.

Ernst C.M. and Schultz P. H. (2005). Investigations of the luminous energy and luminous efficiency of experimental impacts into particulate targets. *Lunar Planet. Sci.*, 36, no. 1475.

Ernst C.M., Schultz P.H., A'Hearn M.F. and the Deep Impact Science Team. (2006). Photometric evolution of the Deep Impact flash. *Lunar Planet. Sci.*, 37, no. 2912.

Fenner N.C. and Daly N.R. (1966). Laser used for mass analyses. *Rev. Sci. Instrum.*, 37, 1068-1072.

Ferris J.P. and Ertem G. (1993). Monmorillonite catalysis of RNA oligomer formation in aqueous solution. A model for prebiotic formation of RNA. *J. Amer. Chem. Soc.*,115, 12270-12275.

Ferris J.P., Joshi P.C., Wang K.-J., Miyakawa S. and Huang W. (2004). Catalysis in prebiotic chemistry: application to the synthesis of RNA oligomers. *Adv. Space Res.*, 33, 100-105.

Ferris J.P., Hill A.R., Liu R. and Orgel L.E. (1996). Synthesis of long prebiotic oligomers on mineral surfaces. *Nature*, 381, 59-61.

Fesenkov V.G. (1949). Atmospheric haziness produced by the fall of the tunguska meteorite on June 30, 1908. *Meteoritika*, 6, 8-12 (in Russian).

Folsome, C. E. (1979). *The Origin of Life*. San Fran cisco: W. H. Freeman.

Fortey R. (1998). *Life*. New York: Knopf.

Fortov V.E., Khrapak A.G., Khrapak S.A., Molotkov V.I. and Petrov O.F. (2004). *Phys.-Usp.*, 47. 447.

Fortov,V.E., Ternovoi,V.Ya., Zhernokletov M.V., Mochalov, M.A., Mikhailov, A.L., Filimonov, A.S., Pyalling, A.A., Mintsev, V.B., Gryaznov, V.K. and Iosilevskii, I.L. (2003). *J. Exp. Theor. Phys.*, 97(2), 259-278.

Fox S.W. and Dose K. (1977). *Molecular evolution and the origin of life (Rev. ed.)*. N.Y.: Marcel Dekker, 370 p.

Fox S. and Nakashima T. (1980). The assembly and properties of protobiological structures: the beginnings of cellular peptide synthesis. *Bio-Systems*, 12, 155-166.

Frank F.C. (1953). On spontaneous asymmetric synthesis. *Biochem. Biophys. Acta.*, 11, 459-463.

Frank-Kamenetskii D.A. (1968). *Lectures on plasma physics*. (2-nd ed.). Moscow: Atomizdat (in Russian).

Friichtenicht J.F. and Slattery J.C. (1963). Ionization associated with hypervelocity impact. *NASA Tech. Note*, D-2091.

Gal'chenko V.F. (2003). Cryptobiosphere of Mars. *Aviakosm. Ekol. Med.*, 5, 15-22 (in Russian).

Gal'chenko V.F. (2004). Cryptolife on Mars and Jovian satellites. *Proc. Vinogradski Inst. Microbiol. Russ. Acad. Sci.*, 12, 64-79 (in Russian).

Galeev A.A. and Sudan R.N. (eds.). (1983). *Basic Plasma Physics I. Volume I of Handbook of Plasma Physics*. Amsterdam: North-Holland Publishing Company, 751 pp.

Galimov E.M. (2001). *Phenomenon of life: between equilibrium and nonlinearity*. Moscow: Editorial URSS, p. 256 (in Russian).

Gerasimov M.V. and Mukhin L.M. (1978). Interstellar molecules and prebiological evolution. *Sov. Phys. – Dokl.*, 23, 777-778.

Gerasimov M.V., Mukhin L.M. and Safonova E.N. (1991). *Izv. Akad. Nauk SSSR, Ser. Geol.*, no. 4, 119-126 (in Russian).

Gerasimov M.V., Safonova E.N. and Dikov Yu.P. (2007). Synthesis of organic molecules during impacts at accretion of the Earth and planets. In: Second International Conference. Biosphere Origin and Evolution/ October 28 – November 2, 2007. Loutraca, Greece. Abstracts, O-14.

Gilbert W. (1986). The RNA world. *Nature*. 319, 618.

Gilichinsky D.A., Vorobyova E.A., Erokhina L.G., Fyodorov-Davydov D.G. and Chaikovskaya N.R. (1992). Long-term preservation of microbial ecosystems in permafrost. *Adv. Space Res.*, 12, 255-263.

Gilichinsky, D.A., Wilson, G.S., Friedmann, C.P., McKay C.P., Sletten, R.S., Rivkina, E.M., Vishnivetskaya, T.A., Erokhina L.G., Ivanushkina, N.E., Kochkina, G.A., Shcherbakova, V.A., Soina, V.S., Spirina E.V., Vorobyova, E.A., Fyodorov-Davydov, D.G., Hallet, B.,

Ozerskaya, S.M., Sorokovikov, V.A., Laurinavichus, K.S., Shatilovich, A.V., Chanton, J.P., Ostroumov, V.E., and Tiedje J.M. (2007). Microbial populations in Antarctic permafrost: biodiversity, state, age, and implication for astrobiology. *Astrobiology*, 7(2), 275-311.

Ginzburg V.L. (1970). *The Propagation of Electromagnetic Waves in Plasmas*. New York: Pergamon.

Gledhill T. and McCall A. (2000). Circular polarization by scattering from spheroidal dust grains. *Mon. Not. R. Astron. Soc.*, 314, 123-137.

Goldanskii V.I. (1979). Facts and hypotheses of molecular chemical tunnelling. *Nature*, 279, 109-115.

Goldanskii V.I. (1993). Revival of the concept of the cold prehistory of life. *Eur. Rev.*, 1(2), 137-147.

Goldanskii V.I., Anikin S.A., Avetisov V.A. and Kuz'min V.V. (1987). *Comm. Mol. Cell. Biophys*., 4, 79-98.

Goldanskii V.I. and Kuz'min V.V. (1989). Spontaneous breaking of mirror symmetry in nature and the origin of life. *Sov. Phys.-Usp.*, 32, 1–29.

Gold T. (1992). The deep hot biosphere. *Proc. Nat. Acad. Sci. USA.*, 89, 6045-6049.

Goldsmith D. and Owen T. (1980). *The search for life in the Universe*. Sasaulito: University Science Books.

Goldsworthy B.J., Burchell M.J., Cole M.J., Green S.F., Leese M.R., McBride N., McDonnell J.A.M., Müller M., Grün E., Srama R., Armes S.P. and Khan M.A. (2002). Laboratory calibration of the Cassini Cosmic Dust Analyzer (CDA) using new, low density projectiles. *Adv. Space Res.* 29(8), 1139-1144.

Göller J. and Grün E. (1989). Calibration of the Galileo/Ulysses dust detectors with different projectile materials and at varying impact angles. *Planet. Space Sci*., 37(10), 1197.

Goresy A. and Donnay G. A. (1968). New allotropic form of carbon from the Ries crater. *Science*, 161, 363-364.

Greenberg J.M., Kouchi A., Niessen W., Irth H., Paradijs J., Groot M. and Hermsen W. (1994). Interstellar dust, chirality, comets and the origins of life: Life from dead stars? *J. Biol. Phys*, 20, 61-70.

Grieve R.A.F. (1980). Impact bombardment and its role in proto-continental growth on the early Earth. (1980). *Precambrian Res.*, 10, 217-247.

Guerrier-Takada C., Gardiner K., Marsh T., Pace N. and Altman S. (1983). The RNA moiety of ribonucleases P is the catalytic subunit of the enzyme. *Cell*, 35, 849-857.

Gurevich L.E. (1940). *Principles of physical kinetics*. Leningrad-Moscow: Gostekhizdat (in Russian).

Guring D. (1973). *Hypervelocity impact from engineering point of view. Hypervelocity impact phenomena//* Moscow: Nauka, p. 468 (in Russian).

Hartmann W.K., Ryder G., Dones L. and Grinspoon D (1990). The time-dependent intense bombardment of the primordial Earth / Moon system. In: *Origin of the Earth and Moon.* Canup R.M. and Righter K. (eds.). Tuscon: Univ. Arizona Press, p. 493-512.

Hendrickson R.A., Arnoldy R.L. and Minckler J.R. (1976). Echo III: the study of electric and magnetic fields with conjugate echoes from artificial electron beams injected into the auroral zone ionosphere. *Geophys. Res. Lett*., 3, 409.

Hess W.N., Trichel M.G., Davis T.N., Beggs W.C., Kraft G.E., Stassinopoulos E. and Maier E.J.R. (1971). Artificial aurora experiment : experiment and principal results. *J. Geophys. Res.*, 76, 6067-6081.

Heymann D., Chibante L.P.F., Brooks R.R., Woldbach W.S., Smith J., Korochantsev A., Nazarov M.A. and Smalley R.E. (1996). Fullerenes of possible wildfire origin in Cretaceous-Tertiary boundary sediments. *The Cretaceous-Tertiary Event and Other Catastrophes in Earth History. The Geological Society of America, Special Paper 307*, p. 453-464.

Hillenkamp F., Unsold E., Kaufmann R. and Nitsche R. (1975). Laser microprobe mass analysis of organic materials. *Nature*, 256, 119.

Hoover R.B. (2003). Comets, carbonaceous meteorites, and the origin of the biosphere. *BioSci*. Discussions, 3, 23-70.

Hoover R.B. and Rozanov A.Yu. (2002). Chemical biomarkers and microfossils in carbonaceous meteorites. *Proc. SPIE.*, 4495, 1-18.

Hornung K., Malama Yu. G. and Thoma K. (1996). Modeling of the very high velocity impact process with respect to in-situ ionization measurements. *Adv. Space Res.*, 17, 77-86.

Hornung K., Malama Yu. and Kestenboim Kh. (2000). Impact vaporization and ionization of cosmic dust particles. *Astrophys. Space Sci.*, 274, 355-363.

Horowitz N.H. (1986). *To Utopia and Back: The Search for Life in the Solar System.* New York: W. H. Freeman.

Hough J., Lucas P.W and Bailey J. (2006). The polarization signature of extra-solar planets. The Scientific Requirements for Extremely Large Telescopes, Proceedings of the 232nd Symposium of the International Astronomical Union, Held in Cape Town, South Africa, November 14-18, 2005. Whitelock P.A., Dennefeld M. and Leibundgut B. (eds.). Cambridge: Cambridge University Press, pp. 350-355.

Huang W. and Ferris J.P. (2003). Synthesis of 35-49 mers of RNA oligomers from unblocked monomers. A simple approach to the RNA world. *Chem. Commun.*, 12, 1458-1459.

Inubushi Y., Nishimura H., Ochiai M., Fujioka S., Johzaki T., Mima K., Kawamura T., Nakazaki S., Kai T., Sakabe S. and Izawa Y. (2006). X-ray line polarization spectroscopy to study hot electron transport in ultra-short laser produced plasma. *J. Quant. Spectrosc. Radiat. Transfer*, 99, 305-313.

Ivanov B.A. (2004). Heating of the lithosphere during meteorite cratering. *Sol. System Res.*, 38, 266.

Ivanov B.A. (2008a). Size-frequency distribution of asteroids and impact craters: estimates of impact rate. In: *Catastrophic Events Caused by Cosmic Objects*. Adushkin V.V. and Nemchinov I.V. (eds.). Berlin: Springer, p. 91-116.

Ivanov B.A. (2008b). Geologic effects of large terrestrial impact crater formation. In: *Catastrophic Events Caused by Cosmic Objects*. Adushkin V.V. and Nemchinov I.V. (eds.). Berlin: Springer, p. 163-206.

Ivanov M.V. and Lein A.Yu. (1991). Methane-producing microorganisms: component of Mars biosphere. *Dokl. Akad. Nauk*, 321, p. 1272-1276 (in Russian). (English translation in: Ivanov M.V. and Lein A.Yu. (1992). Methane-producing microorganisms: component of Mars biosphere.*JPRS Report: Science and Technology. Central Eurasia: Life Sciences*., p. 10).

Johnson A.P., Cleaves H.J., Dworkin J.P., Glavin D.P., Lazcano A. and Bada J.L. (2008). The Miller Volcanic Spark Discharge Experiment. *Science*, 322, 404. Jones W. (2004). *Life in the Solar system and beyond.* Chichester: Praxis.

Joyce G.F. and Orgel L.E. (1993). Prospects for understanding the origin of the RNA world. In: *The RNA World.* Gesteland R.F. and Atkins J.F. (eds.). New York: Cold Spring Harbor Laboratory Press, p. 1-25.

Joyce G.F. and Orgel L.E. (2006). Progress toward understanding the origin of the RNA world. In: *The RNA World*, (3-rd. ed.).. Gesteland R.F., Cech T.R. and Atkins J.F. (eds.). N.Y.: Cold Spring Lab. Press, 2006. p. 23-56.

Joyce G.F., Visser G.M., van Boeckel C.A.A., Van Boom J. H., Orgel L. E. and Van Westrenen J. (1984). Chiral selection in poly(C)-directed synthesis of oligo(G). *Nature*, 310, 602-604.

Juk A.Z., Borodina T.I., Milyavskii V.V., and Fortov V.E. (2000). *Dokl. Akad. Nauk*, 370, 328-331.

Kajander E.O., Bjorklund M. and Ciftcioglu N. (1999). Suggestions from observations on nanobacteria isolated from blood . In: *Size limits of very small microorganisms. Proceedings of a workshop*. Washington: Nat. Acad. Press, p. 50-55.

Kaldor A., Cox D.M. and Reichmann K.C. (1988). Carbon clusters revisited: The ''special'' behavior of C60 and large carbon clusters. *J. Chem. Phys*., 88, 1588-1597.

Karas, M. Gluckmann M. and Schafer J. (2000). Ionization in matrix-assisted laser desorption/ionization: singly charged molecular ions are the lucky survivors. *J. Mass Spectrom.* 35, 1-12.

Kasatochkin V.I., Sladkov A.M., Kudryavtsev Yu.P., Popov N.M. and Korshak V.V. (1967). *Dokl. Akad. Nauk*, 177, 358 (in Russian).

Kawasaki T., Hatase K., Fujii Y., Jo K., Soai K. and Pizzarello S. (2006). The distribution of chiral asymmetry in meteorites: An investigation using asymmetric autocatalytic chiral sensors. *Geochim. Cosmochim. Acta*, 70, 5395-5402.

Kent P.R.C., Towler M.D., Needs R.J. and Rajagopal G. (2000). Carbon clusters near the crossover to fullerene stability. *Phys. Rev. B*, 62, 15394-15397.

Keszthelyi L. (1995). Origin of the homochirality of biomolecules. *Q. Rev. Biophys.*, 28, 473.

Khochenmuss R., Stortelder A., Breuker K. and Zenobi R. (2000). Secondary ion–molecule reactions in matrix-assisted laser desorption/ionization. *J. Mass Spectrom.*, 35, 1237-1245.

Kieffer J.C., J.P. Matte, M. Chaker, Y. Beaudoin, C.Y. Chien, S. Coe, G.Mourou, J. Dubau and M.K. Inal. (1993). X-ray-line polarization spectroscopy in laser-produced plasmas. *Phys. Rev. E*, 48, 4648-4658.

Kieffer, J.C., Matte, J.P., Pépin, H., Chaker, M., Beaudoin, Y., Johnston, T.W., Chien, C.Y., Coe, S., Mourou and G., Dubau, J. (1992). Electron distribution anisotropy in laser-produced plasmas from x-ray line polarization measurements. *Phys. Rev. Lett.* 68, 480-483.

Kim P.H. and Namba S. (1967). *Bull. Amer. Phys. Soc.*, 12, 137.

Kissel J. and Krüger F.R. (1987a). Ion formation by impact of fast dust particles and comparison with related techniques. *Appl. Phys. A*, 42, 69-85.

Kissel J. and Krüger F.R. (1987b). The organic component in dust from comet Halley as measured by the PUMA mass spectrometer on board Vega 1. *Nature*., 326, 755-760.

Kissel J., Sagdeev R.Z., Bertaux J.L., Angarov V.N., Audouze J., Blamont J.E., Buchler K., Evlanov E.N., Fechtig H., Fomenkova M.N., von Hoerner H., Inogamov N.A., Khromov V.N., Knabe W., Krueger F.R., Langevin Y., Leonasv B., Levasseur-Regourd A.C., Managadze G.G., Podkolzin S.N., Shapiro V.D., Tabaldyev S.R. and Zubkov B.V. (1986). Composition of comet Halley dust particles from VEGA observations. *Nature*, 321, 280-282.

Kizel' V. A. (1985). *Physical Causes of the Dissymmetry of Living Systems*. Moscow: Nauka, p. 118 (in Russian).

Kleiman J., Heiman R., Hawken D., and Salansky N.M. (1984). Shock compression and flash heating of graphite/metal mixtures at temperatures up to 3200 K and pressures up to 25 GPa. *J. Appl. Phys.*, 56, 1440-1454.

Klumov B.A., Kim V.V., Lomonosov I.V., Sultanov V.G., Shutov A.V. and Fortov, V. (2005). Deep Impact experiment: possible observable effects. *Phys.-Usp.*, 48, 733-742.

Knabe W. and Kruger F. R. (1982). ion formation from alkali iodide solids by swift dust particle impact. *Z. Naturforsch., A*, 37, 1335-1340.

Knecht W.L. (1965). Initial energies of laser-induced electron emission from W. *Appl. Phys. Lett.*, 6, 99.

Knecht W.L. (1966). Laser-induced spontaneous electron emission from rear side of metal foils. *Appl. Phys. Lett.*, 8, 254.

Kobayashi K. and Saito T. (2000). Energetics for Chemical Evolution on the Primitive Earth. In: *The Role of Radiation in the Origin and Evolution of the Life*, Akabosh M., Full N. and Neverro-Gonzales R., Kyoto: Kyoto Univ. Press, p. 25-38.

Kon'kova G.D., Sil'nikov E.E., Sysoev Aleksei A. and Sysoev Aleksandr A. (2009). A model of pulsed target evaporation and ion generation in laser plasma. *Tech. Phys. Lett.*, 35, 144-146.

Konash P.V. and Lebo I.G. (2006). Simulations of electron-beam scattering by spontaneous magnetic fields in a laser plasma. *Quantum. Electron.*, 36, 767-777.

Kondepudi D.K. and Nelson G.W. (1983). Chiral Symmetry Breaking in Nonequilibrium Systems. *Phys. Rev. Lett.*, 50, 1023-1026.

Kondepudi D.K. and Nelson G.W. (1984). Chiral-symmetry-breaking states and their sensitivity in nonequilibrium chemical systems. *Physica A*, 125, 465.

Kondepudi D.K. and Nelson G.W. (1985). Weak neutral currents and the origin of biomolecular chirality. *Nature*, 314, 438-441.

Korobkin V. V., Motylev S.A., Serov R.V. and Edvards D.F. (1977). Structure of spontaneous magnetic field in a laser plasma. *JETP Lett.*, 25, 497-500.

Kostin V.V, Fortov V.E, Krasyuk I.K, Kunizhev B.I. and Temrokov A.I. (1997). Investigation of shock-wave and destruction processes under conditions of high-velocity impact and laser stimulation of a target of organic material. *High Temp.*, 35, 949.

Kovalev A., Nemchinov I., Shuvalov V., Kosarev I. and Zetzer Yu. (2008). Ionospheric and magnetospheric effects. In: *Catastrophic Events Caused by Cosmic Objects*. Adushkin V.V. and Nemchinov I.V. (eds.). Berlin: Springer, p.313-332.

Kruger K, Grabowski P.J., Zaug A.J., Sands J., Gottschling D.E. and Cech T.R. (1982). Self-splicing RNA: Autoexcision and autocyclization of the ribosomal RNA intervening sequence of Tetrahymena. *Cell*, 31, 147-157.

Ksanfomaliti L.V. (1997). *Planetary alignment*. Moscow: Nauka i Zhizn', p. 260 (in Russian).

Ksanfomaliti L.V. (2005). *Huygens mission to the Saturnian Moon Titan. Nauka i Zhizn'*, Issue 3, 3-8 (in Russian).

Lagow R.J., Kampa J.J., Han-Chao Wei, Battle S.L., Genge J.W., Laude D.A., Harper C.J., Bau R., Stevens R.C., Haw J.F. and Munson E. (1995). Synthesis of linear acetylenic carbon: the "sp" carbon allotrope. *Science*, 267, 362-367.

Laska L., Krasa J., Juha L., Hamplova V. and Soukup L. (1996). Fullerene production driven by long-pulses of near-infrared laser radiation. *Carbon*, 34, 363-368.

Lebedev A.T. (2003). *Mass spectroscopy in organic chemistry*. Moscow: BINOM. Laboratoriya znanii, 493 pp. (in Russian).

Levkin P., Levkina A. and Schurig V. (2006). Combining the enantioselectivities of L-valine diamide and permethylated B-cyclodextrin in one gas cromatographic chiral stationary phase. *Anal. Chem.*, 78, 5143-5148.

Limpoukh I. and Rozanov V.B. (1984). Transverse structures (filaments, spicules, jets) in a laser plasma. *Quantum Electron.*, 14, 955-960.

Lisse C.M. and the Deep Impact Spitzer Science Team. (2006). Spitzer Space Telescope observations of the nucleus and dust of Deep Impact target Comet 9P/Tempel 1. *Lunar Planet. Sci.*, 37, no. 1960.

Löb W. (1906). Studien uber die chemische Wirkung der stillen elektrischen Entladung. *Z. Electrochem.*, 11, 282-316.

Lucas P., Hough J., Bailey J., Chrysostomou A., Gledhill T. and McCall A. (2005). UV Circular polarisation in star formation regions: the origin of homochirality? *Origins Life Evol. Biospheres*, 35, 29-60.

Lyakhov S.B. and Managadze G.G. (1977). Beam-plasma discharge in the vicinity of the rocket in the Zarnitsa-2 experiment. *Sov. J. Plasma Phys.*, 3, 763-769.

Mackie J.C., Colket M.B and Nelson P.F. (1990). Shock tube pyrolysis of pyridine. *J. Phys. Chem.*, 94, 4099-4106.

Malyarova A.M. and Tantsyrev G.D. (1991). Analysis of aqueous solutions of organic substances by secondary ion mass spectrometry. *J. Anal. Chem. USSR*, 46, 1318-1320.

Managadze G.G. (1986). *Time-of-flight ion mass analyzer*. USA Patent no. 4, 611, 118.

Managadze G.G. (1992a). *Universal Multi-Purpose Transportable Mass-Spectrometric Complex*. APTI report, Washington.

Managadze G.G. (1992b). TOF mass-spectrometer. // Patent 1732396 (RF). Priority of invention: 1988. Registered: 1992. Published in: *Russ. Bull. Inventions* No. 17, 1992 (in Russian).

Managadze G.G. (1994). *In situ studies of relict organic molecules on munor bodies of the Solar System*. Preprint. Pr-1885. Moscow: Space Research Institute of the Russian Academy of Sciences. 26 pp (in Russian).

Managadze G.G. (2001). Organic compound synthesis in experiments modeling high-speed meteor impact. Proceedings of 26th General Assembly of the European Geophysical Society. *Geophys. Res. Abstr.*, 3, 7595.

Managadze G.G. (2002a). Molecular synthesis in recombinating impact plasma. In: *Proceedings of 27th General Assembly of the European Geophysical Society, Nice*, Abstract EGS02-A-06871. 2002, p. 334.

Managadze G.G. (2002b). Determination of organic compounds in solid and gas phases using miniaturized axially symmetric time-of-flight mass-reflectrons. *J. Anal. Chem.*, 57, 537-543.

Managadze G.G. (2003). The synthesis of organic molecules in a laser plasma similar to the plasma that emerges in hypervelocity collisions of matter at the early evolutionary stage of the Earth and in interstellar clouds. *J. Exp. Theor. Phys.*, 97, 49-60.

Managadze G.G. (2005a). *Abiogenous synthesis of chiral organic compounds in the plasma generated under the influence of SHVI.* Preprint. Pr-2107. Moscow: Space Research Institute of the Russian Academy of Sciences, 20 pp. (in Russian).

Managadze G.G. (2005b). A universal mechanism of synthesis of organic compounds in the processes of superfast impact on the stage of prebiotic evolution. *Probl. Upravlen. Inf.*, no. 6, 34–47 (in Russian).

Managadze G.G. (2007a). A new universal mechanism of organic compounds synthesis during prebiotic evolution. *Planet. Space Sci.*, 55, 134-140.

Managadze G.G. (2007b). A novel scenario of prebiotic stage of evolution of life based on the universal mechanism of organic compounds synthesis in the plasma torch of meteorite impact. In: Second International Conference. Biosphere Origin and Evolution/ October 28 – November 2, 2007. Loutraca, Greece. Abstracts, O-15.

Managadze G.G. (2009). *Plasma of meteorite impact and prebiotic evolution.* – Moscow: FIZMATLIT. 2009 – 352 pp (in Russian).

Managadze G. (2010a). Plasma and collision processes of hypervelocity meteorite impact in prehistory of life. *Int. J. Astrobiol.* 9, 157-174.

Managadze G.G. (2010b). Plasma and craters of meteorite impact and prehistory of life. In: X European workshop on Astrobiology. Russia, Pushchino, September 6-8, 2010. Abstracts of papers.

Managadze G.G., Brinckerhoff W.B. and Chumikov A.E. (2003a). New mechanism of molecular synthesis in hypervelocity impact plasmas on primitive Earth and in interstellar clouds. Auroral Phenomena and Solar-Terrestrial Relations. In: *Int. Sympos. in memory of Professor Yuri Galperin*. Moscow IKI RAS. February 4-7, SRA RAS.

Managadze G.G., Brinckerhoff W.B. and Chumikov A.E. (2003b). Possible synthesis of organic molecular ions in plasmas similar to those generated in hypervelocity impacts. *Int.J. Impact Eng.*, 29, 449-458.

Managadze G.G., Brinckerhoff W.B. and Chumikov A.E. (2003c). Molecular synthesis in hypervelocity impact plasmas on the primitive Earth and in interstellar clouds. *Geophys. Res. Lett.*, 30, 1247-1251.

Managadze G.G., Brinkerkhoff V., Managadze N.G. and Chumikov A.E. (2006). *Identification of amino acids abiogenically synthesized in the plasma torch simulating a hypervelocity impact torch*. Preprint. Pr- 2126. Moscow: Space Research Institute of the Russian Academy of Sciences, 23 pp. (in Russian).

Managadze G.G. and Eismont N.A. (2009). A low-orbit experiment on modeling superhigh-velocity impact of a meteorite reproducing abiogenic synthesis of complex organic compounds in torch plasma. *Cosmic Res.*, 47, 491-499.

Managadze G.G. and Managadze N.G. (1997). TOF mass spectrometer. Patent 2096861 (RF). Priority of invention: 1994. Registered: 1997. Published in: *Russ. Bull. Inventions* No. 32, 1997 (in Russian).

Managadze G.G. and Managadze N.G. (1999). Quantitative reference-free express analysis of some alloys on a laser time-of-flight mass spectrometer. *Tech. Phys.*, 44, 1253-1257.

Managadze N. G., Managadze G. G. and Ziegler A. (1997). Quantitative analysis of metal alloys, ceramics, minerals without standard samples with the help of compact laser TOF mass-spectrometer. In: Proceedings of the 45th Conference of ASMS, Palm Springs, USA, 1997, p. 1243.

Managadze G.G. and Podgornyi I.M. (1968). Model of the Earth's magnetic field. *Sov. Phys. Dokl.*, 13, 593.

Managadze G.G. and Podgornyi I.M. (1969). Modeling of the Interaction of the Solar Wind with the Geomagnetic Field. *Geomagn. Aeron.,* 8, 496.

Managadze G. G., Riedler W., Balebanov B.M., Friedrich M., Gagua T.I., Klos Z., Laliashvili N.A., Leonov N.A., Lyakhov S.B., Martinson A.A. and Mayorov A.D. (1983). Plasma processes in the region of electron beam injection from a high-altitude payload. In: *Active experiments in space. Proceedings of the International Symposium held at Alpbath, Austria, 24-28 May, 1983. In ESA Active Expts. in Space, July 1983*. ESA SP-195, (SEE N84-14172 05-12), p. 161-169.

Managadze G.G. and Sagdeev R.Z. (1987). *Space mass-spectrometry probe*. Certificate of authorship No. 1190849 (1987). Published in: *Russ. Bull. Inventions* No. 5. P. 279, 1987 (in Russian).

Managadze G.G. and Sagdeev R.Z. (1988). Chemical composition of small bodies of the solar system determined from the effects of solar-wind interaction with their surfaces. *Icarus*, 73, 294-302.

Managadze G.G., Sagdeev R.Z. and Shutyaev I.Yu. (1987). *Remote laser mass-reflectron*. Certificate of authorship No. 1218852 (1987). Published in: *Russ. Bull. Inventions* No. 17, p. 273, 1987 (in Russian).

Managadze G.G. and Shutyaev I.Y. (1993). Exotic instruments and applications of laser ionization mass spectrometry in space research. *Chem. Anal.*, 124, 505-547.

Margulis L. and Sagan C. (1997). *Microcosmos.* Berkley and LA: Univ. of Calif. Press.

Mason S.F. (1991). *Chemical Evolution. Origin of the Elements, Molecules and Living Systems.* Oxford: Clarendon Press.

Matsu T. and Abe Y. (1986). Evolution of an impact-induced atmosphere and magma ocean on the accreting Earth. *Nature*, 319, 303-305.

McFadden L. and A'Hearn M. (2004). Our first look inside a comet: Deep Impact. *Planet. Rep.*, 24, 12-17.

McKay D.S., Gibson E.K., Thomas-Keprta K.L., Vali H., Romanek Ch.S., Clemett S.J., Chillier X.D.F., Maechling C.R. and Zare R.N. (1996). Search for past life on Mars: possible relict biogenic activity in Martian meteorite ALH84001. *Science*, 273, 924-930.

Meierhenrich U.J., Nahon L., Alcaraz C., Bredehöft J.H., Hoffmann S.V., Barbier B. and Brack A. (2005). Asymmetric Vacuum UV photolysis of the Amino Acid Leucine in the Solid State. Angew. Chem., 117, 5774-5779; *Angew. Chem., Int. Ed. Engl.*, 44, , 5630-5634.

Melnichenko V.M., Nikulin Yu.N. and Sladkov A.M. (1985). Layer-chain carbons. *Carbon*, 23, 3-7.

Melosh H. J. and the Deep Impact Team. (2006). Deep Impact: the first second. *Lunar Planet. Sci.*, 37, no. 1165.

Mendel C.W. and Olsen J.N. (1975). Charge-Separation Electric Fields in Laser Plasmas. *Phis. Rev. Lett.*, 34, 859-866.

Mendis D.A. (1988). *Exploration of Halley's Comet*. Growing M., Praderie F. and Reinhard R. (eds.). Berlin: Springer-Verlag, p. 939.

Miller S.L. (1982). Prebiotic syntheses of organic compounds. In: *Mineral Deposits and Involution of the Biosphere*. Holland H.D. and Schidlowski M. (eds.), N.Y.: Springer-Verlag, p. 155-176.

Miller S. and Orgel L. (1974). *The origins of life on Earth*. Englewood Cliffs, NJ: Prentice-Hall.

Miller S.L. and Urey H.C. (1959). Organic compound synthesis on the primitive Earth. *Science*, 130, 245-251.

Mitrofanov I.G., Zuber M.T., Litvak M.L., Demidov N.E., Sanin A.B., Boynton W.V., Gilichinsky D.A., Hamara D., Kozyrev A.S., Saunders R.D., Smith D.E. and Tretyakov V.I. (2007). Water ice permafrost on Mars: Layering structure and subsurface distribution according to HEND/Odyssey and MOLA/MGS data. *Geophys. Res. Lett.*, 34, L18102.

Miyakawa S., Tamura H., Kobayashi K. and Sawaoka A. B. (1997). New Application of a Magneto-Plasma Dynamic Arc-Jet to Amino Acid Synthesis. *Jpn. J. Appl. Phys.*, 36, 4481-4485.

Miyakawa S., Murasawa K., Kobayashi K. and Sawaoka A. B. (1999). Cytosine and uracil synthesis by quenching with high-temperature plasma. *J. Am. Chem. Soc.*, 121, 8144-8145.

Mojzsis S., Harrison M. and Pidgeon R. (2001). Oxygen-isotope evidence from ancient zircons for liquid water at the Earth's surface 4,300 Myr ago. *Nature*, 409, 178-181.

Moroz V.I. (1978). *Physics of planet Mars*. Moscow: Nauka (in Russian).

Moroz V.I. and Mukhin L.M. (1977). Early evolutionary stages in the atmosphere and climate of the terrestrial planets. *Cosmic Res.*, 15, 774-791.

Morozov L.L. (1978). *Dokl. Akad. Nauk SSSR*, 241, 233 (in Russian).

Morozov L.L. (1979). Mirror symmetry breaking in biochemical evolution. *Origins Life*, 9, 187-217.

Morozov L.L. and Goldanskii V.I. (1984). Breaking of chiral symmetry in prebiotic evolution and the physical conditions for the emergence of life. *Vestn. Akad. Nauk SSSR.*, no. 6, p. 54 (in Russian).

Morozov L.L., Kuzmin V.V., Goldanskii V.I.// Sov.Sci.Rev.D (New York, London; Harwood Acad.Publ.) 1982.

Mukhin L.M. (1974). Evolution of organic compounds in volcanic regions. *Nature*, 251, 50-51.

Mukhin L. (1980). *Planets and life*. Moscow: Molodaya Gvardiya (in Russian).

Mukhin L.M., Gerasimov M.V. and Safonova E.N. (1989). Origin of precursors of organic molecules during evaporation of meteorites and mafic terrestrial rocks. *Nature*, 340, 46-48.

Mulyukin A.L., Sorokin V.V., Vorobyova E.A., Suzina N.E., Duda V.I. and El'-Registan G.I. (2002). Detection of microorganisms in the environment and the preliminary appraisal of their physiological state by x-ray microanalysis. *Microbiol.*, 71, 723-734.

Nakamura S., Uchimura T., Katagiri S., Suzuki A., Sawabe M. and Yamamoto Y. Results of the precise orbit determination. experiment with ADEOS-II. In: *Proceedings of the 18 th International Symposium on Space Flight Dynamics. 11-15 October 2004. Munich, Germany*. ESA, SP-548, pp.163-168.

Namba S., Kim P.H. and Mitsuyama A. (1966). Energies of ions produced by laser irradiation. *J. Appl. Phys*. 37, 3300-3301.

Namba S., Kim P.H., Itoh T., Arai T. and Kinoshita N. (1967a). Relation of laser induced ion energy to laser power, *Jpn. J. Appl. Phys*., 6, 273.

Namba S., Kim P.H. and Schwarz H. (1967b). Laser-induced emission of ions from clean surfaces. In: *Phenomena in Ionized Gases*. Proceedings of the Eighth International Conference held August 27 - September 2, 1967, in Vienna, Austria, p.59.

Nemchinov I.V., Svetsov V.V. and Shuvalov V.V. (2008). Main factors of hazards due to comets and asteroid. In: *Catastrophic Events Caused by Cosmic Objects*. Adushkin V.V. and Nemchinov I.V. (eds.). Berlin: Springer, 1-90.

Newkome G.R., Moorefield C.N. and Vogtle F. (1996). *Dendritic Molecules (Concepts, Synthesis, Perspectives)*. N.Y.: Wienheim.

Nicolis G. and Prigogine I. (1981). Symmetry breaking and pattern selection in far-from-equilibrium system. *Proc. Natl. Acad. Sci. USA*, 78, 659-663.

Nicolis G. and Prigogine I. (1989). *Exploring Complexity*. San Francisco: W. H. Freeman and Co..

Nikolaev E.N. and Tantsyrev G.D. (1980). Mechanism of the formation of secondary quasi-molecular ions of polar substances. In: *Secondary ion and ion-photon emission*. – Khar'kov: Khar'kov State University, p. 39-41 (in Russian).

O'Neill P.T. and Williams D.A. (1999). Interstellar water and interstellar ice. *Astrophys. Space Sci*., 266, 539-548.

Oetliker M., Hovestadt D., Klecker B., Collier M.R., Gloeckler G., Hamilton D.C., Ipavich F.M., Bochsler P. and Managadze G.G. (1997). The isotopic composition of iron in the solar wind: first measurements with the mass sensor on the WIND spacecraft. *Astrophys. J.*, 474, L69-L72.

Onstott T.S., Tobin K., Dong H., Deflaun M.F., Fredrickson J.K., Bailey T., Brockman F.J., Kieft Th.L., Peacock A., White D.C., Balkwill D., Phelps T.J. and Boone D.R. (1997). Deep gold mines of South Africa: windows into the subsurface biosphere. Instruments, methods, and missions for the investigation of extraterrestrial microorganisms. *Proc. SPIE*, 3111, 344-357. Instruments, Methods, and Missions for the Investigation of Extraterrestrial Microorganisms, Hoover R.B. (ed.). San Diego, California.

Oparin A.I. (1924). *The origin of life on Earth*. New York: Academic Press.

Opower H. (1967). Kinetic energy of produced from hot laser plasma. *Z. Naturforsch., A*, 1967, 22, 1392-1397.

Orgel L.E. (1968). Evolution of the genetic apparatus. *J. Mol. Biol*., 38, 381-393.

Orgel L.E. (1973). *The Origin of Life: molecules and natural selection*. N.Y.: Wiley, p. 237.

Orgel L.E. (1998). The origin of life – a review of facts and speculations // *Trends in Biochem. Sci.*, 23, 491-495.

Orgel L.E. (2004). Prebiotic chemistry and the origin of the RNA world. Crit. Rev. *Biochem. Mol., Biol*. 39, 99-123.

Oro J. (1961). Mechanism of synthesis of adenine from hydrogen cyanide under plausible primitive earth conditions. *Nature*, 191, 1193-1194.

Ostrovskii V.E. and Kadyshevich E.A. (2007). Generalized hypothesis of the origin of the living-matter simplest elements, transformation of the Archean atmosphere, and the formation of methane-hydrate deposits. *Phys.-Usp.*, 50, 175-196.

Parkes, R.J. and Maxwell, J.R. (1993). Some like it hot (and oily). *Nature*, 365, 694- 695.

Parmon V.N. (2002). The Prebiotic Phase of the Origin of Life. *Herald Russ. Acad. Sci.*, 72, 592–598.

Parmon V.N. (2007). Autocatalytic reactions and natural selection at prebiological steps of the Earth evolution. In: Second International Conference. Biosphere Origin and Evolution/ October 28 – November 2, 2007. Loutraca, Greece. Abstracts, O-8.

Pasteur L. (1884). *Bull, Soc. Chem. France N.S.*, 1215-220.

Pechernikova G.V. and Vityazev A.V. (2008). Impacts and evolution of early Earth. In: *Catastrophic Events Caused by Cosmic Objects*. Adushkin V.V. and Nemchinov I.V. (eds.). Berlin: Springer, p. 333-351.

Pedersen K. (1993). The deep subterranean biosphere. *Earth Sci. Rev.*, 34, 243-260.

Petrovtsev A.V., Politov V.Yu. and Sapozhnikov A.T. (1998). *Preprint of the Russian Federal Nuclear Center - The All-Russian Research Institute of Experimental Physics (RFNC - VNIIEF)* No. 135. Snezhinsk (in Russian).

Phipps C.R. and Dreyfus R.W. (1993). Laser ablation and plasma formation. In: *Laser Ionization Mass Analysis*. Vertes A., Gijbels R. and Adams F. (eds.). N.Y.: J.Wiley and Sons, Inc., p. 369.

Ponnamperuma S. (1972). *The origins of life*. New York: Dutton.

Ponnamperuma C., Lemmon R.M., Mariner R. and Calvin N. (1963). Formation of adenine by electron irradiation of methane, ammonia and water. *Proc. Natl. Acad. Sci. U.S.A.*, 49, 737-740.

Porco C.C., Baker E., Barbara J., Beurle K., Brahic A., Burns J.A., Charnoz S., Cooper N., Dawson D.D., Del Genio A.D., Denk T., Dones L., Dyudina U., Evans M.W., Fussner S., Giese B., Grazier K., Helfenstein P., Ingersoll A.P., Jacobson R.A., Johnson T.V., McEwen A., Murray C.D., Neukum G., Owen W.M., Perry J., Roatsch Th., Spitale J., Squyres S., Thomas P., Tiscareno M., Turtle E.P., Vasavada A.R., Veverka J., Wagner R. and West R. (2005). Imaging of from the Cassini Spacecraft. *Nature*, 434, 159-168.

Porco C.C., Helfenstein P., Thomas P.C., Ingersoll A.P., Wisdom J., West R., Neukum G., Denk T., Wagner R., Roatsch T., Kieffer S., Turtle E., McEwen A., Johnson T.V., Rathbun J., Veverka J., Wilson D., Perry J., Spitale J., Brahic A., Burns J.A., Del Genio A.D., Dones L., Murray C.D. and Squyres S. .(2006). Cassini observes the active South Pole of Enceladus. *Science*, 311, 1393-1401.

Prigogine I. (1967). *Introduction to Thermodynamics of Irreversible Processes*. N.Y.: John Wiley.

Prigogine, I. (1980). *From Being to Becoming: Time and Complexity in the Physical Sciences*. San Francisco: W. H. Freeman and Co.

Prigogine I.O. and Kondepudi D. (1998). *Modern Thermodynamics. From Heat Engines to Dissipative Structures*. New York: Wiley.

Prigogine I. and Stengers I. (1993). *Time, Chaos and the Quantum: Towards the Resolution of the Time Paradox*. New York: Harmony Books.

Raizer Yu.P. (1959). *Zh. Eksp. Teor. Fiz.*, 37, 1741 (in Russian).

Raizer Yu.P. (2009). *Physics of gaseous discharge*. Scientific publication. Dolgoprudnyi: «Intellekt», 736 pp. (in Russian).

Ramendik G.I., Kryuchkova O.I., Derzhiev V.I., Stroganova N.S. and Strel'nikova E.B. (1979). On possibility of mass-spectrometric studies of chemical processes in the plasma. *Dokl. Phys. Chem.*, 245, 336.

Rampino M.R. and Haggerty B.M. (1996). The Cretaceous-Tertiary event and other catastrophes in Earth history. *The Geological Society of America, Special Paper*, no. 307, 11-30.

Ratcliff P.R., Burchell M.J., Cole M.J, Murphy T.H. and Alladfadi F. (1997). Experimental measurements of hypervelocity impact plasma yield and energetics. *Int.J. Impact Eng.*, 20, 663-674.

Ratcliff P. R.; Gogu F., Grun E., Srama R. (1996). Plasma production by secondary impacts: Implications for velocity measurements by in-situ dust detectors. *Adv. Space Res.*, 17, 111-115.

Raup D.M. (1991). *Extinction: Bad Genes or Bad Luck?* NY: Norton.

Reinhard R. (1988). The Giotto mission to Halley's comet. In: *Exploration of Halley's Comet.* Growing M., Praderie F. and Reinhard R. (eds.). Berlin: Springer-Verlag, Berlin, p. 950-958.

Roberts J.A. (1984). Supernovae and Life. *Nature*, 308, 318.

Rode A.V., Gamaly E.G. and Luther-Davies B. (2000). Formation of cluster-assembled carbon nano-foam by high-repetition-rate laser ablation. *Appl. Phys. A*, 70, 135-144.

Rohlfing E.A., Cox D.M. and Kaldor A. (1984). Production and characterization of supersonic carbon cluster beams. *J. Chem. Phys.*, 81, 3322-3330.

Romanovskii Yu.M., Stepanova N.V. and Chernavkii D.S. (1984). *Mathematical biophysics.* Moscow: Nauka (in Russian).

Roybal R., Stein C., M'iglionico C. and Shively J. (1995). Laboratory simulation of hypervelocity debris. *Int.J. Impact Eng.*, 17, 707-718.

Rubenstein E., Bonner W.A., Noyes H.P. and Brown G.S. (1983). Supernovae and Life. *Nature*, 306, 118.

Safronov V.S. (1972). *Evolution of the protoplanetary cloud and formation of the earth and planets*. Jerusalem (Israel): Keter Publishing House, 212pp.

Safronov V.S. and Vityazev A.V. (1983). Origin of the Solar System. *Itogi Nauki Tekh., Ser.: Astron.*, 24, 5-93 (in Russian).

Sagan C. and Khare B. N. (1971). Long-wavelength ultraviolet photoproduction of amino acids on the primitive Earth. *Science*, 173, 417-420.

Sagdeev R.Z. (1988). The Vega mission to Halley's comet. . In: *Exploration of Halley's Comet*. Growing M., Praderie F. and Reinhard R. (eds.). Berlin: Springer-Verlag, Berlin, p. 959-964.

Sagdeev R.Z., Anisimov S.I., Galeev A.A., Shapiro V.D. and Shevchenko V.I. (1982). Dust hazard near Halley Comet in case of the VEGA projec. *Adv. Space Res.*, 2, 133-143.

Sagdeev R.Z., Kambu F. and Zhulin I.A. (1977). Active experiments in the ionosphere and magnetosphere. In: *Nauka i chelovechestvo* (Science and Mankind). Moscow: Znanie, p. 216-233 (in Russian).

Sagdeev R.Z., Kissel J., Evlanov E.N., Fomenkova M.N., Inogamov N.A., Khromov V.N., Managadze G.G., Prilutski O.F., Shapiro V.D., Shutyaev I.Y. and Zubkov B.V (1987). The dependence of mass resolution and sensitivity of the PUMA instrument on the energy spread of ions produced by hypervelocity impacts. *Astron. Astrophys.*, 187, 179-182.

Sagdeev R.Z., Managadze G.G., Tur A.V. Yanovskii V.V. (1986a). In: *Proceedings of the International Workshop on Phobos, Space Research Institute (IKI), USSR Academy of Sciences*, Moscow, p. 129-161 (in Russian).

Sagdeev R.Z., Prokhorov A.M., Managadze G.G., Kissel J., Pellinen R., Balebanov V.M., Tabaldyev S.R., Shutyaev I.Yu., Bondarenko A.L., Timofeev P.P., Ter-Mikaelyan V.M., Terent'e V.I., Natenzon M.Ja., Von Hoerner H., Riedler B., Bergr., Zlatev S., Kynchev

A., Belyakov P., Simeonov L., Silen J., Phironen J., Lehto A., Salmu Ch., Peltonen P., Dul'nev G.N., Pashenin P.P., Malyutin A.A., Il'ichev N.N.,Arumov G.P. and Pershin S.M. (1986b). LIMA-D experiment: Methodology, laboratory tests, PHOBOS: Sci. and Methodol. Aspects of the PHOBOS Study. In: *Proceedings of the International Workshop on Phobos, Space Research Institute (IKI), USSR Academy of Sciences*, Moscow, p. 220-231.

Salama, A. (2004). ISO: highlights of recent results. *Adv. Space Res.*, 34, 528-534.

Schibagaki K., Takada N., Sasaki K. and Kadota K. (2003). Synthetic characteristics of large carbon cluster ions by laser ablation of polymers in vacuum. *J. Appl. Phys.*, 93, 655-661.

Schidlowski M. (1998). Beginnings of terrestrial life: problems of the early record and implications for extraterrestrial scenarious. *Proc. SPIE*, 3441, 149-157.

Schopf J.W. (1993). Microfossils of the Early Archean Apex chert: new evidence of the antiquity of life. *Science*, 260, 640-646.

Schopf J.W. (2000). *The Cradle of Life*. Princeton, N. J: Princeton University Press.

Schultz, P.H. (1996). Effect of impact angle on vaporization. *J. Geophys. Res.,* 101, 21117-21136.

Schultz, P.H. and Anderson J.L.B. (2005). Alternative cratering scenarios for the deep impact collision. *Lunar Planet. Sci.*, 36, no. 1926.

Schultz P.H., Ernst C.M. and Anderson J.L.B. (2005). Expectations for crater size and photometric evolution from the Deep Impact collision. *Space Sci. Rev.*, *117*, 207-239.

Schultz P.H., Ernst C., A'Hearn M.F., Eberhardy C., Sunshine J.M., and the Deep Impact Team. (2006). The deep impact collision: a large-scale oblique impact experiment. *Lunar Planet. Sci.*, 37, no. 2294.

Shapiro R. (1984). The improbability of prebiotic nucleic acid synthesis. *Origins Life Evol. Biosphere*, 14, 565-570.

Shapiro R. (1995). The prebiotic role of adenine: a critical analysis. *Origins Life Evol. Biosphere*. 25, 83-98.

Shklovskii I.S. (1965). *The universe, life, and intelligence*. Moscow: Nauka. 1965. 284 pp. (English translation in: Shklovskii I.S. and Sagan C. (1966). *Intelligent life in the universe. Authorized translation by Paula Fern*. San Francisco: Holden-Day).

Sholin G.V. (1968). The anomalous energy dependence of the polarization of the radiation from atoms excited by a beam of electrons. *Sov. Phys. Dokl.*, 12, 811.

Sholin G.V. and Oks E.A. (1973). Theory of optical polarization measurements of the turbulence spectrum in plasma. *Sov. Phys. Dokl.*, 18, 254 .

Shuvalov V.V., Trubetskaya I.A. and Artem'eva N.A. (2008). Marine target impacts. In: *Catastrophic Events Caused by Cosmic Objects*. Adushkin V.V. and Nemchinov I.V. (eds.). Berlin: Springer, p. 291 - 312.

Simionescu C. and Denes F. *Originea vietii* (Origin of life). Bucuresti: Editura Academii Republicii Socialiste Romania. 1983 (in Romanian).

Smirnov B.M. (1982). *Excited atoms*. Moscow: Energoatomizdat (in Russian).

Snytnikov V.N. (2006). Astrocatalysis as an initial stage of. geobiological processes. Does life create planets? In: *Evolyutsiya biosfery i bioraznoobraziya* (Evolution of the Biosphere and Biodiversity). Moscow: KMK, p.49-59 (in Russian).

Snytnikov V.N. (2007a). Abiogenic synthesis of prebiotic matter for the Earth's biosphere as a stage of self-organization on an astrophysical and paleontological time scale. *Paleontol. J.*, 41, 473-480.

Snytnikov V.N. (2007b). Astrocatalysis. In: Second International Conference. Biosphere Origin and Evolution/ October 28 – November 2, 2007. Loutraca, Greece. Abstracts, PL-4.

Soai K., Shibata T., Morioka H. and Choji K. (1995). Asymmetric autocatalysis and amplification of enantiomeric excess of a chiral molecule. *Nature*, 378, 767-768.

Spirin A.S. (2001). Protein biosynthesis, the RNA world, and the origin of life. *Herald Russ. Acad. Sci.*, 71, 146-153.

Spirin A.S. (2003). Ribonucleic acids: the key link of living matter. *Herald Russ. Acad. Sci.*, 73, 30-39.

Spirin A.S. (2005). Origin, possible forms of being, and size of the primeval organisms. *Paleontol. J.*, 39, 364-371.

Spirin A.S. (2007). When, where, and in what environment could the RNA world appear and evolve? *Paleontol. J.*, 41, 481-488.

Spitzer, L.Jr. (1978). *Physical Processes in the Interstellar Medium*. New York: Wiley.

Srama R. and Grun E. (1997). The dust sensor for Cassini. *Adv. Space Res.*, 20, 1467-1470.

Srama R., Woiwode W., Postberg F., Armes S.P., Fujii S., Dupin D., Ormond-ProutJ.,Sternovsky Z. Kempf S.,Moragas-Klostermeyer G., Mocker A. and Grun E. (2009). Mass spectrometry of hyper-velocity impacts of organic micrograins. *Rapid Commun. Mass Spectrom.*, 23, 3895–3906.

Stamper J.A. (1991). Review on spontaneous magnetic fields in laser-produced plasmas: Phenomena and measurements. *Laser Part. Beams.*, 9, 841-862.

Steele A., Goddard D., Beech I.B., Tapper R.C., Stapleton D. and Smith J.R. (1998). Atomic force microscopy imaging of fragments from the Martian meteorite ALH84001. *J. Microsc.*, 189, 2-7.

Stetter K.O. (1999). Smallest cell sizes within hyperthermophilic Archea (“Archeabacteria”). In: *Size limits of very small microorganisms*. Proceedings of a workshop. Washington: Nat. Acad. Press, p. 68-73.

Stevens T.O. and McKinley J.P. (1995). Lithoautotrophic microbial ecosystems in deep basalt aquifer. *Science*, 270, 450-454.

Stubig M. (2002). New insights in impact ionization and time-of-flight mass spectroscopy with micrometeoroid detectors by improved impact simulations in the laboratory. Dissertation for the Doctor of Natural Sciences. Combined Faculties for the Natural Sci. and for Math. of Ruperto-Carola Univ. of Heidelberg, Germany.

Stubig M., Srama R., Grun E. and Schafer G. (2002). Time–of–flight mass spectra from new projectile materials for calibration of micrometeorite detector CDA. In // *Proc EGS 27th General Assembly*. Abstract EGS02-A-01364, European Geophysical Society, Nice, April, 2002, p. 274.

Stubig M., Schafer G., Ho T.M., Srama R. and Grun E. (2001). Laboratory simulation improvements for hypervelocity meteorite impacts with a new dust particle source. *Planet. Space Sci.*, 49, 853.

Sunyaev R.A. (1989). *Fizika kosmosa: Malen'kaya entsiklopediya*. (Space physics. A small encyclopedia). // Moscow: Sov. Entsiklopediya, p. 415 (in Russian).

Tanaka K.A., Hara M., Ozaki N., Sasatani Y. and Anisimov S. (2000). Multi-layered flyer accelerated by laser induced shock waves. *Phys. Plasmas.*, 7, 676-678.

Tantsyrev G.D. and Nikolaev E.N. (1971). Formation of clusters in ion bombardment of films of frozen polar substances. *JETP Lett.*, 13, 337-339.

Tielens A.G.G.M., McKee C.F., Seab G.G., Hollenbach D.J. (1994). The physics of grain-grain collisions and gas-grain sputtering in interstellar shocks.// *Astrophys. J.*, 431, 321-340.

Tonon G.F. (1965). *Academ. Science*, 262, 1413-1417.

Trubnikov B.A. (1969). *Introduction into plasma theory*. Parts 1-3. Moscow: MIFI.

Urey H.C. (1952). On the early chemical history of the Earth and the origin of life. *Proc. Natl. Acad. Sci.*, 38, 351-363.

Vainshtein M.B. and Kudryashova E.B. (2000). Nanobacteria. *Mikrobiologiya*, 69, 163-174 (in Russian).

Visser G.M. (1986). Thesis D. Sci. — Leiden Univ..

Visser G.M., van Westrenen J., van Boeckel C.A.A. and van Boom J.H. (1984). Synthesis of the mirror image of β-D-riboguanosine 5'-phosphate a substrate to study chiral selection in non-enzymatic RNA synthesis. *Recl. Trav. Chim. Pays-Bas*, 103, 141-142.

Vityazev A.V., Pechernikova G.V. and Safronov V.S. (1990). *Terrestrial planets: origin and early evolution.* // Moscow: Nauka, 296pp (in Russian).

Vlasov G.P., Pavlov G.M., Bayanova N.V., Korneeva E.V., Ebel C., Khodorkovskii M.A. and Artamonova T.O. (2004). Dendrimers based on α-amino acids: synthesis and hydrodynamic characteristics. *Dokl. Phys. Chem.*, 399, 290-292.

Vlasov G.P., Tarasenko I.I., Valueva S.V., Kipper A.I., Tarabukina E.B., Filippov A.P., Skvortsova E.V. and Vorob'ev V.I. (2005). *Polym. Sci. Series A*, 47, 422-429.

Vorobyova E.A., Soina V.S., Bolshakova A.V. and Yaminsky I.V. (2001). Astrobiology strategies, life strategies, and methodologies applied. In: *Proceedings of the First Workshop on Exo // Astrobiology, Frascati, 21-23 May*, ESA SP-496, August 2001, 419-422.

Vorobyova E., Soina V., Gorlenko M., Minkovskaya N., Zalinova N., Mamukelashvili A., Gilichinsky D., Rivkina E. and Vishnivetskaya T. (1997). The deep cold biosphere: facts and hypothesis. *FEMS Microbiol. Rev.*, 20, 277-290.

Vorobyova E.A., Soina V.S., Mamukelashvili A.S., Bolshakova A., Yaminsky I.V. and Mulyukin A.L. (2005). Living cells in permafrost as models for astrobiology research. In: *Life in Ancient Ice*, Castello J.D. and Roger S.O. (eds.). Priceton: Princeton University Press, pp. 227-288.

Vorobyova E.A., Soina V.S., Zvyagintsev D.G. and Gilichinskii D.A. (2002). Viable ecosystems in the cryolithosphere. In: *Bacterial paleontology*. Paleontological Institute of the Russian Academy of Sciences. Rozanov A.Yu. (ed.). Moscow, P. 155-168 (in Russian).

Voshchinnikov N.V. (1986). Interstellar dust. *Itogi Nauki Tekh., Ser.: Issled. Kosm. Prostranstva*, 25 (in Russian).

Wachtershauser G. (1992). Groundworks for en evolutionary biochemistry: the iron-sulfur world. *Prog. Biophys. Mol. Biol.*, 58, 85-201.

Whittker A.G., Neudorffer M.E. and Watts E.J. (1983). Carbon: a rhombohedral carbyne form. *Carbon*, 21, 597-599.

Wilde S.A., Valley J.W., Peck W.H. and Graham C.M. (2001). Evidence from detrital zircons for the Existence of continental crust and oceans on the Earth 4.4 Gyr ago. *Nature*, 409, 175-178.

Woese C. (1967). The evolution of the genetic code. In: *The Genetic Code*. New York: HarperandRow, p. 179-195.

Woese S.R. (1998). The universal ancestor. *Proc. Natl. Acad. Sci. U.S.A.*, 95, 6854-6859.

Yarus M. (2002). Primordial genetics: phenotype of the ribocyte. *Ann. Rev. Genet*, 36, 125-151.

Zavarzin G.A. (2006). The evolution of the biosphere: The view of geologists and biologists. *Herald Russ. Acad. Sci.*, 76, 97-99.

Zavarzin G.A. (2006). Does evolution make the essence of biology? *Herald Russ. Acad. Sci.*, 76, 292-302.

Zel'dovich Ya.B and Raizer Yu.P. (2002). *Physics of Shock Waves and High-Temperature Hydrodynamic Phenomena*. Mineola, NY: Dover Publications.

Zel'dovich Ya.B. and Saakyan D.B. (1980). Asymmetry of breakup of optically active molecules by longitudinally polarized relativistic electrons. *J. Exp. Theor. Phys.*, 51, 1118.

Zenobi R., Khochenmuss R., (1998). Ion formation in MALDI mass spectrometry. *Mass Spectrom. Rev.*, 17, 337-366.

Zhang R., Achiba Y., Fisher J. K., Gadd G.E., Hopwood F.G., Ishigaki Toshinobu, Smith D.R., Suzuki Shinzo and Willett G.D. (1999). Laser ablation mass spectrometry of pyrolyzed koppers coal-tar pitch: a precursor for fullerenes and metallofullerenes. *J. Phys. Chem. B*, 103, 9450-9458.

Zvyagintsev D.G., Gilichinskii D.A., Blagodatskii S.A., Vorobyova E.A., Khlebnikova G.M., Arkhangelova A.A. and Kudryavtseva N.N. (1985). Duration of preservation of microorganisms in frozen sediments and buried soils. *Mikrobiol.*, 54, 155-161 (in Russian).

INDEX

A

B

C

D

E

F

G

H

I

J

K

L

M

N

O

P

Q

R

S

T

U

V

W

Y